Analysis and Control of Coupled Neural Networks with Reaction-Diffusion Terms

Jin-Liang Wang · Huai-Ning Wu
Tingwen Huang · Shun-Yan Ren

Analysis and Control of Coupled Neural Networks with Reaction-Diffusion Terms

Springer

Jin-Liang Wang
Tianjin Polytechnic University
Tianjin
China

Tingwen Huang
Texas A&M University at Qatar
Doha
Qatar

Huai-Ning Wu
Beihang University
Beijing
China

Shun-Yan Ren
Tianjin Polytechnic University
Tianjin
China

ISBN 978-981-10-4906-4 ISBN 978-981-10-4907-1 (eBook)
DOI 10.1007/978-981-10-4907-1

Library of Congress Control Number: 2017940609

Printed on acid-free paper

This Springer imprint is published by Springer Nature
The registered company is Springer Nature Singapore Pte Ltd.
The registered company address is: 152 Beach Road, #21-01/04 Gateway East, Singapore 189721, Singapore

Preface

Up to date, many researchers have devoted much effort to dynamical behaviors for coupled neural networks (CNNs) because of their wide applications in different fields. For instance, the CNNs have been triumphantly applied to harmonic oscillation generation, chaos generators design, secure communication, the electronic circuits, and memorizing and reproducing complex oscillatory patterns. Moreover, the research about synchronization of CNNs is a significant step to comprehend brain science. On the other hand, it is well known that neural networks are implemented by electric circuits, and the diffusion phenomena inevitably appear in electric circuits once electrons transport in a nonuniform electromagnetic field. Obviously, it is extremely necessary to consider the diffusion phenomena in coupled neural networks. Therefore, the investigation of dynamical behaviors about coupled reaction-diffusion neural networks (CRDNNs) has both practical and theoretical significance.

The aim of this book is to introduce recent research work on analysis and control of the passivity and synchronization for CRDNNs. This book is organized as follows:

Chapter 1: The background of reaction-diffusion neural network and CRDNNs is introduced as well as the organization of this book, and some important definitions, useful lemmas, and some basic knowledge about graph are also provided in this chapter.

Chapter 2: Two types of CRDNNs are proposed in this chapter. In the first one, the nodes are coupled through their states. In the second one, the nodes are coupled through the spatial diffusion terms. The synchronization of the first model is investigated by utilizing Lyapunov functional method and pinning control technique. In addition, considering that the theoretical coupling strength required for synchronization may be much larger than the needed value, we propose an adaptive strategy to adjust the coupling strength for achieving suitable value. For the latter, we establish a criterion for synchronization by using the designed pinning controllers. It is found that the CRDNNs with state coupling under the given linear feedback pinning controllers can realize synchronization when the coupling strength is very large, which is contrary to CRDNNs with spatial diffusion coupling.

Moreover, a general criterion for ensuring network synchronization is derived by pinning a small fraction of nodes with adaptive feedback controllers.

Chapter 3: This chapter is concerned with a directed CRDNNs. Based on the Lyapunov functional method and the pinning control technique, some sufficient conditions are obtained to guarantee the synchronization of the proposed network model. In addition, an adaptive strategy is proposed to obtain appropriate coupling strength for achieving network synchronization. Furthermore, the pinning adaptive synchronization problem is also investigated in this chapter, and a general criterion for ensuring network synchronization is established.

Chapter 4: This chapter applies the impulsive control method to achieve the synchronization of CRDNNs with time-varying delay. By combining the Lyapunov functional method with the impulsive delay differential inequality and comparison principle, a few sufficient conditions are derived to guarantee the global exponential synchronization of coupled neural networks with reaction-diffusion terms. Especially, the estimate for the exponential convergence rate is also given, which relies on time delay, system parameters, and impulsive interval.

Chapter 5: Two types of CRDNNs are proposed and the adaptive synchronization of these two types of CRDNNs is, respectively, investigated in this chapter. Based on local information of node dynamics, some novel adaptive strategies to tune the coupling strengths among network nodes are designed. By constructing appropriate Lyapunov functionals and utilizing inequality techniques, several sufficient conditions are given for reaching synchronization by using the designed adaptive laws.

Chapter 6: This chapter proposes a model of CRDNNs with hybrid coupling, which is composed of spatial diffusion coupling and state coupling. By utilizing the Lyapunov functional method combined with the inequality techniques, a sufficient condition is given to ensure that the proposed network model is synchronized. In addition, when the external disturbances appear in the network, a criterion is obtained to guarantee the $\mathcal{H}_\infty$ synchronization of the network. Moreover, some adaptive strategies to tune the coupling strengths among network nodes are designed for reaching synchronization and $\mathcal{H}_\infty$ synchronization. Some criteria for synchronization and $\mathcal{H}_\infty$ synchronization are derived by using the designed adaptive laws.

Chapter 7: A CRDNNs consisting of N linearly and diffusively coupled identical RDNNs is proposed in this chapter. By utilizing some inequality techniques, a sufficient condition ensuring the output strict passivity is derived for the proposed network model. Then, we reveal the relationship between output strict passivity and synchronization of the proposed network model. Moreover, based on the obtained passivity result and the relationship between output strict passivity and synchronization, a criterion for synchronization is established.

Chapter 8: This chapter is concerned with the passivity problem of a model of CRDNNs with adaptive coupling. In order to ensure the passivity of the CRDNNs, some adaptive strategies to tune the coupling strengths among network nodes are designed. By utilizing some inequality techniques and the designed adaptive laws, several sufficient conditions ensuring passivity are obtained. In addition, we reveal the relationship between passivity and synchronization of the CRDNNs. Based on the obtained passivity results and the relationship between passivity and synchronization, a global synchronization criterion is established.

Chapter 9: This chapter considers two CRDNNs with different dimensions of input and output. The only difference between them is whether time-varying delay is incorporated in the mathematical model of network. We, respectively, analyze dissipativity and passivity of these CRDNNs. Firstly, for the systems with different dimensions of input and output vectors, two new passivity definitions are proposed. Then, by exploiting some inequality techniques, several dissipativity and passivity criteria for these CRDNNs are established. Furthermore, we analyze stability of passive CRDNNs.

Chapter 10: This chapter studies a CRDNNs consisting of N identical neural networks with reaction-diffusion terms. Firstly, several passivity definitions for the systems with different dimensions of input and output are given. By utilizing some inequality techniques, several criteria are presented, ensuring passivity of the CRDNNs under the designed adaptive law. Then, we discuss the relationship between synchronization and output strict passivity of the proposed network model. Furthermore, these results are extended to the case when topological structure of the network is undirected.

Acknowledgement: This book was supported by the National Natural Science Foundation for Distinguished Young Scholars of China under Grant 61625302, the National Natural Science Foundation of China under Grants 61403275, 61473011, and 61421063, the Natural Science Foundation of Tianjin, China, under Grant 15JCQNJC04100, and the National Priorities Research Program (NPRP) from the Qatar National Research Fund (a member of Qatar Foundation) under Grant NPRP 9-166-1-031. I'd like to begin by acknowledging my postgraduates Bei-Bei Xu, Pu-Chong Wei, Shu-Xue Wang, and Meng Xu who have unselfishly given their valuable time in arranging these raw materials into something I'm proud of. Thanks to Zhen Qin, Xiao-Xiao Zhang, Wei-Zhong Chen, and Shui-Han Qiu for your scrupulous check and review, and your eagle eyes spared me from sweating the details.

Tianjin, China — Jin-Liang Wang
Beijing, China — Huai-Ning Wu
Doha, Qatar — Tingwen Huang
Tianjin, China — Shun-Yan Ren
January 2017

Contents

1 Introduction 1
1.1 Background 1
1.1.1 Reaction-Diffusion Neural Networks 1
1.1.2 Coupled Neural Networks with Reaction-Diffusion Terms 2
1.2 Book Organization 4
1.3 Some Definitions 5
1.4 Graph Theory 7
1.5 Some Lemmas 7
1.6 Assumption 12

2 Pinning Control Strategies for Synchronization of CRDNNs 13
2.1 Introduction 13
2.2 Pinning Control of CRDNNs with State Coupling 15
2.2.1 Network Model 15
2.2.2 Pinning Synchronization of CRDNNs 17
2.2.3 Pinning Synchronization of CRDNNs with Adaptive Coupling Strength 21
2.3 Pinning Control of CRDNNs with Spatial Diffusion Coupling 24
2.3.1 Network Model 25
2.3.2 Pinning Synchronization of CRDNNs 26
2.3.3 Pinning Adaptive Synchronization of CRDNNs 30
2.4 Numerical Examples 33
2.5 Conclusions 38

3 Pinning Control for Synchronization of CRDNNs with Directed Topologies 39
3.1 Introduction 39
3.2 Pinning Synchronization of a CRDNNs with Directed Topologies 40

3.2.1 Network Model 40
3.2.2 Pinning Synchronization Criteria 42
3.2.3 Adaptive Coupling Strength 51
3.2.4 Pinning Adaptive Synchronization Criterion 54
3.3 Numerical Example 57
3.4 Conclusions 61

4 Impulsive Control for the Synchronization of CRDNNs 63
4.1 Introduction 63
4.2 Main Results 64
4.3 Numerical Examples 77
4.4 Conclusions 80

5 Novel Adaptive Strategies for Synchronization of CRDNNs 81
5.1 Introduction 81
5.2 Adaptive Synchronization in CRDNNs 82
5.2.1 Network Model 83
5.2.2 Adaptive Strategy for Synchronization of CRDNNs . . . 84
5.2.3 Adaptive Pinning Strategy for Synchronization of CRDNNs 89
5.3 Adaptive Synchronization in CRDNNs with Time-Varying Coupling Delay 91
5.3.1 Network Model 91
5.3.2 Adaptive Strategy for Synchronization of CRDNNs . . . 92
5.3.3 Adaptive Pinning Strategy for Synchronization of CRDNNs 95
5.4 Numerical Examples 97
5.5 Conclusions 103

6 Synchronization and Adaptive Control of CRDNNs with Hybrid Coupling 105
6.1 Introduction 105
6.2 Synchronization in CRDNNs with Hybrid Coupling 106
6.2.1 Network Model 106
6.2.2 Synchronization of CRDNNs 107
6.2.3 $\mathcal{H}_\infty$ Synchronization of CRDNNs 111
6.3 Adaptive Synchronization of CRDNNs with Hybrid Coupling 114
6.3.1 Network Model 114
6.3.2 Adaptive Strategies for Synchronization of CRDNNs 114
6.3.3 Adaptive Strategies for $\mathcal{H}_\infty$ Synchronization of CRDNNs 120
6.4 Numerical Examples 124
6.5 Conclusions 128

7 Passivity-Based Synchronization of CRDNNs with Time-Varying Delay 129
7.1 Introduction 129
7.2 Network Model. 130
7.3 Passivity-Based Synchronization of CRDNNs 131
7.4 Numerical Example 139
7.5 Conclusion 141

8 Passivity and Synchronization of CRDNNs with Adaptive Coupling. 143
8.1 Introduction 143
8.2 Network Model. 144
8.3 Passivity and Synchronization of CRDNNs 146
8.3.1 Passivity Analysis of CRDNNs. 146
8.3.2 The Relationship Between Passivity and Synchronization of CRDNNs 152
8.3.3 Synchronization of CRDNNs 155
8.4 Numerical Example 155
8.5 Conclusion 159

9 Passivity Analysis of CRDNNs with Dirichlet Boundary Conditions 161
9.1 Introduction 161
9.2 Passivity of CRDNNs. 162
9.2.1 Network Model 162
9.2.2 Passivity Analysis of CRDNNs. 163
9.2.3 The Stability of Passive CRDNNs. 168
9.3 Passivity of CRDNNs with Time-Varying Delay 172
9.3.1 Network Model 172
9.3.2 Passivity Analysis of CRDNNs with Time-Varying Delay. 172
9.3.3 The Stability of Passive Delayed CRDNNs. 176
9.4 Numerical Examples. 176
9.5 Conclusion 184

10 Passivity of Directed and Undirected CRDNNs with Adaptive Coupling Weights 185
10.1 Introduction 185
10.2 Network Model. 187
10.3 Passivity of CRDNNs with Directed Topology and Adaptive Coupling Weights. 188
10.3.1 Passivity Analysis of CRDNNs. 190
10.3.2 The Relationship Between Output Strict Passivity and Synchronization of CRDNNs 195

10.4 Passivity of CRDNNs with Undirected Topology and Adaptive Coupling Weights 197
10.4.1 Passivity Analysis of CRDNNs 198
10.4.2 The Relationship Between Output Strict Passivity and Synchronization of CRDNNs 201
10.5 Numerical Examples 202
10.6 Conclusion 209

References 211

Symbols and Acronyms

$\mathbb{N}$	1, 2, 3, $\cdots$
$\mathbb{R}$	field of real numbers
$\mathbb{R}^n$	n-dimensional real Euclidean space
$\mathbb{R}^{m\times n}$	space of all $m \times n$ real matrices
I_n	$n \times n$ real identity matrix
$P > 0$	symmetric positive definite
$P \geqslant 0$	symmetric positive semi-definite
$P < 0$	symmetric negative definite
$P \leqslant 0$	symmetric negative semi-definite
B^{T}	transpose of matrix B
B^{-1}	inverse of matrix B
$\otimes$	Kronecker product of two matrices
$\lambda_M(A)$	maximum eigenvalue of matrix A
$\lambda_m(A)$	minimum eigenvalue of matrix A
$\mathrm{diag}\{\cdots\}$	block-diagonal matrix
$\|\cdot\|$	Euclidean norm of a vector and its induced norm of a matrix
sup	supremum
inf	infimum
Ω	$\{x = (x_1, x_2, \cdots, x_q)^T \,\|\, \|x_k\| < l_k, k = 1, 2, \cdots, q\}$ is an open bounded domain in $\mathbb{R}^q$ with smooth boundary $\partial\Omega$
$\mathrm{PC}[[-\tau, +\infty), \mathbb{R}]$	$\{\phi : [-\tau, +\infty) \to \mathbb{R}, \phi(t)$ is continuous everywhere except for the points $t_k, k \in \mathbb{N}$ at which $\phi(t_k^+) = \phi(t_k)$ and $\phi(t_k^-)$ exist$\}$
T_{k-1}	$t_k - t_{k-1}$
$T_{\min}$	$\inf_{k\in\mathbb{N}}\{T_{k-1}\}$
$T_{\max}$	$\sup_{k\in\mathbb{N}}\{T_{k-1}\}$
PDEs	Partial Differential Equations
DPDE	Delay Partial Differential Equation
CNNs	Coupled Neural Networks
RDNNs	Reaction-Diffusion Neural Networks
CRDNNs	Coupled Reaction-Diffusion Neural Networks

Chapter 1
Introduction

1.1 Background

1.1.1 Reaction-Diffusion Neural Networks

Recently, neural networks have found successful applications in diverse areas such as pattern classification, parallel computation, signal processing, and nonlinear functions approximation. These applications heavily depend on the dissipativity, stability, asymptotic stability, passivity, output strict passivity, and synchronization of neural networks. Therefore, many interesting results on the dynamical behaviors have been presented for fuzzy, stochastic, complex-valued, and impulsive network models [1–21]. In [1], the authors investigated the dissipativity problem of discrete-time fuzzy neural network with parameter uncertainties based on interval type-2 fuzzy model. Zhang et al. [2] discussed the stochastic stability of delayed neural networks with local impulses. In [10], a delay-dependent global exponential stability condition was established for complex-valued neural networks.

However, the reaction-diffusion effect is not taken into consideration in these existing works [1–11, 22, 23]. It is well known that neural networks are implemented by electric circuits, and the diffusion phenomena inevitably appear in electric circuits once electrons transport in a nonuniform electromagnetic field [24–33]. Therefore, it is essential to take the reaction-diffusion effect into consideration when we investigate neural networks [24–32]. In [29], the authors gave a new passivity definition for the systems with time and space-dependent input and output. By exploiting some inequality techniques, several stability and passivity criteria for RDNNs were established [29]. Ma et al. [32] studied the asymptotic stability of RDNNs with time-varying delays.

J.-L. Wang et al., *Analysis and Control of Coupled Neural Networks with Reaction-Diffusion Terms*, DOI 10.1007/978-981-10-4907-1_1

1.1.2 *Coupled Neural Networks with Reaction-Diffusion Terms*

Many systems in nature and society, such as food webs, communication networks, social networks, power grids, cellular networks, World Wide Web, metabolic systems, disease transmission networks, etc. [34, 35], can be modeled as complex networks. Therefore, analysis and control of dynamical behaviors in complex networks have become a focal research topic in recent years. Especially, as two most significant and interesting dynamical properties of complex networks, synchronization and passivity have received much attention. Recently, many interesting results on synchronization [36–71] and passivity [72–75] were derived for various complex networks associated with models being time invariant, time-varying, discrete, and impulsive. Cheng et al. [36] studied the adaptive pinning synchronization of delayed complex networks with nonlinear coupling by employing Lyapunov stability theory. In [39], several criteria were derived to guarantee the synchronization of complex network model with fractional order chaotic nodes by using the LaSalle invariance principle. Wang et al. [74] investigated input passivity and output passivity for a complex network with non-linear, time-varying, non-symmetric and delayed coupling. By constructing some suitable Lyapunov functionals, several sufficient conditions ensuring input passivity and output passivity were derived for complex dynamical networks. In [75], the authors considered the passivity problem for a class of Markovian switching complex dynamic networks with multiple time-varying delays and stochastic perturbations. By constructing appropriate stochastic Lyapunov–Krasovskii functional and employing stochastic theory, linear matrix inequality technique and properties of Weiner process, some sufficient conditions were obtained to guarantee that the complex dynamic networks with multiple time-varying delays and stochastic perturbations under Markovian switching are passive in the sense of expectation. However, in these existing works [36–75], the node state is only dependent on time. But, in many circumstances, the node state is not only dependent on time, but also intensively dependent on space variable.

As a special class of complex networks, coupled neural networks have attracted the attention of researchers from different fields. The main reason for this is that it has many applications in different areas. For instance, Hoppensteadt and Izhikevich [76] proposed an architecture of coupled neural networks to store and retrieve complex oscillatory patterns as synchronization states. In [77], a secure communication system based on coupled cellular neural networks was presented. In addition, the study of synchronization of coupled neural networks is an important step for understanding brain science [78, 79]. Therefore, it is interesting to study the dynamical behaviors of arrays of coupled neural networks [80–90]. In [80, 91], Feng et al. considered the stochastic synchronization for an array of neural networks with hybrid coupling. Wang and Song [81] discussed a class of coupled stochastic discrete-time neural networks with discrete and distributed time-delays. By utilizing an adaptive control method and some stochastic analysis techniques, several sufficient conditions were derived to ensure the synchronization [82] and robust synchronization [83] of coupled delayed neural networks. Tang et al. [84] proposed a controllable probabilistic particle swarm optimization algorithm. In [85, 86], some sufficient conditions were derived for global exponential synchronization of coupled neural networks with constant and delayed coupling by using the Lyapunov functional method and Kronecker product technique. In [87], the authors investigated the synchronization control problem of a class of switched linearly coupled delayed neural networks by using the Lyapunov functional method and linear matrix inequality approach. Lu et al. [88] considered the global exponential synchronization for an array of delayed coupled neural networks with an asymmetric coupling configuration matrix. In [89], the global exponential synchronization was studied for uncertain coupled neural networks with both constant coupling and distributed-delay coupling. Li et al. [90] discussed the global exponential synchronization for an array of coupled discrete-time Cohen–Grossberg neural networks with time-varying delay. Li and Cao [91] discussed the passivity and robust synchronization problems of switched coupled neural networks with uncertain parameters.

Unfortunately, in these existing works [80–91], the diffusion effects have not been considered. To our knowledge, very few researchers have investigated the synchronization and passivity for an array of coupled neural networks with reaction-diffusion terms [92–103]. In [97], Liu discussed the μ-synchronization and pinning control problems for a class of CRDNNs with the Dirichlet boundary conditions and unbounded time delays. Wang et al. [98] discussed the adaptive synchronization in an array of linearly coupled neural networks with reaction-diffusion terms and time delays. Based on the LaSalle invariant principle of functional differential equations and the adaptive feedback control technique, some sufficient conditions for ensuring network synchronization were established in [98]. In [99], the authors investigated the synchronization problem of two kinds of CRDNNs using edge-based adaptive strategy. Yang et al. [101] studied the global exponential synchronization of a class of CRDNNs with time-varying delay by adding impulsive controller to a small fraction of nodes. Wang et al. [100] considered a general array model of CRDNNs with hybrid coupling, which is composed of spatial diffusion coupling and state coupling. In [95, 96], some passivity criteria were derived for CRDNNs with fixed and adaptive coupling weights. Moreover, they analyzed the synchronization of passive CRDNNs. Unfortunately, the input vector has the same dimension as the output vector in the network models considered in [29, 43, 50, 74, 95, 96]. Practically, input and output vectors in some networks have different dimensions. But, for the systems with different dimensions of input and output vectors, we can not deal with by traditional passivity definitions. However, very few authors have discussed passivity of CRDNNs with different dimensions of input and output [93, 94]. Wang et al. [93] proposed two CRDNNs with different dimensions of input and output, and respectively analyzed the dissipativity and passivity of these CRDNNs. Furthermore, they analyzed the stability of passive CRDNNs. In [94], several criteria ensuring passivity of CRDNNs under the designed adaptive law were presented by utilizing some inequality techniques. Moreover, the authors discussed the relationship between synchronization and output strict passivity of the proposed network model.

1.2 Book Organization

Recently, the dynamical behaviors of CNNs have been extensively investigated by researchers, and many important results on analysis and control of dynamical behaviors have been established for various CNNs. Although research on CNNs has attracted so much attention, little of that has been devoted to the dynamical behaviors of CNNs with reaction-diffusion terms. Thus, the main aim of this book is to introduce some recent results on analysis and control of the dynamical behaviors for CRDNNs, and this book can serve as a stepping stone to study the dynamical behaviors of CRDNNs. The rest of this book is organized as follows:

Chapter 2 proposes two types of CRDNNs respectively coupled through states and the spatial diffusion terms [103], and investigates the synchronization problems for these proposed networks by utilizing pinning control strategies.

Chapter 3 presents a directed CRDNNs consisting of N linearly and diffusively coupled identical RDNNs [102]. By exploiting the pinning control method combined with adaptive control technique, several sufficient conditions are obtained to guarantee the synchronization of the proposed network model.

Chapter 4 applies the impulsive control method to investigate the synchronization of CRDNNs with time-varying delay [92], and a few sufficient conditions are derived to guarantee the global exponential synchronization of the proposed network as well as the estimation for the exponential convergence rate.

Chapter 5 gives two types of CRDNNs [99], and respectively investigate the synchronization of these two types of CRDNNs using edge-based adaptive strategy.

Chapter 6 discusses a CRDNNs with hybrid coupling, which is composed of spatial diffusion coupling and state coupling [100], and the synchronization, adaptive synchronization, $\mathcal{H}_\infty$ synchronization and adaptive $\mathcal{H}_\infty$ synchronization of the proposed network model are considered respectively.

Chapter 7 puts forward a CRDNNs consisting of N linearly and diffusively coupled identical RDNNs [95], and the passivity problem for the proposed network model is investigated and the relationship between output strict passivity and synchronization is revealed. Moreover, a passivity-based synchronization criterion is also established.

Chapter 8 is concerned with CRDNNs with adaptive coupling [96], and the passivity of the proposed network is investigated by designing some adaptive strategies to tune the coupling strengths among network nodes. In addition, we reveal the relationship between passivity and synchronization of the proposed network.

Chapter 9 considers two CRDNNs with different dimensions of input and output [93], and dissipativity and passivity of these CRDNNs are analyzed respectively based on two new passivity definitions.

Chapter 10 studies a CRDNNs consisting of N identical RDNNs [94], establishes several criteria to guarantee passivity of the CRDNNs under the designed adaptive laws, and the relationship between synchronization and output strict passivity of the proposed network model is also revealed. In addition, these results are extended to the undirected topological case.

1.3 Some Definitions

Definition 1.1 For $e(x,t) = (e_1(x,t), e_2(x,t), \ldots, e_n(x,t))^T \in \mathbb{R}^n$, $(x,t) \in \Omega \times \mathbb{R}$, we define

$$\|e(\cdot,t)\|_2 = \sqrt{\int_{\Omega} \sum_{i=1}^{n} e_i^2(x,t)dx},$$

$$\|e(\cdot,t)\|_{\tau} = \sup_{-\tau \leqslant \theta \leqslant 0} \|e(\cdot, t+\theta)\|_2.$$

Definition 1.2 (*see* [104]) The matrix A of order n is called *reducible* if there is a permutation matrix $P \in \mathbb{R}^{n\times n}$ such that

$$PAP^T = \begin{pmatrix} A_1 & 0 \\ A_{21} & A_2 \end{pmatrix},$$

where A_1 and A_2 are square matrices of order at least one. If A is not reducible, then A is called *irreducible*. Notice that a matrix of order 1 is irreducible.

Definition 1.3 (*see* [105]) A system with supply rate ϑ is said to be dissipative if there exists a nonnegative function $S : \mathbb{R}^+ \to \mathbb{R}^+$, called the storage function, such that

$$\int_{t_0}^{t_p} \vartheta(u,y)dt \geqslant S(t_p) - S(t_0)$$

for any $t_p, t_0 \in \mathbb{R}^+$ and $t_p \geqslant t_0$, where $u(x,t) \in \mathbb{R}^n$ and $y(x,t) \in \mathbb{R}^n$ are the input and output of the system at time t and in space x, respectively, $(x,t) \in \Omega \times \mathbb{R}^+$.

Definition 1.4 (*see* [29, 105–108]) A system is said to be passive if it is dissipative with respect to

$$\vartheta(u,y) = \int_{\Omega} y^T(x,t)u(x,t)dx,$$

where $u(x,t) \in \mathbb{R}^n$ and $y(x,t) \in \mathbb{R}^n$ are the input and output of the system at time t and in space x, respectively, $(x,t) \in \Omega \times \mathbb{R}^+$.

Definition 1.5 (*see* [29, 105–108]) A system is said to be strictly passive if it is dissipative with respect to

$$\vartheta(u,y) = \int_{\Omega} y^T(x,t)u(x,t)dx - \gamma_1 \int_{\Omega} u^T(x,t)u(x,t)dx - \gamma_2 \int_{\Omega} y^T(x,t)y(x,t)dx$$

for $\gamma_1 \geqslant 0, \gamma_2 \geqslant 0, \gamma_1 + \gamma_2 > 0$, where $u(x,t) \in \mathbb{R}^n$ and $y(x,t) \in \mathbb{R}^n$ are the input and output of the system at time t and in space x, respectively, $(x,t) \in \Omega \times \mathbb{R}^+$.

The system is said to be input-strictly passive if $\gamma_1 > 0$ and output-strictly passive if $\gamma_2 > 0$.

Remark 1.6 In recent years, many authors have studied the passivity of ordinary differential equation systems, in which the input and output variables are only dependent on the time. However, the input and output variables in many systems are varied with the time and space variables. But this case can not be dealt with by traditional passivity theory. In this paper, we give several concepts of passivity (Definitions 1.4–1.5) for the case where input and output variables are varied with the time and space variables, which can be obtained by Definition 1.3 through an appropriate choice of the supply rate. Obviously, the Definitions 1.4–1.5 generalize some existing passivity definitions.

Definition 1.7 (*see* [105]) A system with supply rate ϑ is said to be dissipative if there exists a nonnegative function $S : \mathbb{R}^+ \to \mathbb{R}^+$, called the storage function, such that

$$\int_{t_0}^{t_p} \vartheta(u,y)dt \geqslant S(t_p) - S(t_0)$$

for any $t_p, t_0 \in \mathbb{R}^+$ and $t_p \geqslant t_0$, where $u(x,t) \in \mathbb{R}^p$ and $y(x,t) \in \mathbb{R}^m$ are the input and output of the system at time t and in space x, respectively, $(x,t) \in \Omega \times \mathbb{R}^+$.

Definition 1.8 A system is said to be passive if it is dissipative with respect to

$$\vartheta(u,y) = \int_\Omega y^T(x,t)Pu(x,t)dx,$$

where $P \in \mathbb{R}^{m\times p}$ is a constant matrix, $u(x,t) \in \mathbb{R}^p$ and $y(x,t) \in \mathbb{R}^m$ are the input and output of the system at time t and in space x, respectively, $(x,t) \in \Omega \times \mathbb{R}^+$.

Definition 1.9 A system with input $u(x,t) \in \mathbb{R}^p$ and output $y(x,t) \in \mathbb{R}^m((x,t) \in \Omega \times [0,+\infty))$ is said to be strictly passive if it is dissipative with respect to

$$\begin{aligned}\vartheta(u,y) = &\int_\Omega y^T(x,t)Pu(x,t)dx - \int_\Omega y^T(x,t)Qy(x,t)dx \\ &- \int_\Omega u^T(x,t)Wu(x,t)dx\end{aligned}$$

for $0 \leqslant W \in \mathbb{R}^{p\times p}, 0 \leqslant Q \in \mathbb{R}^{m\times m}, P \in \mathbb{R}^{m\times p}$ and $\lambda_m(W) + \lambda_m(Q) > 0$.

The system is said to be input-strictly passive if $W > 0$ and output-strictly passive if $Q > 0$.

Remark 1.10 In [29, 43, 50, 74, 95, 96], the authors respectively studied the passivity of ordinary and partial differential equation network models. But the input vector has the same dimension as the output vector in the network models considered in [29, 43, 50, 74, 95, 96]. Unfortunately, in many systems, the input and output vectors have different dimensions. However, this case can not be dealt with by traditional passivity theory. As a natural extension of the existing concepts of passivity, we give two passivity definitions (see Definitions 1.8 and 1.9) for the case where the input and output vectors have different dimensions. Letting $W = \gamma_1 I_m$, $Q = \gamma_2 I_m$ and $P = I_m$, then these traditional passivity definitions in [29, 43, 50, 74, 95, 96] can be derived by Definitions 1.8 and 1.9.

1.4 Graph Theory

Denote $\mathcal{G} = \{\mathcal{V}, \mathcal{E}, G\}$ as a weighted directed network composed of a set of vertices $\mathcal{V} = \{1, 2, \ldots, N\}$, set of directed edges $\mathcal{E} \subseteq \mathcal{V} \times \mathcal{V}$, and a weighted adjacency matrix $G = (G_{ij})_{N\times N}$. A directed edge $\mathcal{E}_{ij}$ in network $\mathcal{G}$ is represented by the ordered pair of vertices (i, j), where i is called the initial vertex and j is called the terminal vertex of $\mathcal{E}_{ij}$. If there is an edge from vertex i to vertex j in $\mathcal{G}$, then $G_{ij} > 0$; otherwise, $G_{ij} = 0$. The outdegree of vertex i is the number of edges leaving vertex i. The indegree of vertex i is the number of edges entering vertex i. A sequence $(i_1, i_2), (i_2, i_3), \ldots, (i_{k-1}, i_k)$ of edges in network $\mathcal{G}$ with distinct vertices $i_s (s = 1, 2, \ldots, k)$ is called a directed path from vertex i_1 to vertex i_k. The weighted directed network $\mathcal{G}$ is strongly connected if there is a directed path from every vertex to every other vertex.

1.5 Some Lemmas

Lemma 1.11 (see Theorem 3.2.1 of [104, 109]) *Let G be a matrix of order n. Then G is irreducible if and only if its corresponding network $\mathcal{G}$ is strongly connected.*

Lemma 1.12 (see Theorem 3.2.4 of [104]) *Let A be a matrix of order n. Then there exists a permutation matrix P of order n and an integer $m \geqslant 1$ such that*

$$PAP^T = \begin{pmatrix} A_1 & A_{12} & \ldots & A_{1m} \\ 0 & A_2 & \ldots & A_{2m} \\ \vdots & \vdots & \ddots & \vdots \\ 0 & 0 & \ldots & A_m \end{pmatrix}, \tag{1.1}$$

where $A_1, A_2, \ldots, A_m$ are square irreducible matrices. The matrices $A_1, A_2, \ldots, A_m$ that occur as diagonal blocks in (1.1) are uniquely determined to within simultaneous permutation of their lines, but their ordering in (1.1) is not necessarily unique.

The form in (1.1) is called the Frobenius normal form of the square matrix A.

Lemma 1.13 (see [63, 110]) *Let $\alpha \in \mathbb{R}$, A, B, C, D be matrices with suitable dimensions. Then the Kronecker product has the following properties:*

$$
\begin{aligned}
&(1)\quad (A \otimes B)^T = A^T \otimes B^T;\\
&(2)\quad (\alpha A) \otimes B = A \otimes (\alpha B);\\
&(3)\quad (A + B) \otimes C = A \otimes C + B \otimes C;\\
&(4)\quad (A \otimes B)(C \otimes D) = (AC) \otimes (BD).
\end{aligned}
$$

Lemma 1.14 *For any vectors $x, y \in \mathbb{R}^n$ and matrix $0 < P \in \mathbb{R}^{n\times n}$, the following inequality holds:*

$$x^T y + y^T x \leqslant x^T P x + y^T P^{-1} y.$$

Lemma 1.15 (Boyd et al. [111]) *The linear matrix inequality (LMI)*

$$\begin{pmatrix} Q & S \\ S^T & B \end{pmatrix} > 0,$$

where $Q = Q^T$, $B = B^T$, is equivalent to

$$B > 0, \;\; Q - SB^{-1}S^T > 0.$$

Lemma 1.16 ([112]) *Suppose that $G = (G_{ij})_{N\times N}$ is a real symmetric and irreducible matrix, where*

$$G_{ij} \geqslant 0 \;\; (i \neq j), \quad G_{ii} = -\sum_{\substack{j=1\\ j\neq i}}^{N} G_{ij}.$$

Then,

(1) 0 is an eigenvalue of matrix G with multiplicity 1, associated with the eigenvector $(1, 1, \ldots, 1)^T$.
(2) all the other eigenvalues of G are strictly negative.
(3) there exists a unitary matrix $\Phi = (\phi_1, \phi_2, \ldots, \phi_N)$ such that

$$G\phi_i = \lambda_i \phi_i, i = 1, 2, \ldots, N,$$

where $0 = \lambda_1 > \lambda_2 \geqslant \lambda_3 \geqslant \cdots \geqslant \lambda_N$ are the eigenvalues of G.

Lemma 1.17 (see [113]) *Suppose that $G = (G_{ij})_{N\times N}$ is a real symmetric and irreducible matrix, where*

$$G_{ij} \geqslant 0 \ (i \neq j), \quad G_{ii} = -\sum_{\substack{j=1 \\ j \neq i}}^{N} G_{ij}.$$

Then,

(1) The largest nonzero eigenvalue $\lambda_2(G)$ of the matrix G satisfies

$$\lambda_2(G) = \max_{\substack{x \neq 0, \\ x^T \mathbf{1}_N = 0}} \frac{x^T G x}{x^T x}$$

where $\mathbf{1}_N = (1, 1, \ldots, 1)^T \in \mathbb{R}^N$.

(2) For any $\eta = (\eta_1, \eta_2, \ldots, \eta_N)^T \in \mathbb{R}^N$,

$$\eta^T G \eta = -\frac{1}{2} \sum_{i=1}^{N} \sum_{j=1}^{N} G_{ij} (\eta_i - \eta_j)^2.$$

Lemma 1.18 *Suppose that $G = (G_{ij})_{N \times N} (N \geqslant 2)$ is an irreducible matrix, where*

$$G_{ij} \geqslant 0 \ (i \neq j), \quad G_{ii} \leqslant -\sum_{\substack{j=1 \\ j \neq i}}^{N} G_{ij}.$$

Then, there exists a positive definite diagonal matrix $\Xi = \mathrm{diag}(\Xi_1, \Xi_2, \ldots, \Xi_N) \in \mathbb{R}^{N \times N}$ such that

$$\Xi G + G^T \Xi \leqslant 0.$$

Proof Let $U_{ij} = G_{ij} (i \neq j), U_{ii} = -\sum_{\substack{j=1 \\ j \neq i}}^{N} U_{ij}, H_{ii} = G_{ii} - U_{ii} \leqslant 0$. Then, we have

$$G = U + H,$$

where $U = (U_{ij})_{N \times N}, H = \mathrm{diag}(H_{11}, H_{22}, \ldots, H_{NN})$. By the similar proof of Lemma 2.12 in [109], we can find the positive definite diagonal matrix $\Xi = \mathrm{diag}(\Xi_1, \Xi_2, \ldots, \Xi_N)$ satisfying

$$\Xi U + U^T \Xi \leqslant 0.$$

Since $H \leqslant 0$, we can easily derive

$$\Xi U + \Xi H + U^T \Xi + H \Xi = \Xi G + G^T \Xi \leqslant 0.$$

The proof is completed.

Lemma 1.19 (see [109, 114, 115]) *Suppose that $G = (G_{ij})_{N\times N}$ is irreducible and satisfies $\sum_{j=1}^{N} G_{ij} = 0$ with $G_{ij} \geqslant 0 (i \neq j)$. Then, there exists a positive vector $\theta = (\theta_1, \theta_2, \ldots, \theta_N)^T \in \mathbb{R}^N$ such that*

(1) $G^T\theta = 0$;

(2) $\hat{G} = \Xi G + G^T \Xi$ is symmetric and

$$\sum_{j=1}^{N} \hat{G}_{ij} = \sum_{j=1}^{N} \hat{G}_{ji} = 0 \text{ for all } i = 1, 2, \ldots, N,$$

where $\Xi = \mathrm{diag}(\theta_1, \theta_2, \ldots, \theta_N)$.

Lemma 1.20 (see [116]) *Let $0 \leqslant \tau_i(t) \leqslant \tau_i, \sigma > 0, m_i \geqslant 0, m \in \mathbb{R}, i = 1, 2, \ldots, n$. Assume that $m + \frac{\ln\sigma}{\rho} + \sigma^{\mathrm{sgn}(\ln\sigma)} \sum_{i=1}^{n} m_i < 0$ and $u(t) \in \mathrm{PC}[[-\tau, +\infty), \mathbb{R}^+]$ satisfies*

$$\begin{cases} D^+u(t) \leqslant mu(t) + \sum_{i=1}^{n} m_i u(t - \tau_i(t)), & t \geqslant 0, \\ u(t_k) \leqslant \sigma u(t_k^-), & k \in \mathbb{N}, \\ u(t) = \phi(t), & -\tau \leqslant t \leqslant 0, \end{cases}$$

in which $\tau = \max_{i=1,2,\ldots,n}\{\tau_i\}$, and $\phi(t)$ is bounded and continuous on $[-\tau, 0]$, then

$$u(t) \leqslant \sigma^{\mathrm{sgn}(\ln\sigma)} e^{-\lambda t} \sup_{-\tau\leqslant s\leqslant 0} \phi(s), \quad t \geqslant 0,$$

where $\lambda > 0$ is a unique solution of

$$\lambda + m + \frac{\ln\sigma}{\rho} + \sigma^{\mathrm{sgn}(\ln\sigma)} \sum_{i=1}^{n} m_i e^{\lambda\tau_i} = 0,$$

in which $\rho \geqslant T_{\max}$ if $\sigma < 1$, otherwise, $\rho \leqslant T_{\min}$;

$$\mathrm{sgn}(\ln\sigma) = \begin{cases} -1 & \text{if } \sigma < 1, \\ 0 & \text{if } \sigma = 1, \\ 1 & \text{if } \sigma > 1. \end{cases}$$

Remark 1.21 We always assume $0 < T_{\min} \leqslant T_{\max} < +\infty$. Lemma 1.20 is very significant for us to analyze the global exponential synchronization of CRDNNs via impulsive control.

Lemma 1.22 (see [117]) *Let Ω be a cube $|x_k| < l_k (k = 1, 2, \ldots, q)$ and let $h(x)$ be a real-valued function belonging to $C^1(\Omega)$ which vanishes on the boundary $\partial\Omega$ of Ω, i.e., $h(x)|_{\partial\Omega} = 0$. Then*

$$\int_{\Omega} h^2(x)dx \leqslant l_k^2 \int_{\Omega} \left(\frac{\partial h}{\partial x_k}\right)^2 dx,$$

where $x = (x_1, x_2, \ldots, x_q)^T$.

Consider the following delay partial differential equation (DPDE) system

$$\begin{cases} \frac{\partial \varphi(x,t)}{\partial t} = D\Delta\varphi(x,t) + g(\varphi(x,t-\tau(t))) + f(\varphi(x,t)), \\ \varphi(x,t) = \phi(x,t), \quad (x,t) \in \Omega \times [-\tau, 0], \\ \varphi(x,t) = 0, \quad (x,t) \in \partial\Omega \times [-\tau, +\infty), \end{cases} \tag{1.2}$$

where $\varphi(x,t) = (\varphi_1(x,t), \varphi_2(x,t), \ldots, \varphi_n(x,t))^T \in \mathbb{R}^n$ is the state vector of system at time t and in space x; $x = (x_1, x_2, \ldots, x_q)^T \in \Omega \subset \mathbb{R}^q$; $\Delta = \sum_{k=1}^{q} \frac{\partial^2}{\partial x_k^2}$ is the Laplace operator on Ω; $\tau(t)$ is the time-varying delay with $0 \leqslant \tau(t) \leqslant \tau$; $D = \mathrm{diag}(d_1, d_2, \ldots, d_n) > 0$; $\phi(x,t) \in \mathbb{R}^n$ is bounded and continuous on $\Omega \times [-\tau, 0]$; the functions $f(\cdot) \in \mathbb{R}^n$ and $g(\cdot) \in \mathbb{R}^n$ satisfy $f(0) = 0$, $g(0) = 0$, and the following assumption:

The functions $f(\cdot)$ and $g(\cdot)$ are continuous, and there exist two positive constants α_1 and α_2 such that

$$|f(\xi_1) - f(\xi_2)| \leqslant \alpha_1 |\xi_1 - \xi_2|,$$
$$|g(\xi_1) - g(\xi_2)| \leqslant \alpha_2 |\xi_1 - \xi_2|$$

for any $\xi_1, \xi_2 \in \mathbb{R}^n$, where $|\cdot|$ is the Euclidean norm.

Let $\varphi(x,t,\phi)$ be the state trajectory of system (1.2) with initial condition $\phi(x,t)$. The equilibrium solution $\varphi^* = 0$ of system (1.2) is said to be asymptotically stable if for any $\epsilon > 0$, there exists $\delta(\epsilon)$ such that

$$\|\phi(\cdot, 0)\|_{\tau} < \delta(\epsilon)$$

implies

$$\|\varphi(\cdot, t, \phi)\|_2 < \epsilon$$

for $t \geqslant 0$, and there is a $b_0 > 0$ such that

$$\|\phi(\cdot, 0)\|_{\tau} < b_0$$

implies

$$\|\varphi(\cdot, t, \phi)\|_2 \to 0 \text{ as } t \to +\infty.$$

To our knowledge, there are very few works on the asymptotic stability of DPDE systems. It is well known that Lyapunov functional is a very convenient tool to analyze the asymptotic stability of DODE systems. In the following, we give a sufficient condition for the asymptotic stability of the equilibrium solution $\varphi^* = 0$ of system (1.2).

Lemma 1.23 (see [118]) *Suppose that* $v_1, v_2, v_3 : \mathbb{R}^+ \to \mathbb{R}^+$ *are continuous and strictly monotonically nondecreasing functions,* $v_1(s), v_2(s), v_3(s)$ *are positive for* $s > 0$ *with* $v_1(0) = v_2(0) = 0$. *If there is a continuous functional* $V : \mathbb{R}^+ \times C \to \mathbb{R}^+$ *such that*

$$v_1(\|\varphi(\cdot, t)\|_2) \leqslant V(t, \varphi_t(x, \theta)) \leqslant v_2(\|\varphi(\cdot, t)\|_\tau),$$
$$\dot{V}(t, \varphi_t(x, \theta)) \leqslant -v_3(\|\varphi(\cdot, t)\|_2),$$

where $\varphi_t(x, \theta) \triangleq \varphi(x, t + \theta), -\tau \leqslant \theta \leqslant 0$, $\dot{V}$ *is the derivative of* V *along the solution of system* (1.2), *then the equilibrium solution* $\varphi^* = 0$ *of system* (1.2) *is asymptotically stable.*

1.6 Assumption

Throughout this book, the function $f_j(\cdot)(j = 1, 2, \ldots, n)$ satisfies the Lipschitz condition, that is, there exists positive constant ρ_j such that

$$|f_j(\xi_1) - f_j(\xi_2)| \leqslant \rho_j|\xi_1 - \xi_2|$$

for any $\xi_1, \xi_2 \in \mathbb{R}$, where $|\cdot|$ is the Euclidean norm.

Chapter 2
Pinning Control Strategies for Synchronization of CRDNNs

2.1 Introduction

As one of the most significant and interesting dynamical properties of the complex networks, synchronization has been extensively studied by the researchers [41, 42, 45, 47, 53, 57, 63, 66, 119–121]. To our knowledge, in most existing works on the synchronization of complex networks, three kinds of coupling forms (named respectively state coupling, output coupling and derivative coupling) were considered in network models. So far, a great many important results on synchronization have been obtained for various complex networks with state coupling, see ([41, 45, 47, 53, 66, 119] and relevant references therein). In [57], Chen proposed a complex dynamical network model, in which nodes are connected by measured outputs experiencing the random sensor delay. Synchronization in the proposed network model was analyzed by the stochastic stability theory. Considering that the node state in complex networks is difficult to be observed or measured, some researchers investigated the output synchronization of complex delayed dynamical networks with output coupling [42, 63]. In [120, 121], the synchronization was studied for complex dynamical networks with non-derivative and derivative coupling.

It should be noticed that the above mentioned works are based on the network models with time-varying state variables. However, in reality, the node state is not only dependent on the time, but also intensively dependent on space variable in many circumstances. As a special class of complex networks, coupled neural networks have attracted much attention in recent years. Especially, the synchronization problem of coupled neural networks has stirred much research interest due to its fruitful applications in various fields [76–79]. It is well known that the diffusion phenomena can not be ignored in neural networks and electric circuits once electrons transport in a nonuniform electromagnetic field [28–30]. Therefore, we must consider the diffusion effects in neural networks. Obviously, in CRDNNs, the state variable of node is seriously dependent on the time and space.

More recently, researchers have investigated the synchronization problem of coupled neural networks with reaction-diffusion terms, and some interesting results

J.-L. Wang et al., *Analysis and Control of Coupled Neural Networks with Reaction-Diffusion Terms*, DOI 10.1007/978-981-10-4907-1_2

have also been established, e.g., see also [97–99, 101]. Yang et al. [101] studied the global exponential synchronization of a class of CRDNNs with time-varying delay by adding impulsive controller to a small fraction of nodes. In [99], Wang et al. respectively investigated the synchronization problem of two kinds of linearly coupled neural networks with reaction-diffusion terms using edge-based adaptive strategy. Unfortunately, most of the existing results of synchronization for CRDNNs are concerned about state coupling. But, in reaction-diffusion networks, different diffusion of node may cause different changes of other nodes [122]. For example, as is well known, different diffusion of species may cause different movements of other species in food webs [123, 124]. Therefore, it is also interesting to study the coupled neural networks with reaction-diffusion terms and spatial diffusion coupling. To our knowledge, very few researchers have investigated the synchronization of CRDNNs with spatial diffusion coupling [100]. In [100], the authors proposed a general model of an array of N linearly coupled RDNNs with spatial diffusion coupling, and respectively investigated the synchronization and $\mathcal{H}_\infty$ synchronization of the proposed network model.

Motivated by the above discussions, in this chapter, we propose two kinds of CRDNNs. In the first one, the nodes are coupled through their states. In the second one, the nodes are coupled through the spatial diffusion terms. In many circumstances, CRDNNs can not be synchronized by themselves, thus some control strategies should be adopted to achieve synchronization. Considering that it is difficult to apply control actions to all nodes in a large-scale network, some authors developed several pinning control schemes for complex networks [125–127]. For instance, Tang et al. [125] investigated the pinning distributed synchronization problem of a class of nonlinear dynamical networks with multiple stochastic disturbances using fixed pinning and switching pinning schemes. In [126], distributed robust pinning synchronization was investigated for a class of complex networks with parameter uncertainties and stochastic coupling. In these existing works [125–127], the node state is only dependent on the time. Obviously, it is also beneficial to apply the pinning control technique to study the synchronization problem of the CRDNNs. To our knowledge, very few researchers have investigated the pinning control of the CRDNNs [97, 101]. Therefore, the objective of this chapter is to design some pinning control strategies such that all nodes in the CRDNNs can synchronize onto a desired state. It is well known that the topological structure and the coupling strength are two key factors impacting the synchronization in CRDNNs. Therefore, it is natural to raise the following problem: Does the coupling form play an important role in the synchronization of the CRDNNs? This chapter also analyzes the relationship among pinning synchronization, the coupling form, the coupling strength, and the topological structure in CRDNNs.

The main contributions of this chapter are as follows. First, several sufficient conditions are established to guarantee the synchronization of the CRDNNs with state coupling by using the designed pinning controllers. Second, an effective adaptive strategy to adjust the coupling strength of the CRDNNs with state coupling is designed. Third, a sufficient condition ensuring synchronization of the CRDNNs with spatial diffusion coupling is obtained by using the designed pinning controllers, and an adaptive strategy is proposed to obtain appropriate pinning feedback gains for achieving network synchronization.

2.2 Pinning Control of CRDNNs with State Coupling

In this section, we consider a CRDNNs consisting of N identical nodes with state coupling, in which each node is an n-dimensional reaction-diffusion neural network. By using Lyapunov functional method and pinning control technique, some sufficient conditions are established to ensure that the CRDNNs is synchronized. In addition, an adaptive strategy to tune the coupling strength is proposed, and a general criterion for synchronization is obtained by using the designed adaptive law.

2.2.1 Network Model

To facilitate the readers, the CRDNNs model is presented in a step-by-step format.

A single reaction-diffusion neural network with Dirichlet boundary conditions is described by the following PDEs:

$$\frac{\partial w_i(x,t)}{\partial t} = d_i \Delta w_i(x,t) - a_i w_i(x,t) + J_i + \sum_{j=1}^{n} b_{ij} f_j(w_j(x,t)), \tag{2.1}$$

where $i = 1, 2, \ldots, n$, n is the number of neurons in the network; $x = (x_1, x_2, \ldots, x_q)^T \in \Omega \subset \mathbb{R}^q$; $w_i(x,t) \in \mathbb{R}$ is the state of the ith neuron at time t and in space x; $\Delta = \sum_{k=1}^{q} \frac{\partial^2}{\partial x_k^2}$ is the Laplace diffusion operator on Ω; $d_i > 0$ represents the transmission diffusion coefficient along the ith neuron; $a_i > 0$ represents the rate with which the ith neuron will reset its potential to the resting state when disconnected from the network and external input; b_{ij} denotes the strength of the jth neuron on the ith neuron; $f_j(\cdot)$ denotes the activation function of the jth neuron; J_i is a constant external input.

The initial value and boundary value conditions associated with system (2.1) are given in the form

$$w_i(x,0) = \phi_i(x), \quad x \in \Omega, \tag{2.2}$$

$$w_i(x,t) = 0, \quad (x,t) \in \partial\Omega \times [0, +\infty), \tag{2.3}$$

where $\phi_i(x) (i = 1, 2, \ldots, n)$ is bounded and continuous on Ω.

We can rewrite system (2.1) in a compact form as follows:

$$\frac{\partial w(x,t)}{\partial t} = D\Delta w(x,t) - Aw(x,t) + J + Bf(w(x,t)), \tag{2.4}$$

where $D = \text{diag}(d_1, d_2, \ldots, d_n)$, $B = (b_{ij})_{n\times n}$, $J = (J_1, J_2, \ldots, J_n)^T$, $A = \text{diag}(a_1, a_2, \ldots, a_n)$, $f(w(x,t)) = (f_1(w_1(x,t)), f_2(w_2(x,t)), \ldots, f_n(w_n(x,t)))^T$, $w(x,t) = (w_1(x,t), w_2(x,t), \ldots, w_n(x,t))^T$.

N mutually coupled RDNNs (2.4) can result in a CRDNNs, which is described by

$$\frac{\partial z_i(x,t)}{\partial t} = D\Delta z_i(x,t) - Az_i(x,t) + J + Bf(z_i(x,t)) + c\sum_{j=1}^{N} G_{ij}\Gamma z_j(x,t), \tag{2.5}$$

where $i = 1, 2, \ldots, N$, N is the number of nodes in the network; $z_i(x,t) = (z_{i1}(x,t), z_{i2}(x,t), \ldots, z_{in}(x,t))^T \in \mathbb{R}^n$ is the state vector of node i; c is a positive real number, which represents the overall coupling strength; $\Gamma \in \mathbb{R}^{n\times n} > 0$ is inner coupling matrix; $G = (G_{ij})_{N\times N}$ is the coupling configuration matrix representing the topological structure of the network, where G_{ij} is defined as follows: if there exists a connection from node i to node j, then $G_{ij} > 0$; otherwise, $G_{ij} = 0 (i \neq j)$; and the diagonal elements of matrix G are defined by

$$G_{ii} = -\sum_{\substack{j=1\\ j\neq i}}^{N} G_{ij}, i = 1, 2, \ldots, N.$$

In this section, we always assume that CRDNNs (2.5) is strongly connected. The initial value and boundary value conditions associated with network (2.5) are given in the form

$$z_i(x,0) = \Phi_i(x) \in \mathbb{R}^n, \quad x \in \Omega, \tag{2.6}$$
$$z_i(x,t) = 0, \quad (x,t) \in \partial\Omega \times [0, +\infty), \tag{2.7}$$

where $\Phi_i(x)$ is bounded and continuous on Ω.

Suppose $w^*(x,t) = (w_1^*(x,t), w_2^*(x,t), \ldots, w_n^*(x,t))^T$ is an arbitrary desired solution of the system (2.1), then it satisfies (2.3) and

$$\frac{\partial w^*(x,t)}{\partial t} = D\Delta w^*(x,t) - Aw^*(x,t) + J + Bf(w^*(x,t)). \tag{2.8}$$

The objective of this section is to design some pinning control strategies such that the solution of the controlled network (2.5) can achieve synchronization in the sense that

$$\lim_{t\to+\infty} \|z_i(\cdot,t) - w^*(\cdot,t)\|_2 = 0, \ i = 1, 2, \ldots, N.$$

2.2.2 Pinning Synchronization of CRDNNs

Without loss of generality, rearrange the order of all nodes and let the first $l(1 \leqslant l < N)$ nodes be selected to be pinned. Thus, the pinning controlled network can be described by

$$\begin{aligned}\frac{\partial z_i(x,t)}{\partial t} &= D\Delta z_i(x,t) - Az_i(x,t) + J + Bf(z_i(x,t)) + c\sum_{j=1}^{N} G_{ij}\Gamma z_j(x,t) \\ &\quad + u_i,\ i = 1, 2, \ldots, l, \\ \frac{\partial z_i(x,t)}{\partial t} &= D\Delta z_i(x,t) - Az_i(x,t) + J + Bf(z_i(x,t)) + c\sum_{j=1}^{N} G_{ij}\Gamma z_j(x,t), \\ &\quad i = l+1, \ldots, N, \end{aligned} \tag{2.9}$$

where

$$u_i = -ck_i\Gamma(z_i(x,t) - w^*(x,t)),\ i = 1, 2, \ldots, l \tag{2.10}$$

are n-dimensional linear feedback controllers with all the control gains $k_i > 0$. Defining $e_i(x,t) = z_i(x,t) - w^*(x,t)$, then the dynamics of the error vector $e_i(x,t)$ is governed by the following equation:

$$\begin{aligned}\frac{\partial e_i(x,t)}{\partial t} &= D\Delta e_i(x,t) - Ae_i(x,t) + Bf(z_i(x,t)) - Bf(w^*(x,t)) \\ &\quad + c\sum_{j=1}^{N} G_{ij}\Gamma e_j(x,t) - ck_i\Gamma e_i(x,t), \end{aligned} \tag{2.11}$$

where $i = 1, 2, \ldots, N$, and $k_i = 0$ for $i = l+1, l+2, \ldots, N$.

For the convenience, we denote

$$\begin{aligned} \tilde{D} &= \sum_{k=1}^{q} \frac{D}{l_k^2}, \quad \Theta = \mathrm{diag}(\rho_1^2, \rho_2^2, \ldots, \rho_n^2), \\ K &= \mathrm{diag}(k_1, k_2, \ldots, k_N), \\ \Upsilon &= -\tilde{D} - A + \frac{\Theta}{2} + \frac{BB^T}{2}. \end{aligned}$$

Theorem 2.1 *If there exists a positive definite diagonal matrix* $\Xi = \mathrm{diag}(\eta_1, \eta_2, \ldots, \eta_N) \in \mathbb{R}^{N\times N}$ *such that*

$$\Xi \otimes \Upsilon - c[(\Xi K)\otimes \Gamma] + c\left(\frac{\Xi G + G^T\Xi}{2} \otimes \Gamma\right) < 0, \tag{2.12}$$

then the pinning controlled network (2.9) *is synchronized.*

Proof Define the following Lyapunov functional for the system (2.11):

$$V_1(t) = \frac{1}{2}\sum_{i=1}^{N}\eta_i \int_{\Omega} e_i^T(x,t)e_i(x,t)dx. \tag{2.13}$$

In the following, we calculate the time derivative $\dot{V}_1(t)$ along the trajectory of system (2.11)

$$\begin{aligned}\dot{V}_1(t) &= \sum_{i=1}^{N}\eta_i \int_{\Omega} e_i^T(x,t)\frac{\partial e_i(x,t)}{\partial t}dx\\ &= \sum_{i=1}^{N}\int_{\Omega}\eta_i e_i^T(x,t)\big(D\Delta e_i(x,t) - Ae_i(x,t) + Bf(z_i(x,t))\\ &\quad + c\sum_{j=1}^{N}G_{ij}\Gamma e_j(x,t) - Bf(w^*(x,t)) - ck_i\Gamma e_i(x,t)\big)dx.\end{aligned} \tag{2.14}$$

From Green's formula and the boundary condition, we have

$$\int_{\Omega} e_{is}(x,t)\Delta e_{is}(x,t)dx = -\sum_{k=1}^{q}\int_{\Omega}\left(\frac{\partial e_{is}(x,t)}{\partial x_k}\right)^2 dx,$$

where $e_i(x,t) = (e_{i1}(x,t), e_{i2}(x,t), \ldots, e_{in}(x,t))^T$, $s = 1, 2, \ldots, n$. According to Lemma 1.22, we can obtain

$$\begin{aligned}\int_{\Omega} e_i^T(x,t)D\Delta e_i(x,t)dx &= \sum_{s=1}^{n}\int_{\Omega} d_s e_{is}(x,t)\Delta e_{is}(x,t)dx\\ &= -\sum_{k=1}^{q}\sum_{s=1}^{n}\int_{\Omega} d_s\left(\frac{\partial e_{is}(x,t)}{\partial x_k}\right)^2 dx\\ &\leqslant -\sum_{k=1}^{q}\frac{1}{l_k^2}\sum_{s=1}^{n}\int_{\Omega} d_s e_{is}^2(x,t)dx\\ &= -\int_{\Omega} e_i^T(x,t)\tilde{D}e_i(x,t)dx.\end{aligned} \tag{2.15}$$

Furthermore, we can easily derive

$$\begin{aligned}&e_i^T(x,t)B[f(z_i(x,t)) - f(w^*(x,t))]\\ &\leqslant \frac{1}{2}e_i^T(x,t)BB^Te_i(x,t) + \frac{1}{2}e_i^T(x,t)\Theta e_i(x,t).\end{aligned} \tag{2.16}$$

It follows from (2.14) to (2.16) that

$$\begin{aligned}
\dot{V}_1(t) &\leqslant \sum_{i=1}^{N} \int_{\Omega} \eta_i e_i^T(x,t)(-\tilde{D} - A + \frac{BB^T}{2} + \frac{\Theta}{2} - ck_i\Gamma)e_i(x,t)dx \\
&\quad + c\sum_{i=1}^{N}\sum_{j=1}^{N} \int_{\Omega} \eta_i G_{ij} e_i^T(x,t)\Gamma e_j(x,t)dx \\
&= \int_{\Omega} e^T(x,t)\left[\Xi \otimes \Upsilon + c\left(\frac{\Xi G + G^T\Xi}{2} \otimes \Gamma\right) - c((\Xi K)\otimes \Gamma)\right] e(x,t)dx \\
&\leqslant \gamma_1 \|e(\cdot,t)\|_2^2, \qquad (2.17)
\end{aligned}$$

where $e(x,t) = (e_1^T(x,t), e_2^T(x,t), \ldots, e_N^T(x,t))^T$, $\gamma_1 = \lambda_M(\Xi \otimes \Upsilon + c(\frac{\Xi G + G^T\Xi}{2} \otimes \Gamma) - c(\Xi K \otimes \Gamma)) < 0$. By the definition of $V_1(t)$, we have

$$\gamma_2 \|e(\cdot,t)\|_2^2 \leqslant V_1(t) \leqslant \gamma_3 \|e(\cdot,t)\|_2^2, \qquad (2.18)$$

where $\gamma_2 = \min_{i=1,2,\ldots,N}\{\frac{\eta_i}{2}\}$, $\gamma_3 = \max_{i=1,2,\ldots,N}\{\frac{\eta_i}{2}\}$. Thus, by (2.17) and (2.18), we can get

$$\dot{V}_1(t) \leqslant \frac{\gamma_1}{\gamma_3} V_1(t). \qquad (2.19)$$

Then, we can derive from (2.18) and (2.19) that

$$\|e(\cdot,t)\|_2 \leqslant \sqrt{\frac{\gamma_3}{\gamma_2}} e^{\frac{\gamma_1}{2\gamma_3}t} \|e(\cdot,0)\|_2.$$

Obviously, the pinning controlled network (2.9) is synchronized. The proof is completed.

According to Lemmas 1.11, 1.16 and 1.19, there obviously exists a positive definite diagonal matrix $\Xi = \mathrm{diag}(\eta_1, \eta_2, \ldots, \eta_N)$ such that

$$c\left(\frac{\Xi G + G^T\Xi}{2} \otimes \Gamma\right) \leqslant 0. \qquad (2.20)$$

When $\Upsilon < 0$, we can derive from (2.20) that

$$\Xi \otimes \Upsilon + c\left(\frac{\Xi G + G^T\Xi}{2} \otimes \Gamma\right) < 0.$$

From Theorem 2.1, CRDNNs (2.5) can synchronize by itself. Therefore, in this section, we always assume $\lambda_M(\Upsilon) \geqslant 0$.

By applying Theorem 2.1, we can easily obtain the following conclusion.

Corollary 2.2 *If there exists a positive definite diagonal matrix* $\Xi = \mathrm{diag}(\eta_1, \eta_2, \ldots, \eta_N) \in \mathbb{R}^{N\times N}$ *such that*

$$c > \frac{-\lambda_M(\Xi \otimes \Upsilon)}{\lambda_M\left(\frac{\Xi(G-K)+(G-K)^T\Xi}{2} \otimes \Gamma\right)}, \tag{2.21}$$

where $\Xi(G-K) + (G-K)^T\Xi < 0$, *then the pinning controlled network* (2.9) *is synchronized.*

Remark 2.3 According to Corollary 2.2, there exists a critical coupling strength c^* for given Ξ such that the pinning controlled network (2.9) will synchronize if $c > c^*$. Therefore, if

$$\Xi(G-K) + (G-K)^T\Xi < 0$$

is satisfied, then the pinning controlled network (2.9) can synchronize as long as the coupling strength c is large enough.

Remark 2.4 From Lemma 1.19, there is a positive definite diagonal matrix $\Xi = \mathrm{diag}(\eta_1, \eta_2, \ldots, \eta_N) \in \mathbb{R}^{N\times N}$ such that the sum of the entries in each row of matrix $\Xi G + G^T\Xi$ is zero. In addition, it is obvious that matrix $\Xi G + G^T\Xi$ is symmetric and irreducible. Then, by utilizing Lemma 1.16, we can easily obtain the eigenvalues of $\Xi G + G^T\Xi$ are real-valued and strictly negative except an eigenvalue 0 with multiplicity 1. Thus, we have that for any $y = (y_1, y_2, \ldots, y_N)^T \neq 0 \in \mathbb{R}^N$,

$$y^T(\Xi G + G^T\Xi)y = 0 \iff y_1 = y_2 = \cdots = y_N \neq 0.$$

Then, we can get

$$y^T[\Xi(G-K) + (G-K)^T\Xi]y < 0$$

for any $y = (y_1, y_2, \ldots, y_N)^T \neq 0 \in \mathbb{R}^N$, namely,

$$\Xi(G-K) + (G-K)^T\Xi < 0.$$

Therefore, for any given matrices G and K, we can always find the positive definite diagonal matrix Ξ satisfying

$$\Xi(G-K) + (G-K)^T\Xi < 0.$$

2.2.3 Pinning Synchronization of CRDNNs with Adaptive Coupling Strength

The coupling strength c given in (2.21) is very conservative, usually much larger than the needed value. Therefore, an adaptive strategy to tune the coupling strength c is designed in this subsection.

The pinning controlled network (2.9) with adaptive coupling strength can be described by

$$\frac{\partial z_i(x,t)}{\partial t} = D\Delta z_i(x,t) - Az_i(x,t) + J + Bf(z_i(x,t)) + c(t)\sum_{j=1}^{N} G_{ij}\Gamma z_j(x,t) + u_i,\ i = 1, 2, \ldots, l,$$

$$\frac{\partial z_i(x,t)}{\partial t} = D\Delta z_i(x,t) - Az_i(x,t) + J + Bf(z_i(x,t)) + c(t)\sum_{j=1}^{N} G_{ij}\Gamma z_j(x,t),\ i = l+1, \ldots, N, \tag{2.22}$$

where

$$u_i = -c(t)k_i\Gamma(z_i(x,t) - w^*(x,t)),\ i = 1, 2, \ldots, l \tag{2.23}$$

are n-dimensional linear feedback controllers with all the control gains $k_i > 0$. Let $e_i(x,t) = (e_{i1}(x,t), e_{i2}(x,t), \ldots, e_{in}(x,t))^T = z_i(x,t) - w^*(x,t)$. Then, the dynamics of the error vector $e_i(x,t)$ is governed by the following equation:

$$\frac{\partial e_i(x,t)}{\partial t} = D\Delta e_i(x,t) - Ae_i(x,t) + Bf(z_i(x,t)) - Bf(w^*(x,t)) + c(t)\sum_{j=1}^{N} G_{ij}\Gamma e_j(x,t) - c(t)k_i\Gamma e_i(x,t), \tag{2.24}$$

where $i = 1, 2, \ldots, N$, and $k_i = 0$ for $i = l+1, l+2, \ldots, N$.

Theorem 2.5 *If there exists a positive definite diagonal matrix* $\Xi = \mathrm{diag}(\eta_1, \eta_2, \ldots, \eta_N) \in \mathbb{R}^{N\times N}$ *such that*

$$\Xi(G-K) + (G-K)^T\Xi < 0, \tag{2.25}$$

where $K = \mathrm{diag}(k_1, k_2, \ldots, k_N)$*, then the pinning controlled network* (2.9) *is synchronized under the following adaptive law:*

$$\dot{c}(t) = \beta\sum_{i=1}^{N}\eta_i\int_{\Omega}(z_i(x,t) - w^*(x,t))^T\Gamma(z_i(x,t) - w^*(x,t))dx, \tag{2.26}$$

where $c(0) > 0$, β *is a positive real number.*

Proof Firstly, according to (2.25), there obviously exists a positive constant r_1 such that

$$\Xi(G-K)+(G-K)^T\Xi+2r_1\Xi<0.$$

Construct a Lyapunov functional for the system (2.24) as follows:

$$V_2(t)=\frac{1}{2}\sum_{i=1}^{N}\eta_i\int_{\Omega}e_i^T(x,t)e_i(x,t)dx+\frac{r_1}{2\beta}(c(t)-\tilde{c})^2, \tag{2.27}$$

where $\tilde{c}$ is a positive real number.

Calculating the time derivative of $V_2(t)$ along the trajectory of system (2.24), we can get

$$\begin{aligned}
\dot{V}_2(t)&=\sum_{i=1}^{N}\eta_i\int_{\Omega}e_i^T(x,t)\frac{\partial e_i(x,t)}{\partial t}dx+r_1(c(t)-\tilde{c})\sum_{i=1}^{N}\eta_i\int_{\Omega}e_i^T(x,t)\Gamma e_i(x,t)dx\\
&=\sum_{i=1}^{N}\int_{\Omega}\eta_i e_i^T(x,t)\Big[D\Delta e_i(x,t)-Ae_i(x,t)+Bf(z_i(x,t))\\
&\quad+c(t)\sum_{j=1}^{N}G_{ij}\Gamma e_j(x,t)-Bf(w^*(x,t))-c(t)k_i\Gamma e_i(x,t)\Big]dx\\
&\quad+r_1(c(t)-\tilde{c})\sum_{i=1}^{N}\eta_i\int_{\Omega}e_i^T(x,t)\Gamma e_i(x,t)dx\\
&\leqslant\int_{\Omega}e^T(x,t)[\Xi\otimes(-\tilde{D}-A+\frac{BB^T}{2}+\frac{\Theta}{2}-r_1\tilde{c}\Gamma)]e(x,t)dx\\
&\quad+\frac{c(t)}{2}\int_{\Omega}e^T(x,t)\Big\{\big[\Xi(G-K)+(G-K)^T\Xi\\
&\quad+2r_1\Xi\big]\otimes\Gamma\Big\}e(x,t)dx\\
&\leqslant\int_{\Omega}e^T(x,t)\big[\Xi\otimes(-\tilde{D}-A+\frac{BB^T}{2}+\frac{\Theta}{2}-r_1\tilde{c}\Gamma)\big]e(x,t)dx,
\end{aligned} \tag{2.28}$$

where $e(x,t)=(e_1^T(x,t),e_2^T(x,t),\ldots,e_N^T(x,t))^T$, $\Theta=\mathrm{diag}(\rho_1^2,\rho_2^2,\ldots,\rho_n^2)$, $\tilde{D}=\sum_{k=1}^{q}\frac{D}{l_k^2}$.

By selecting $\tilde{c}$ sufficiently large such that

$$-\tilde{D}-A+\frac{BB^T}{2}+\frac{\Theta}{2}-r_1\tilde{c}\Gamma<0,$$

one obtains

$$\dot{V}_2(t) \leqslant -\varrho \|e(\cdot, t)\|_2^2, \tag{2.29}$$

where $\varrho = \lambda_m(\Xi \otimes (\tilde{D} + A - \frac{BB^T}{2} - \frac{\Theta}{2} + r_1\tilde{c}\Gamma)) > 0$.

Obviously, $V_2(t)$ is non-increasing, and each term of $V_2(t)$ is bounded. Consequently, $c(t)$ is bounded, and $\lim_{t\to+\infty} V_2(t)$ exists and is a non-negative real number. Since $c(t)$ is monotonically increasing, one can conclude that $c(t)$ asymptotically converges to a finite positive value. Therefore, by the definition of $V_2(t)$, we can derive that

$$\lim_{t\to+\infty} \sum_{i=1}^{N} \eta_i \int_\Omega e_i^T(x,t)e_i(x,t)dx$$

exists and is a non-negative real number. In what follows, we shall prove that

$$\lim_{t\to+\infty} \sum_{i=1}^{N} \eta_i \int_\Omega e_i^T(x,t)e_i(x,t)dx = 0.$$

If this is not true, we have

$$\lim_{t\to+\infty} \sum_{i=1}^{N} \eta_i \int_\Omega e_i^T(x,t)e_i(x,t)dx = \mu > 0.$$

Then, there obviously exists a real number $M > 0$ such that

$$\sum_{i=1}^{N} \eta_i \int_\Omega e_i^T(x,t)e_i(x,t)dx > \frac{\mu}{2}$$

for $t \geqslant M$. Therefore,

$$\|e(\cdot, t)\|_2^2 > \frac{\mu}{2\delta}, \quad t \geqslant M, \tag{2.30}$$

where $\delta = \max_{i=1,2,\ldots,N}\{\eta_i\}$. From (2.29) and (2.30), we can get

$$\dot{V}_2(t) < -\frac{\varrho\mu}{2\delta}, \quad t \geqslant M. \tag{2.31}$$

By integrating (2.31) with respect to t over the time period M to $+\infty$, we can obtain

$$\begin{aligned}
-V_2(M) \leqslant V_2(+\infty) - V_2(M) &= \int_M^{+\infty} \dot{V}_2(t)dt \\
&< -\int_M^{+\infty} \frac{\varrho\mu}{2\delta}dt \\
&= -\infty.
\end{aligned}$$

This yields a contradiction, and so

$$\lim_{t\to+\infty}\sum_{i=1}^{N}\eta_i\int_{\Omega}e_i^T(x,t)e_i(x,t)dx=0.$$

Then, we can easily obtain

$$\lim_{t\to+\infty}\|e(\cdot,t)\|_2=0.$$

Therefore, the pinning controlled network (2.9) is synchronized under the designed adaptive law. The proof is completed.

Remark 2.6 To our knowledge, in many real-world networks, the coupling strength is adaptively adjusted according to changes in the environment or the network itself (for instance, wireless sensor networks, biological networks, neural networks) [128]. Therefore, it is important and interesting to study the pinning synchronization of CRDNNs with adaptive coupling strength. In Theorem 2.5, a sufficient condition is obtained to guarantee the synchronization of the pinning controlled network (2.9) by using the designed adaptive law.

Remark 2.7 Obviously, for any given matrices G and K, we can always find the positive definite diagonal matrix Ξ satisfying

$$\Xi(G-K)+(G-K)^T\Xi<0.$$

Therefore, the CRDNNs (2.5) can realize synchronization under any pinning controllers in the form of (2.10) if the coupling strength c is adjusted according to the designed adaptive law.

Remark 2.8 It follows from Corollary 2.2 that the CRDNNs (2.5) can realize synchronization by controlling only one node. But it requires a very large coupling strength c, which may not be very practical [47]. According to Theorem 2.5, the CRDNNs (2.5) is synchronized by pinning only one node if the coupling strength is adjusted according to the adaptive law (2.26). In this case, $c(t)$ may also converge to a very large positive real number. Therefore, we need to find a good balance between the number of pinned nodes and the coupling strength such that they are as small as possible and acceptable for practical use. Obviously, this is an important and interesting problem and will become our future investigative direction.

2.3 Pinning Control of CRDNNs with Spatial Diffusion Coupling

In this section, we consider a CRDNNs consisting of N identical nodes with spatial diffusion coupling, in which each node is an n-dimensional reaction-diffusion neural network. Based on the Lyapunov functional method and the pinning control technique, a sufficient condition is obtained to guarantee the synchronization of the CRDNNs. In addition, we also investigate the pinning adaptive synchronization of the CRDNNs, and a general criterion for ensuring network synchronization is established.

2.3.1 *Network Model*

In this section, we consider a CRDNNs consisting of N identical RDNNs (2.4) with spatial diffusion coupling. The mathematical model of the CRDNNs can be described as follows:

$$\begin{aligned}\frac{\partial z_i(x,t)}{\partial t} &= D\Delta z_i(x,t) - Az_i(x,t) + J + Bf(z_i(x,t)) \\ &\quad + \hat{c}\sum_{j=1}^{N}\hat{G}_{ij}\hat{\Gamma}\Delta z_j(x,t),\end{aligned} \tag{2.32}$$

where $i = 1, 2, \ldots, N$, N is the number of nodes in the network; $z_i(x,t) = (z_{i1}(x,t), z_{i2}(x,t), \ldots, z_{in}(x,t))^T \in \mathbb{R}^n$ is the state vector of node i; $\hat{c}$ is a positive real number, which represents the overall coupling strength; $\hat{\Gamma} = (\hat{\Gamma}_{ij})_{n\times n} \in \mathbb{R}^{n\times n} > 0$ is inner coupling matrix; $\hat{G} = (\hat{G}_{ij})_{N\times N}$ is the coupling configuration matrix representing the topological structure of the network, where $\hat{G}_{ij}$ is defined as follows: if there exists a connection from node i to node j, then $\hat{G}_{ij} > 0$; otherwise, $\hat{G}_{ij} = 0 (i \neq j)$; and the diagonal elements of matrix $\hat{G}$ are defined by

$$\hat{G}_{ii} = -\sum_{\substack{j=1\\ j\neq i}}^{N}\hat{G}_{ij}, i = 1, 2, \ldots, N.$$

In this section, we always assume that CRDNNs (2.32) is strongly connected. The initial value and boundary value conditions associated with network (2.32) are given in the form

$$z_i(x,0) = \hat{\Phi}_i(x) \in \mathbb{R}^n, \quad x \in \Omega, \tag{2.33}$$
$$z_i(x,t) = 0, \quad (x,t) \in \partial\Omega \times [0, +\infty), \tag{2.34}$$

where $\hat{\Phi}_i(x)$ is bounded and continuous on Ω.

The objective of this section is to design some pinning control strategies such that the solutions of the controlled network (2.32) can achieve synchronization in the sense that

$$\lim_{t \to +\infty} \|z_i(\cdot, t) - w^*(\cdot, t)\|_2 = 0, \ i = 1, 2, \ldots, N,$$

where $w^*(x, t) = (w_1^*(x, t), w_2^*(x, t), \ldots, w_n^*(x, t))^T$ is an arbitrary desired solution of the system (2.1).

2.3.2 *Pinning Synchronization of CRDNNs*

Without loss of generality, rearrange the order of all nodes and let the first $l(1 \leqslant l < N)$ nodes be selected to be pinned. Thus, the pinning controlled network can be described by

$$\begin{aligned}
\frac{\partial z_i(x,t)}{\partial t} &= D\Delta z_i(x,t) - Az_i(x,t) + J + Bf(z_i(x,t)) + \hat{c}\sum_{j=1}^{N} \hat{G}_{ij}\hat{\Gamma}\Delta z_j(x,t) \\
&\quad + u_i, \ i = 1, 2, \ldots, l, \\
\frac{\partial z_i(x,t)}{\partial t} &= D\Delta z_i(x,t) - Az_i(x,t) + J + Bf(z_i(x,t)) + \hat{c}\sum_{j=1}^{N} \hat{G}_{ij}\hat{\Gamma}\Delta z_j(x,t), \\
&\quad i = l+1, \ldots, N,
\end{aligned} \tag{2.35}$$

where

$$u_i = -\hat{c}k_i\hat{\Gamma}(z_i(x,t) - w^*(x,t)), \ i = 1, 2, \ldots, l \tag{2.36}$$

are n-dimensional linear feedback controllers with all the control gains $k_i > 0$. Defining $e_i(x,t) = z_i(x,t) - w^*(x,t)$, then the dynamics of the error vector $e_i(x,t)$ is governed by the following equation:

$$\begin{aligned}
\frac{\partial e_i(x,t)}{\partial t} &= D\Delta e_i(x,t) - Ae_i(x,t) + Bf(z_i(x,t)) - Bf(w^*(x,t)) \\
&\quad + \hat{c}\sum_{j=1}^{N} \hat{G}_{ij}\hat{\Gamma}\Delta e_j(x,t) - \hat{c}k_i\hat{\Gamma}e_i(x,t),
\end{aligned} \tag{2.37}$$

where $i = 1, 2, \ldots, N$, and $k_i = 0$ for $i = l+1, l+2, \ldots, N$.

For the convenience, we denote

$$\begin{aligned}
\tilde{D} &= \sum_{k=1}^{q} \frac{D}{l_k^2}, \ \Theta = \mathrm{diag}(\rho_1^2, \rho_2^2, \ldots, \rho_n^2), \\
K &= \mathrm{diag}(k_1, k_2, \ldots, k_N), \\
\Upsilon &= -\tilde{D} - A + \frac{\Theta}{2} + \frac{BB^T}{2}.
\end{aligned}$$

Theorem 2.9 *If there exists a positive definite diagonal matrix* $\Xi = \mathrm{diag}(\eta_1, \eta_2, \ldots, \eta_N) \in \mathbb{R}^{N\times N}$ *such that*

$$\Xi \otimes D + \frac{\hat{c}\Xi\hat{G} + \hat{c}\hat{G}^T\Xi}{2} \otimes \hat{\Gamma} \geqslant 0, \tag{2.38}$$

$$\Xi \otimes \Upsilon - (\hat{c}\Xi K) \otimes \hat{\Gamma} - \sum_{k=1}^{q} \frac{\hat{c}}{l_k^2}\left(\frac{\Xi\hat{G} + \hat{G}^T\Xi}{2} \otimes \hat{\Gamma}\right) < 0, \tag{2.39}$$

then the pinning controlled network (2.35) *is synchronized.*

Proof Take the same Lyapunov functional $V_1(t)$ as in Theorem 2.1, that is,

$$V_1(t) = \frac{1}{2}\sum_{i=1}^{N}\eta_i \int_{\Omega} e_i^T(x,t)e_i(x,t)dx.$$

In the following, we calculate the time derivative $\dot{V}_1(t)$ along the trajectory of system (2.37)

$$\begin{aligned}
\dot{V}_1(t) &= \sum_{i=1}^{N}\eta_i \int_{\Omega} e_i^T(x,t)\frac{\partial e_i(x,t)}{\partial t}dx \\
&= \sum_{i=1}^{N}\eta_i \int_{\Omega} e_i^T(x,t)\big(D\Delta e_i(x,t) - Ae_i(x,t) + Bf(z_i(x,t)) - Bf(w^*(x,t)) \\
&\quad - \hat{c}k_i\hat{\Gamma}e_i(x,t) + \hat{c}\sum_{j=1}^{N}\hat{G}_{ij}\hat{\Gamma}\Delta e_j(x,t)\big)dx \\
&\leqslant \int_{\Omega} e^T(x,t)[\Xi \otimes D + (\hat{c}\Xi\hat{G}) \otimes \hat{\Gamma}]\Delta e(x,t)dx \\
&\quad + \int_{\Omega} e^T(x,t)\left[\Xi \otimes \left(-A + \frac{\Theta}{2} + \frac{BB^T}{2}\right) - (\hat{c}\Xi K) \otimes \hat{\Gamma}\right]e(x,t)dx,
\end{aligned}$$

where $e(x,t) = (e_1^T(x,t), e_2^T(x,t), \ldots, e_N^T(x,t))^T$.

From Green's formula and the boundary condition, we then have

$$\begin{aligned}
&\int_{\Omega} e^T(x,t)(\Xi \otimes D)\Delta e(x,t)dx \\
&= \sum_{i=1}^{N}\eta_i \int_{\Omega} e_i^T(x,t)D\Delta e_i(x,t)dx \\
&= \sum_{i=1}^{N}\eta_i \sum_{l=1}^{n} d_l \int_{\Omega} e_{il}(x,t)\Delta e_{il}(x,t)dx
\end{aligned}$$

$$
\begin{aligned}
&= -\sum_{k=1}^{q}\sum_{i=1}^{N}\eta_i\sum_{l=1}^{n}d_l\int_{\Omega}\left(\frac{\partial e_{il}(x,t)}{\partial x_k}\right)^2 dx \\
&= -\sum_{k=1}^{q}\int_{\Omega}\left(\frac{\partial e(x,t)}{\partial x_k}\right)^T (\Xi\otimes D)\frac{\partial e(x,t)}{\partial x_k}dx, \\
&\int_{\Omega} e^T(x,t)[(\hat{c}\Xi\hat{G})\otimes\hat{\Gamma}]\Delta e(x,t)dx \\
&= \sum_{i=1}^{N}\sum_{j=1}^{N}\hat{c}\eta_i\hat{G}_{ij}\int_{\Omega} e_i^T(x,t)\hat{\Gamma}\Delta e_j(x,t)dx \\
&= \sum_{i=1}^{N}\sum_{j=1}^{N}\hat{c}\eta_i\hat{G}_{ij}\sum_{l=1}^{n}\sum_{s=1}^{n}\hat{\Gamma}_{ls}\int_{\Omega} e_{il}(x,t)\Delta e_{js}(x,t)dx \\
&= -\sum_{k=1}^{q}\sum_{i=1}^{N}\sum_{j=1}^{N}\hat{c}\eta_i\hat{G}_{ij}\sum_{l=1}^{n}\sum_{s=1}^{n}\hat{\Gamma}_{ls}\int_{\Omega}\frac{\partial e_{il}(x,t)}{\partial x_k}\frac{\partial e_{js}(x,t)}{\partial x_k}dx \\
&= -\sum_{k=1}^{q}\int_{\Omega}\left(\frac{\partial e(x,t)}{\partial x_k}\right)^T [(\hat{c}\Xi\hat{G})\otimes\hat{\Gamma}]\frac{\partial e(x,t)}{\partial x_k}dx, \\
&\int_{\Omega}(\Delta e(x,t))^T[(\hat{c}\hat{G}^T\Xi)\otimes\hat{\Gamma}]e(x,t)dx \\
&= \sum_{i=1}^{N}\sum_{j=1}^{N}\hat{c}\eta_j\hat{G}_{ji}\int_{\Omega}(\Delta e_i(x,t))^T\hat{\Gamma}e_j(x,t)dx \\
&= \sum_{i=1}^{N}\sum_{j=1}^{N}\hat{c}\eta_j\hat{G}_{ji}\sum_{l=1}^{n}\sum_{s=1}^{n}\hat{\Gamma}_{ls}\int_{\Omega}\Delta e_{il}(x,t)e_{js}(x,t)dx \\
&= -\sum_{k=1}^{q}\int_{\Omega}\left(\frac{\partial e(x,t)}{\partial x_k}\right)^T [(\hat{c}\hat{G}^T\Xi)\otimes\hat{\Gamma}]\frac{\partial e(x,t)}{\partial x_k}dx.
\end{aligned}
\tag{2.40}
$$

By (2.40), we can get

$$
\begin{aligned}
&\int_{\Omega} e^T(x,t)[\Xi\otimes D+(\hat{c}\Xi\hat{G})\otimes\hat{\Gamma}]\Delta e(x,t)dx \\
&= \int_{\Omega} e^T(x,t)\left(\Xi\otimes D+\frac{\hat{c}\Xi\hat{G}}{2}\otimes\hat{\Gamma}\right)\Delta e(x,t)dx \\
&+ \int_{\Omega}(\Delta e(x,t))^T\left(\frac{\hat{c}\hat{G}^T\Xi}{2}\otimes\hat{\Gamma}\right)e(x,t)dx \\
&= -\sum_{k=1}^{q}\int_{\Omega}\left(\frac{\partial e(x,t)}{\partial x_k}\right)^T\left(\Xi\otimes D+\frac{\hat{c}\Xi\hat{G}+\hat{c}\hat{G}^T\Xi}{2}\otimes\hat{\Gamma}\right)\frac{\partial e(x,t)}{\partial x_k}dx.
\end{aligned}
$$

According to (2.38), there exists a real square matrix Q such that

$$\Xi \otimes D + \frac{\hat{c}\Xi\hat{G} + \hat{c}\hat{G}^T\Xi}{2} \otimes \hat{\Gamma} = Q^T Q.$$

Then, we can easily derive

$$\left(\frac{\partial e(x,t)}{\partial x_k}\right)^T \left(\Xi \otimes D + \frac{\hat{c}\Xi\hat{G} + \hat{c}\hat{G}^T\Xi}{2} \otimes \hat{\Gamma}\right) \frac{\partial e(x,t)}{\partial x_k}$$
$$= \left(\frac{\partial (Qe(x,t))}{\partial x_k}\right)^T \frac{\partial (Qe(x,t))}{\partial x_k}.$$

Let $y(x,t) = Qe(x,t)$, for $(x,t) \in \partial\Omega \times [0,+\infty)$ from the boundary condition (2.34), we have $y(x,t) = Qe(x,t) = 0$. In view of Lemma 1.22, one has

$$\sum_{k=1}^{q} \int_{\Omega} \left(\frac{\partial y(x,t)}{\partial x_k}\right)^T \frac{\partial y(x,t)}{\partial x_k} dx$$
$$\geqslant \sum_{k=1}^{q} \frac{1}{l_k^2} \int_{\Omega} e^T(x,t) \left(\Xi \otimes D + \frac{\hat{c}\Xi\hat{G} + \hat{c}\hat{G}^T\Xi}{2} \otimes \hat{\Gamma}\right) e(x,t) dx. \quad (2.41)$$

Therefore,

$$\dot{V}_1(t) \leqslant \int_{\Omega} e^T(x,t) \left[\Xi \otimes \Upsilon - \sum_{k=1}^{q} \frac{\hat{c}}{l_k^2} \left(\frac{\Xi\hat{G} + \hat{G}^T\Xi}{2} \otimes \hat{\Gamma}\right) - (\hat{c}\Xi K) \otimes \hat{\Gamma}\right] e(x,t) dx.$$

Then, following similar arguments as in the proof of Theorem 2.1, we can obtain the desired result immediately.

Remark 2.10 By using the properties of Kronecker product, we can get

$$\Xi \otimes D + \frac{\hat{c}\Xi\hat{G} + \hat{c}\hat{G}^T\Xi}{2} \otimes \hat{\Gamma} < \left(\kappa\Xi + \frac{\hat{c}\Xi\hat{G} + \hat{c}\hat{G}^T\Xi}{2}\right) \otimes \hat{\Gamma},$$

where $\kappa = \frac{2\lambda_M(D)}{\lambda_m(\hat{\Gamma})}$. When $\hat{c} > \max_{i=1,2,\ldots,N}\left\{\frac{\kappa}{|\hat{G}_{ii}|}\right\}$, the diagonal elements of matrix $\kappa\Xi + \frac{\hat{c}\Xi\hat{G} + \hat{c}\hat{G}^T\Xi}{2}$ are strictly negative. In this case, $\kappa\Xi + \frac{\hat{c}\Xi\hat{G} + \hat{c}\hat{G}^T\Xi}{2}$ has at least one negative eigenvalue. Therefore, if

$$\hat{c} > \max_{i=1,2,\ldots,N}\left\{\frac{\kappa}{|\hat{G}_{ii}|}\right\},$$

$\Xi \otimes D + \frac{\hat{c}\Xi\hat{G} + \hat{c}\hat{G}^T\Xi}{2} \otimes \hat{\Gamma}$ always has a negative eigenvalue for any matrix Ξ.

Remark 2.11 In the past few years, some pinning control schemes for the CRDNNs with state coupling have been developed [97, 101]. Obviously, it is also beneficial to apply the pinning control technique to study the synchronization of the CRDNNs with spatial diffusion coupling. To the best of our knowledge, this is the first to consider the pinning control problem of the CRDNNs with spatial diffusion coupling, which is a very important and challenging problem.

Remark 2.12 In Sect. 2.2, we consider a CRDNNs consisting of N identical RDNNs (2.4) with state coupling, and prove that the CRDNNs (2.5) under the pinning controllers (2.10) can synchronize as long as the coupling strength c is large enough [see Corollary 2.2]. In this section, we investigate the pinning synchronization of a CRDNNs with spatial diffusion coupling, and a criterion for reaching synchronization is established by using the designed pinning controllers (2.36) [see Theorem 2.9]. If the coupling strength $\hat{c}$ is very large, it is impossible to find matrix Ξ to satisfy the condition (2.38) in Theorem 2.9 [see Remark 2.10]. In such a case, the CRDNNs (2.32) under the pinning controllers (2.36) may not be synchronized. These results show that coupling form has a strong influence on the dynamic behavior of the CRDNNs.

2.3.3 *Pinning Adaptive Synchronization of CRDNNs*

Obviously, it is desirable to make the pinning feedback gains $k_i (i = 1, 2, \ldots, l)$ as small as possible. Therefore, an effective adaptive strategy to tune the feedback gains k_i is designed in this subsection.

The pinning controlled network (2.35) with adaptive feedback gains can be described by

$$
\begin{aligned}
\frac{\partial z_i(x,t)}{\partial t} &= D\Delta z_i(x,t) - Az_i(x,t) + J + Bf(z_i(x,t)) + \hat{c}\sum_{j=1}^{N} \hat{G}_{ij}\hat{\Gamma}\Delta z_j(x,t) \\
&\quad - \hat{c}k_i(t)\hat{\Gamma}(z_i(x,t) - w^*(x,t)), \ i = 1, 2, \ldots, l, \\
\frac{\partial z_i(x,t)}{\partial t} &= D\Delta z_i(x,t) - Az_i(x,t) + J + Bf(z_i(x,t)) + \hat{c}\sum_{j=1}^{N} \hat{G}_{ij}\hat{\Gamma}\Delta z_j(x,t), \\
&\quad i = l+1, \ldots, N, \qquad (2.42) \\
\dot{k}_i(t) &= \hat{\beta}_i \int_{\Omega} (z_i(x,t) - w^*(x,t))^T \hat{\Gamma}(z_i(x,t) - w^*(x,t))dx, \\
&\quad i = 1, 2, \ldots, l, \qquad (2.43)
\end{aligned}
$$

where $\hat{\beta}_i$ and $k_i(0)$ are positive constants. Let $e_i(x,t) = z_i(x,t) - w^*(x,t)$. Then, the dynamics of the error vector $e_i(x,t)$ is governed by the following equation:

$$\frac{\partial e_i(x,t)}{\partial t} = D\Delta e_i(x,t) - Ae_i(x,t) + Bf(z_i(x,t)) - Bf(w^*(x,t))$$
$$+ \hat{c}\sum_{j=1}^{N} \hat{G}_{ij}\hat{\Gamma}\Delta e_j(x,t) - \hat{c}k_i(t)\hat{\Gamma}e_i(x,t),$$
$$\dot{k}_i(t) = \hat{\beta}_i \int_{\Omega} e_i^T(x,t)\hat{\Gamma}e_i(x,t)dx, i = 1,\ldots,l, \tag{2.44}$$

where $i = 1, 2, \ldots, N$, and $k_i(t) \equiv 0$ for $i = l+1, l+2, \ldots, N$.

Theorem 2.13 *If there exist matrices* $\Xi = \text{diag}(\eta_1, \eta_2, \ldots, \eta_N) \in \mathbb{R}^{N\times N} > 0$ *and* $\hat{K} = \text{diag}(\underbrace{\hat{k}_1, \hat{k}_2, \ldots, \hat{k}_l}_{l}, \underbrace{0, \ldots, 0}_{N-l}) \in \mathbb{R}^{N\times N}$ *such that*

$$\Xi \otimes D + \frac{\hat{c}\Xi\hat{G} + \hat{c}\hat{G}^T\Xi}{2} \otimes \hat{\Gamma} \geqslant 0, \tag{2.45}$$
$$\Xi \otimes \Upsilon - (\hat{c}\Xi\hat{K}) \otimes \hat{\Gamma} - \sum_{k=1}^{q} \frac{\hat{c}}{l_k^2}\left(\frac{\Xi\hat{G} + \hat{G}^T\Xi}{2} \otimes \hat{\Gamma}\right) < 0, \tag{2.46}$$

where $\hat{k}_i > 0, i = 1, 2, \ldots, l$, $\tilde{D} = \sum_{k=1}^{q} \frac{D}{l_k^2}$, $\Theta = \text{diag}(\rho_1^2, \rho_2^2, \ldots, \rho_n^2)$, $\Upsilon = -\tilde{D} - A + \frac{\Theta}{2} + \frac{BB^T}{2}$, *then the controlled network* (2.42) *is synchronized under the adaptive law* (2.43).

Proof Construct a Lyapunov functional for the system (2.44) as follows:

$$V_3(t) = \frac{1}{2}\sum_{i=1}^{N} \eta_i \int_{\Omega} e_i^T(x,t)e_i(x,t)dx + \sum_{i=1}^{l} \frac{\hat{c}\eta_i}{2\hat{\beta}_i}(k_i(t) - \hat{k}_i)^2. \tag{2.47}$$

Calculating the time derivative of $V_3(t)$ along the trajectory of system (2.44), we can get

$$\dot{V}_3(t) = \sum_{i=1}^{N} \eta_i \int_{\Omega} e_i^T(x,t)\frac{\partial e_i(x,t)}{\partial t}dx$$
$$+ \sum_{i=1}^{l} \hat{c}\eta_i(k_i(t) - \hat{k}_i) \int_{\Omega} e_i^T(x,t)\hat{\Gamma}e_i(x,t)dx$$
$$= \sum_{i=1}^{N} \eta_i \int_{\Omega} e_i^T(x,t)\Big[D\Delta e_i(x,t) - Ae_i(x,t) - \hat{c}k_i(t)\hat{\Gamma}e_i(x,t)$$
$$+ Bf(z_i(x,t)) - Bf(w^*(x,t)) + \hat{c}\sum_{j=1}^{N} \hat{G}_{ij}\hat{\Gamma}\Delta e_j(x,t)\Big]dx$$

$$
\begin{aligned}
&+\sum_{i=1}^{l}\hat{c}\eta_i(k_i(t)-\hat{k}_i)\int_{\Omega}e_i^T(x,t)\hat{\Gamma}e_i(x,t)dx\\
&\leqslant\int_{\Omega}e^T(x,t)[\Xi\otimes D+(\hat{c}\Xi\hat{G})\otimes\hat{\Gamma}]\Delta e(x,t)dx\\
&+\int_{\Omega}e^T(x,t)\left[\Xi\otimes\left(-A+\frac{BB^T}{2}+\frac{\Theta}{2}\right)-(\hat{c}\Xi\hat{K})\otimes\hat{\Gamma}\right]e(x,t)dx\\
&\leqslant\int_{\Omega}e^T(x,t)\Bigg[\Xi\otimes\left(-\tilde{D}-A+\frac{BB^T}{2}+\frac{\Theta}{2}\right)-(\hat{c}\Xi\hat{K})\otimes\hat{\Gamma}\\
&-\sum_{k=1}^{q}\frac{\hat{c}}{l_k^2}\left(\frac{\Xi\hat{G}+\hat{G}^T\Xi}{2}\otimes\hat{\Gamma}\right)\Bigg]e(x,t)dx\\
&\leqslant\hat{\varrho}\|e(\cdot,t)\|_2^2,
\end{aligned}
\tag{2.48}
$$

where $e(x,t)=(e_1^T(x,t),e_2^T(x,t),\ldots,e_N^T(x,t))^T$, $\hat{\varrho}=\lambda_M(\Xi\otimes\Upsilon-(\hat{c}\Xi\hat{K})\otimes\hat{\Gamma}-\sum_{k=1}^{q}\frac{\hat{c}}{l_k^2}(\frac{\Xi\hat{G}+\hat{G}^T\Xi}{2}\otimes\hat{\Gamma}))<0$.

Obviously, $V_3(t)$ is non-increasing, and each term of $V_3(t)$ is bounded. Therefore, $k_i(t), i=1,2,\ldots,l$, are bounded, and $\lim_{t\to+\infty}V_3(t)$ exists and is a non-negative real number. Because $k_i(t)$ is monotonically increasing [see (2.43)], one can conclude that $k_i(t)(i=1,2,\ldots,l)$ asymptotically converges to a finite positive value. Therefore, by the definition of $V_3(t)$, we can derive that

$$
\lim_{t\to+\infty}\sum_{i=1}^{N}\eta_i\int_{\Omega}e_i^T(x,t)e_i(x,t)dx
$$

exists and is a non-negative real number. Then, by the similar proof of Theorem 2.5, we can obtain

$$
\lim_{t\to+\infty}\|e(\cdot,t)\|_2=0.
$$

Therefore, the controlled network (2.42) is synchronized under the adaptive law (2.43). The proof is completed.

Remark 2.14 To our knowledge, very few researchers have discussed the adaptive synchronization of the CRDNNs [98–100]. In [98], adaptive method was applied to design controller feedback gains, and some sufficient conditions for adaptive synchronization were obtained. It should be noticed that a network model with state coupling was considered in [98]. In [99, 100], the authors investigated the synchronization problem of the CRDNNs using edge-based adaptive strategy, and some adaptive strategies to tune all (or a small fraction of) the coupling weights were designed. In this subsection, we study the synchronization of the CRDNNs with spatial diffusion coupling by pinning a small fraction of nodes with adaptive feedback controllers, and a criterion is obtained to guarantee the synchronization of the network (2.32).

Remark 2.15 In this chapter, some adaptive strategies to tune the coupling strength c and pinning feedback gains k_i are designed [see (2.26) and (2.43)]. Adaptive technique is a very effective method to tune the coupling strength and pinning feedback gains. Firstly, coupling strength and pinning feedback gains can be adjusted very quickly by utilizing the adaptive laws defined by the synchronization errors. Another important advantage of adaptive technique is to adjust coupling strength and pinning feedback gains for achieving suitable values in practice [129].

2.4 Numerical Examples

As an application of the above obtained theoretical results, two representative examples are given in this section.

Example 2.16 Consider the following 3-dimensional reaction-diffusion neural network

$$\begin{aligned}\frac{\partial w_i(x,t)}{\partial t} &= d_i\frac{\partial^2 w_i(x,t)}{\partial x^2} - a_i w_i(x,t) + J_i + \sum_{j=1}^{3} b_{ij} f_j(w_j(x,t)),\\ w_i(x,t) &= 0,\ (x,t)\in\partial\Omega\times[0,+\infty),\end{aligned} \tag{2.49}$$

where $i = 1, 2, 3$, $\Omega = \{x \mid -0.5 < x < 0.5\}$, $f_j(\xi) = \frac{|\xi+1|-|\xi-1|}{2}$, $d_1 = 0.2$, $d_2 = 0.2$, $d_3 = 0.3$, $a_1 = 0.3$, $a_2 = 0.2$, $a_3 = 0.3$, $J_1 = J_2 = J_3 = 0$, and the matrix $B = (b_{ij})_{3\times 3}$ is chosen as

$$B = \begin{pmatrix} 2 & -0.3 & -0.2 \\ -2.5 & 3 & -0.6 \\ -3 & -2 & 4 \end{pmatrix}.$$

Obviously, $(0, 0, 0)^T \in \mathbb{R}^3$ is an equilibrium solution of the network (2.49), and $f_j(\cdot)(j = 1, 2, 3)$ satisfies the Lipschitz condition with $\rho_j = 1$.

Now we consider a CRDNNs consisting of five linearly coupled identical model (2.49) with state coupling. The state equations of the entire network are

$$\begin{aligned}\frac{\partial z_i(x,t)}{\partial t} &= D\frac{\partial^2 z_i(x,t)}{\partial x^2} - Az_i(x,t) + Bf(z_i(x,t)) + c\sum_{j=1}^{5} G_{ij}\Gamma z_j(x,t),\\ &\quad i = 1, 2, \ldots, 5,\end{aligned} \tag{2.50}$$

where

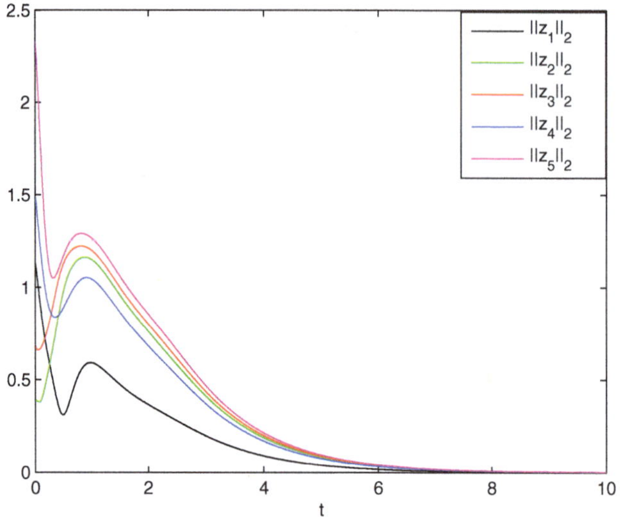

Fig. 2.1 The change processes of $\|z_i(\cdot,t)\|_2$, $i=1,2,\ldots,5$($\|z_1(\cdot,0)\|_2 = 1.1336$, $\|z_2(\cdot,0)\|_2 = 0.3873$, $\|z_3(\cdot,0)\|_2 = 0.6819$, $\|z_4(\cdot,0)\|_2 = 1.4933$, $\|z_5(\cdot,0)\|_2 = 2.3335$)

$$\Gamma = \mathrm{diag}(0.6, 0.8, 0.5),$$
$$G = \begin{pmatrix} -0.2 & 0 & 0.2 & 0 & 0 \\ 0 & -0.6 & 0 & 0.6 & 0 \\ 0 & 0 & -0.4 & 0.4 & 0 \\ 0.3 & 0 & 0 & -0.6 & 0.3 \\ 0 & 0.5 & 0 & 0 & -0.5 \end{pmatrix}.$$

The control objective here is to design appropriate control strategies such that all nodes in the CRDNNs (2.50) can synchronize onto $(0,0,0)^T \in \mathbb{R}^3$.

We choose the node 1 as pinned node. Select the parameters as follows: $k_1 = 0.3$, $\eta_1 = 0.3$, $\eta_2 = 0.1$, $\eta_3 = 0.15$, $\eta_4 = 0.2$, $\eta_5 = 0.12$. It is easy to verify that the condition (2.25) in Theorem 2.5 is satisfied. According to Theorem 2.5, the CRDNNs (2.50) under pinning control and adaptive law is synchronized. The simulation results are shown in Figs. 2.1 and 2.2.

Example 2.17 Consider the following 3-dimensional reaction-diffusion neural network

$$\begin{aligned} \frac{\partial w_i(x,t)}{\partial t} &= d_i \frac{\partial^2 w_i(x,t)}{\partial x^2} - a_i w_i(x,t) + J_i + \sum_{j=1}^{3} b_{ij} f_j(w_j(x,t)), \\ w_i(x,t) &= 0, \ (x,t) \in \partial\Omega \times [0,+\infty), \end{aligned} \tag{2.51}$$

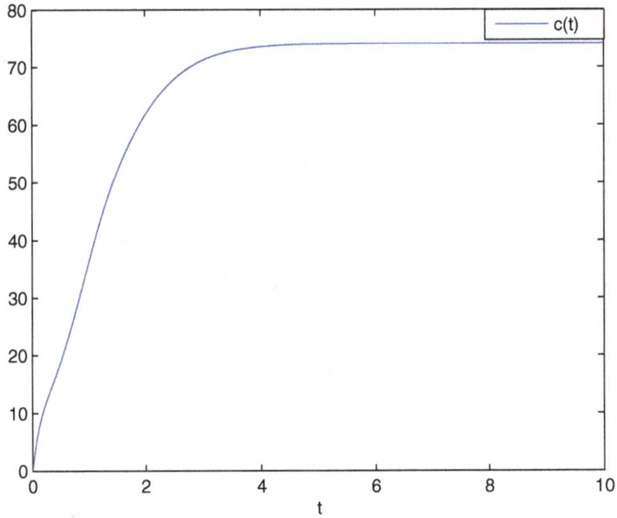

Fig. 2.2 Adaptive coupling strength ($c(0) = 0.1$)

where $i = 1, 2, 3$, $\Omega = \{x \mid -0.5 < x < 0.5\}$, $f_j(\xi) = \frac{|\xi+1|-|\xi-1|}{2}$, $d_1 = 0.6$, $d_2 = 0.7$, $d_3 = 0.5$, $a_1 = 0.6$, $a_2 = 0.8$, $a_3 = 0.4$, $J_1 = J_2 = J_3 = 0$, and the matrix $B = (b_{ij})_{3\times 3}$ is chosen as

$$B = \begin{pmatrix} 0.7 & 0.3 & 0.4 \\ 0.6 & 0.4 & 0.2 \\ 0.5 & 0.6 & 0.3 \end{pmatrix}.$$

Obviously, $(0, 0, 0)^T \in \mathbb{R}^3$ is an equilibrium solution of the network (2.51), and $f_j(\cdot)(j = 1, 2, 3)$ satisfies the Lipschitz condition with $\rho_j = 1$.

Now we consider a CRDNNs consisting of five linearly coupled identical model (2.51) with spatial diffusion coupling. The state equations of the entire network are

$$\frac{\partial z_i(x,t)}{\partial t} = D\frac{\partial^2 z_i(x,t)}{\partial x^2} - Az_i(x,t) + Bf(z_i(x,t)) + \hat{c}\sum_{j=1}^{5}\hat{G}_{ij}\hat{\Gamma}\frac{\partial^2 z_j(x,t)}{\partial x^2},$$
$$i = 1, 2, \ldots, 5, \tag{2.52}$$

where

$$\hat{\Gamma} = \mathrm{diag}(0.5, 0.7, 0.6),\ \hat{c} = 0.5,$$

$$\hat{G} = \begin{pmatrix} -0.1 & 0 & 0 & 0.1 & 0 \\ 0.2 & -0.2 & 0 & 0 & 0 \\ 0 & 0.3 & -0.5 & 0 & 0.2 \\ 0 & 0 & 0.4 & -0.4 & 0 \\ 0.1 & 0 & 0 & 0 & -0.1 \end{pmatrix}.$$

The control objective here is to design appropriate pinning adaptive controllers such that all nodes in the CRDNNs (2.52) can synchronize onto $(0, 0, 0)^T \in \mathbb{R}^3$.

We can find the following matrices

$$\hat{K} = \mathrm{diag}(0.6, 0.7, 0, 0, 0.6),$$
$$\Xi = \mathrm{diag}(0.4, 0.3, 0.5, 0.6, 0.4)$$

satisfying (2.45) and (2.46). That is, we only need to control the nodes 1, 2 and 5 for realizing network synchronization. From Theorem 2.13, the CRDNNs (2.52) can realize synchronization by using the designed pinning adaptive controllers. The simulation results are shown in Figs. 2.3 and 2.4.

According to the change processes of $\|z_i(\cdot, t)\|_2, i = 1, 2, \ldots, 5$, in Figs. 2.1 and 2.3, we clearly see that the CRDNNs is synchronized. Figures 2.2 and 2.4 visualize the change processes of $c(t)$ and $k_i(t)(i = 1, 2, 5)$ in time interval [0, 10]. The numerical results clearly show that $c(t)$ and $k_i(t)(i = 1, 2, 5)$ asymptotically converge to some positive real numbers.

Remark 2.18 Finite-difference method [130], as an effective numerical method, in the past ten years, has been widely used to simulate the reaction-diffusion systems. In this section, we plot the curves of the synchronization errors, adaptive coupling strength and adaptive feedback gains by employing the finite difference method. For instance, by utilizing the finite-difference method, $\frac{\partial^2 z_i(x,t)}{\partial x^2}$ and $\frac{\partial z_i(x,t)}{\partial t}$ can be approximated by

$$\frac{\partial^2 z_i(x_k, t_j)}{\partial x^2} \approx \frac{z_i(x_k, t_j) - 2z_i(x_{k-1}, t_j) + z_i(x_{k-2}, t_j)}{x_{\mathrm{sample}}^2},$$
$$\frac{\partial z_i(x_k, t_j)}{\partial t} \approx \frac{z_i(x_k, t_j) - z_i(x_k, t_{j-1})}{t_{\mathrm{sample}}},$$

where $x_0 = -0.5, x_{N_x} = 0.5, t_0 = 0, t_{N_t} = 10, k \in \mathcal{N}_x, j \in \mathcal{N}_t, x_k - x_{k-1} = x_{\mathrm{sample}} = \frac{x_{N_x} - x_0}{N_x}, \mathcal{N}_x = \{0, 1, \ldots, N_x\}, t_j - t_{j-1} = t_{\mathrm{sample}} = \frac{t_{N_t} - t_0}{N_t}, \mathcal{N}_t = \{0, 1, \ldots, N_t\}$. Similarly, $\dot{c}(t)$ and $\dot{k}_i(t)$ can be discretized.

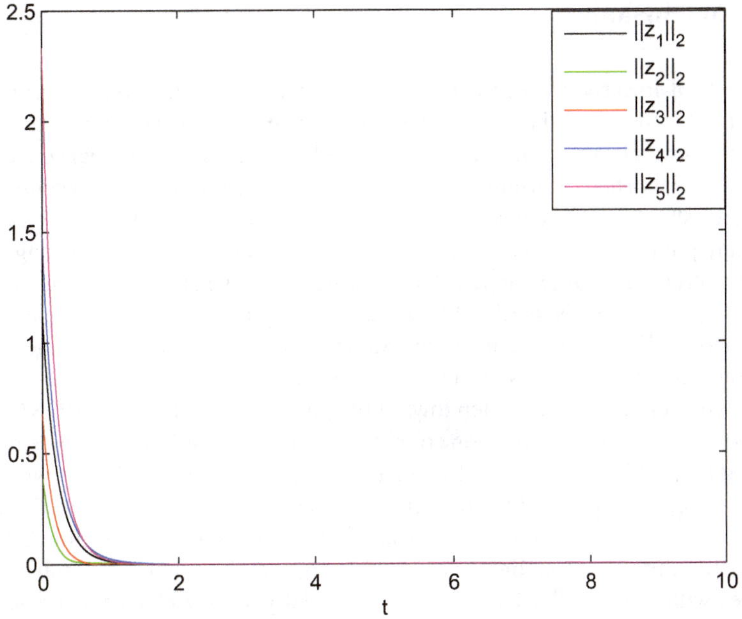

Fig. 2.3 The change processes of $\|z_i(\cdot,t)\|_2$, $i = 1, 2, \ldots, 5$($\|z_1(\cdot,0)\|_2 = 1.1336$, $\|z_2(\cdot,0)\|_2 = 0.3873$, $\|z_3(\cdot,0)\|_2 = 0.6819$, $\|z_4(\cdot,0)\|_2 = 1.4933$, $\|z_5(\cdot,0)\|_2 = 2.3335$)

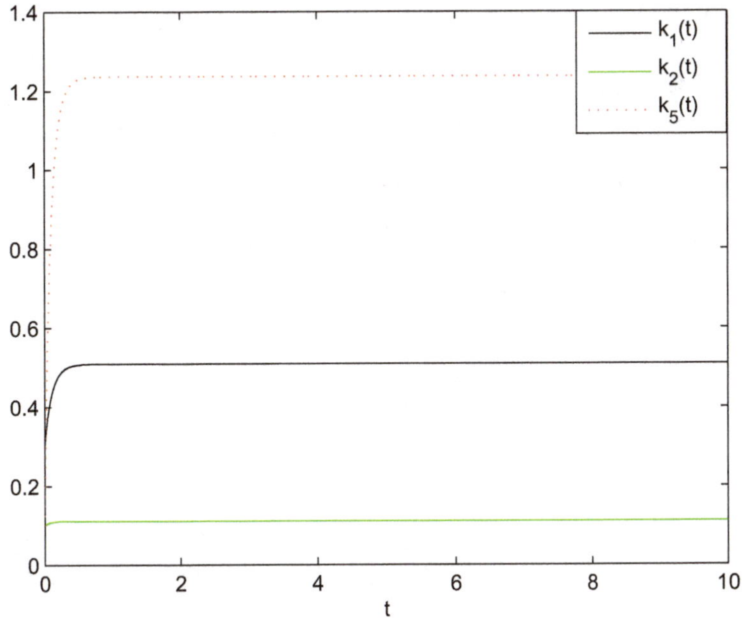

Fig. 2.4 Adaptive feedback gains ($k_1(0) = 0.3$, $k_2(0) = 0.1$, $k_5(0) = 0.1$)

2.5 Conclusions

Two kinds of linearly coupled neural networks with reaction-diffusion terms have been introduced, which have different coupling forms. We have investigated the pinning synchronization of the proposed network models. Some sufficient conditions have been established to ensure that the proposed network models are synchronized. It has been shown that, if the network is strongly connected, then the CRDNNs (2.5) under any pinning controllers in the form of (2.10) can synchronize as long as the coupling strength is large enough. On the other hand, when the coupling strength is very large, the CRDNNs (2.32) under the pinning controllers (2.36) may not be synchronized. Finally, two numerical examples have been provided to verify the correctness and effectiveness of the obtained results.

This chapter is only a first step toward the pinning control of the CRDNNs with state coupling and spatial diffusion coupling, and there are still some interesting and challenging problems deserving further investigation. For example, (1) what kind of nodes should be pinned? (2) how many nodes are needed to be pinned? In this chapter, some sufficient conditions ensuring the synchronization of the CRDNNs with spatial diffusion coupling have been established by pinning a small fraction of nodes with state feedback controllers. Practically, it may be more interesting to consider the case that the external control u_i be defined also by spatial diffusion coupling.

Chapter 3
Pinning Control for Synchronization of CRDNNs with Directed Topologies

3.1 Introduction

In many circumstances, complex dynamical networks can not be synchronized by themselves, thus some control strategies should be adopted to achieve synchronization. Considering that it is difficult to apply control actions to all nodes in a large-scale network, some authors developed several pinning control schemes for the traditional complex networks without reaction-diffusion terms [47, 115, 131–133]. To our knowledge, there are few studies on the pinning synchronization of coupled neural networks with reaction-diffusion terms [97, 101]. In [97], Liu investigated the μ-synchronization for a class of linearly coupled RDNNs by using the pinning control method. In [101], globally exponential synchronization in an array of CRDNNs was studied via pinning-impulsive control strategy. Therefore, it is interesting to further investigate the pinning synchronization of arrays of CRDNNs.

Adaptive control, as an important control technique, in the past ten years, has been widely used to synchronize complex dynamical networks. On the one hand, adaptive control has been developed as an effective method to design controller feedback gains [67, 68, 120, 134–137]. On the other hand, adaptive control method can also be utilized to tune the strengths of the interconnections among network nodes [41, 138–140]. In these existing works [41, 67, 68, 120, 134–140], the node state is only dependent on the time. Obviously, it is also beneficial to apply the adaptive strategy [141–145] to study the synchronization problem of CRDNNs. Therefore, in this chapter, some adaptive strategies are proposed to obtain appropriate coupling strength and feedback gains for achieving network synchronization.

Motivated by the above discussions, the objective of this chapter is to investigate the pinning synchronization of coupled neural networks with reaction-diffusion terms. The main contributions of this chapter are as follows. First, we give a sufficient condition for the asymptotic stability of error system which generalizes the second method of Lyapunov for ordinary differential equation systems. Second, we study the synchronization control problem of a class of CRDNNs by pinning a small

J.-L. Wang et al., *Analysis and Control of Coupled Neural Networks with Reaction-Diffusion Terms*, DOI 10.1007/978-981-10-4907-1_3

fraction of vertices with negative feedback controllers. With the help of Lyapunov functional and some inequality techniques, several network synchronization criteria are established. Third, the theoretical coupling strength required for synchronization may be much larger than the needed value, so an adaptive strategy to tune the coupling strength is designed in this chapter. Fourth, we propose a pinning adaptive control scheme to make the feedback gains as small as possible, and a sufficient condition is given for reaching synchronization by using the designed pinning adaptive controllers.

3.2 Pinning Synchronization of a CRDNNs with Directed Topologies

3.2.1 Network Model

The CRDNNs to be studied in this chapter is described by

$$\begin{aligned}\frac{\partial z_i(x,t)}{\partial t} &= D\Delta z_i(x,t) - Az_i(x,t) + J + Bf(z_i(x,t)) \\ &\quad + c\sum_{j=1}^{N} G_{ij}\Gamma z_j(x,t),\end{aligned} \tag{3.1}$$

where $i = 1, 2, \ldots, N$, N is the number of neural networks in the network; $x = (x_1, x_2, \ldots, x_q)^T \in \Omega \subset \mathbb{R}^q$; $\Delta = \sum_{k=1}^{q} \frac{\partial^2}{\partial x_k^2}$ is the Laplace diffusion operator on Ω; $z_i(x,t) = (z_{i1}(x,t), z_{i2}(x,t), \ldots, z_{in}(x,t))^T \in \mathbb{R}^n$ is the state vector of the ith neural network at time t and in space x; $D = \mathrm{diag}(d_1, d_2, \ldots, d_n) > 0$, and d_i is the transmission diffusion coefficient along the ith neuron; $A = \mathrm{diag}(a_1, a_2, \ldots, a_n) > 0$, and a_i is the rate with which the ith neuron will reset its potential to the resting state when disconnected from the neural network and external input; $J = (J_1, J_2, \ldots, J_n)^T$ is a constant external input vector; $B = (b_{ij})_{n\times n} \in \mathbb{R}^{n\times n}$, and b_{ij} is the strength of the jth neuron on the ith neuron; $f(z_i(x,t)) = (f_1(z_{i1}(x,t)), f_2(z_{i2}(x,t)), \ldots, f_n(z_{in}(x,t)))^T$, and $f_j(\cdot)$ is the activation function of the jth neuron; c is a positive real number, which represents the overall coupling strength; $0 < \Gamma \in \mathbb{R}^{n\times n}$ is inner coupling matrix; $G = (G_{ij})_{N\times N}$ is the coupling configuration matrix representing the coupling strength and topological structure of the network, where G_{ij} is defined as follows: if there exists a connection from neural network i to neural network j, then $G_{ij} > 0$; otherwise, $G_{ij} = 0 (i \neq j)$; and the diagonal elements of matrix G are defined by $G_{ii} = -\sum_{\substack{j=1\\ j\neq i}}^{N} G_{ij}, i = 1, 2, \ldots, N$.

The initial value and boundary value conditions associated with network (3.1) are given in the form

$$z_i(x,0) = \Phi_i(x) \in \mathbb{R}^n, x \in \Omega, \tag{3.2}$$

$$z_i(x,t) = 0, (x,t) \in \partial\Omega \times [0,+\infty), \tag{3.3}$$

where $\Phi_i(x)$ is bounded and continuous on Ω.

Remark 3.1 In this chapter, we do not require that the CRDNNs (3.1) is connected. In other words, the coupling configuration matrix G may be reducible. In addition, the matrix G is not restricted to be symmetric in network model (3.1).

Suppose that $w^*(x) = (w_1^*(x), w_2^*(x), \ldots, w_n^*(x))^T$ is an equilibrium solution of the following system

$$\frac{\partial w(x,t)}{\partial t} = D\Delta w(x,t) - Aw(x,t) + J + Bf(w(x,t)), \tag{3.4}$$

$$w(x,t) = 0, \qquad (x,t) \in \partial\Omega \times [0,+\infty). \tag{3.5}$$

Then, $w^*(x)$ satisfies (3.5) and

$$0 = D\Delta w^*(x) - Aw^*(x) + J + Bf(w^*(x)).$$

Definition 3.2 The CRDNNs (3.1) is said to achieve synchronization if

$$\lim_{t\to+\infty} \|z_i(\cdot,t) - w^*(\cdot)\|_2 = 0, \text{ for all } i = 1,2,\ldots,N.$$

In order to realize synchronization, pinning strategy is applied to a small fraction of the vertices in CRDNNs (3.1). Suppose that the vertices $q_1, q_2, \ldots, q_l (1 \leqslant l < N)$ are selected to be pinned. Then, the pinning controlled network can be described by

$$\begin{aligned}
\frac{\partial z_{q_i}(x,t)}{\partial t} &= D\Delta z_{q_i}(x,t) - Az_{q_i}(x,t) + Bf(z_{q_i}(x,t)) + c\sum_{j=1}^{N} G_{q_i j}\Gamma z_j(x,t) \\
&\quad + J + u_{q_i}(x,t),\ i = 1,2,\ldots,l \\
\frac{\partial z_{q_i}(x,t)}{\partial t} &= D\Delta z_{q_i}(x,t) - Az_{q_i}(x,t) + Bf(z_{q_i}(x,t)) + c\sum_{j=1}^{N} G_{q_i j}\Gamma z_j(x,t) \\
&\quad + J,\ i = l+1, l+2, \ldots, N,
\end{aligned} \tag{3.6}$$

where

$$u_{q_i}(x,t) = -ck_i\Gamma(z_{q_i}(x,t) - w^*(x)),\ i = 1,2,\ldots,l \tag{3.7}$$

are n-dimensional linear feedback controllers with all the control gains $k_i > 0$.

Remark 3.3 Recently, some researchers have investigated the synchronization problem of complex dynamical networks by utilizing the pinning control method [47, 131]. However, in most existing works on the pinning synchronization of complex dynamical networks, they always assume that the first $l(1 \leqslant l < N)$ vertices are selected to be controlled. In this chapter, we remove this restriction and choose the vertices $q_1, q_2, \ldots, q_l$ as pinned vertices. In this case, the controlled CRDNNs can be described by (3.6).

By applying controllers (3.7) into system (3.6), the resulting closed-loop dynamic system is of the form:

$$\begin{aligned}\frac{\partial z_i(x,t)}{\partial t} &= D\Delta z_i(x,t) - Az_i(x,t) + J + Bf(z_i(x,t)) + c\sum_{j=1}^{N} G_{ij}\Gamma z_j(x,t)\\ &\quad - c\hat{k}_i\Gamma(z_i(x,t) - w^*(x)),\ i = 1, 2, \ldots, N, \qquad (3.8)\end{aligned}$$

where

$$\begin{aligned}K &= \mathrm{diag}(\hat{k}_1, \hat{k}_2, \ldots, \hat{k}_N)\\ &= \mathrm{diag}(0, \ldots, 0, \underbrace{k_1}_{q_1}, 0, \ldots, 0, \underbrace{k_2}_{q_2}, 0, \ldots, 0, \underbrace{k_l}_{q_l}, \cdots).\end{aligned}$$

Let $e_i(x,t) = (e_{i1}(x,t), e_{i2}(x,t), \ldots, e_{in}(x,t))^T = z_i(x,t) - w^*(x)$. Then, we have

$$\begin{aligned}\frac{\partial e_i(x,t)}{\partial t} &= D\Delta e_i(x,t) - Ae_i(x,t) + Bf(z_i(x,t)) - Bf(w^*(x))\\ &\quad + c\sum_{j=1}^{N} \hat{G}_{ij}\Gamma e_j(x,t), \qquad (3.9)\end{aligned}$$

where $i = 1, 2, \ldots, N$, $\hat{G} = (\hat{G}_{ij})_{N\times N} = G - K$.

For the convenience, we denote

$$\tilde{D} = \sum_{k=1}^{q} \frac{D}{l_k^2}, \quad \Theta = \mathrm{diag}(\rho_1^2, \rho_2^2, \ldots, \rho_n^2),$$

$$\xi = -\frac{\lambda_m(\tilde{D})}{\lambda_M(\Gamma)} - \frac{\lambda_m(A)}{\lambda_M(\Gamma)} + \frac{\lambda_M(\frac{\Theta}{2})}{\lambda_m(\Gamma)} + \frac{\lambda_M(\frac{BB^T}{2})}{\lambda_m(\Gamma)}.$$

3.2.2 Pinning Synchronization Criteria

Obviously, the pinning controlled network (3.6) can achieve synchronization if the system (3.9) is asymptotically stable.

Next, we give a rigorous mathematical definition for asymptotic stability of the system (3.9).

Definition 3.4 Let $e(x,t,\phi) = (e_1^T(x,t,\phi), e_2^T(x,t,\phi), \ldots, e_N^T(x,t,\phi))^T$ be the state trajectory of system (3.9) with initial condition $\phi(x) \in \mathbb{R}^{Nn}$. The system (3.9) is said to be asymptotically stable if for any $\epsilon > 0$, there exists $\delta(\epsilon)$ such that

$$\|\phi(\cdot)\|_2 < \delta(\epsilon)$$

implies

$$\|e(\cdot,t,\phi)\|_2 < \epsilon \text{ for } t \geqslant 0,$$

and there is a $b_0 > 0$ such that

$$\|\phi(\cdot)\|_2 < b_0$$

implies

$$\|e(\cdot,t,\phi)\|_2 \to 0 \text{ as } t \to +\infty.$$

In the following, a new sufficient condition is presented for the asymptotic stability of the system (3.9).

Theorem 3.5 *Suppose that $\upsilon_1, \upsilon_2, \upsilon_3 : \mathbb{R}^+ \to \mathbb{R}^+$ are continuous and strictly monotonically nondecreasing functions, $\upsilon_1(s), \upsilon_2(s), \upsilon_3(s)$ are positive for $s > 0$ with $\upsilon_1(0) = \upsilon_2(0) = 0$. If there is a continuous functional $V : \mathbb{R}^+ \times \mathbb{R}^{Nn} \to \mathbb{R}^+$ such that*

$$\upsilon_1(\|e(\cdot,t)\|_2) \leqslant V(t, e(x,t)) \leqslant \upsilon_2(\|e(\cdot,t)\|_2),$$
$$\dot{V}(t, e(x,t)) \leqslant -\upsilon_3(\|e(\cdot,t)\|_2),$$

where $e(x,t) = (e_1^T(x,t), e_2^T(x,t), \ldots, e_N^T(x,t))^T$, $\dot{V}$ is the derivative of V along the solution of system (3.9), then the system (3.9) is asymptotically stable.

Proof For any $\epsilon > 0$, there is a $\delta = \delta(\epsilon), 0 < \delta < \epsilon$, such that $\upsilon_2(\delta) < \upsilon_1(\epsilon)$. If $\|\phi(\cdot)\|_2 < \delta$, then we have

$$\upsilon_1(\|e(\cdot,t,\phi)\|_2) \leqslant V(t, e(x,t,\phi)) \leqslant V(0,\phi) \leqslant \upsilon_2(\delta) < \upsilon_1(\epsilon), t \geqslant 0. \quad (3.10)$$

Therefore, $\|e(\cdot,t,\phi)\|_2 < \epsilon, t \geqslant 0$. This proves stability.

For $\epsilon = 1$, choose $b_0 = \delta(1)$ as the above constant for stability. For any $\epsilon > 0$ and $\|\phi(\cdot)\|_2 < b_0$, we shall prove that there exists $T(\phi,\epsilon) > 0$ such that

$$\|e(\cdot,t,\phi)\|_2 < \epsilon$$

for all $t \geqslant T(\phi,\epsilon)$. If this is not true, then there obviously exist ϵ^*, $\hat{\phi}(x)$ and sequence $\{t_k\}$ such that

$$\|e(\cdot, t_k, \hat{\phi})\|_2 \geqslant \delta(\epsilon^*), \quad M_{k-1} + \sigma < t_k < M_k, \quad k \in \mathbb{N},$$

where $M_0 = 0$ and $\lim_{k\to+\infty} M_k = +\infty$, $\delta(\epsilon^*)$ is the above constant for stability, σ is a positive constant and $\|\hat{\phi}(\cdot)\|_2 < b_0$.

From Green's formula and the Dirichlet boundary condition, we can obtain

$$\begin{aligned}
&2\sum_{i=1}^{N}\int_{\Omega} e_i^T(x,t,\hat{\phi})\frac{\partial e_i(x,t,\hat{\phi})}{\partial t}dx \\
&= 2\sum_{i=1}^{N}\int_{\Omega} e_i^T(x,t,\hat{\phi})\Big[D\Delta e_i(x,t,\hat{\phi}) - Ae_i(x,t,\hat{\phi}) + B\tilde{f}(e_i(x,t,\hat{\phi})) \\
&\quad + c\sum_{j=1}^{N}\hat{G}_{ij}\Gamma e_j(x,t,\hat{\phi})\Big]dx \\
&\leqslant -2\sum_{k=1}^{q}\sum_{i=1}^{N}\int_{\Omega}\left(\frac{\partial e_i(x,t,\hat{\phi})}{\partial x_k}\right)^T D\frac{\partial e_i(x,t,\hat{\phi})}{\partial x_k}dx \\
&\quad + \sum_{i=1}^{N}\int_{\Omega} e_i^T(x,t,\hat{\phi})(BB^T + \Theta - 2A)e_i(x,t,\hat{\phi})dx \\
&\quad + 2c\sum_{i=1}^{N}\sum_{j=1}^{N}\int_{\Omega}\hat{G}_{ij}e_i^T(x,t,\hat{\phi})\Gamma e_j(x,t,\hat{\phi})dx \\
&\leqslant \int_{\Omega} e^T(x,t,\hat{\phi})\Big[I_N\otimes(BB^T + \Theta - 2A) + c(\hat{G} + \hat{G}^T)\otimes\Gamma\Big]e(x,t,\hat{\phi})dx \\
&\leqslant \varrho\|e(\cdot,t,\hat{\phi})\|_2^2
\end{aligned} \tag{3.11}$$

where $\varrho = \max\Big\{1, \lambda_M\big(I_N\otimes(BB^T+\Theta-2A)+c(\hat{G}+\hat{G}^T)\otimes\Gamma\big)\Big\}$, $\tilde{f}(e_i(x,t,\hat{\phi})) = f(z_i(x,t)) - f(w^*(x))$. Then, we can derive from (3.11) that

$$\frac{d\int_{\Omega} e^T(x,t,\hat{\phi})e(x,t,\hat{\phi})dx}{dt} < \varrho \tag{3.12}$$

for all $t \geqslant 0$. In what follows, we shall prove that

$$\|e(\cdot,t,\hat{\phi})\|_2 > \frac{\delta(\epsilon^*)}{2}, \quad t_k - \frac{\delta^2(\epsilon^*)}{2\varrho} \leqslant t \leqslant t_k. \tag{3.13}$$

If this is not true, then there exists a $t^* \in \Big[t_k - \frac{\delta^2(\epsilon^*)}{2\varrho}, t_k\Big]$ satisfying

$$\|e(\cdot,t_k,\hat{\phi})\|_2^2 - \|e(\cdot,t^*,\hat{\phi})\|_2^2 \geqslant \frac{3\delta^2(\epsilon^*)}{4}. \tag{3.14}$$

On the other hand, by (3.12), we can easily derive

$$\|e(\cdot, t_k, \hat{\phi})\|_2^2 - \|e(\cdot, t^*, \hat{\phi})\|_2^2 \leqslant \varrho \times \frac{\delta^2(\epsilon^*)}{2\varrho} = \frac{\delta^2(\epsilon^*)}{2}.$$

This contradicts (3.14), and so the estimate (3.13) holds. Then, we can get

$$\dot{V}(t, e(x, t, \hat{\phi})) \leqslant -\upsilon_3\left(\frac{\delta(\epsilon^*)}{2}\right), \quad t_k - \frac{\delta^2(\epsilon^*)}{2\varrho} \leqslant t \leqslant t_k.$$

Let $\sigma = \frac{\delta^2(\epsilon^*)}{\varrho}$. It is obvious that these intervals do not overlap, and thus

$$V(t_k, e(x, t_k, \hat{\phi})) - V(0, \hat{\phi}) \leqslant -\upsilon_3\left(\frac{\delta(\epsilon^*)}{2}\right)\frac{\delta^2(\epsilon^*) \times k}{2\varrho}.$$

Let $K(\epsilon^*, \varrho)$ be the smallest integer $\geqslant \frac{\upsilon_2(b_0)}{\upsilon_3(\frac{\delta(\epsilon^*)}{2})\frac{\delta^2(\epsilon^*)}{2\varrho}}$. If $k > K(\epsilon^*, \varrho) + 1$, then

$$\begin{aligned} V(t_k, e(x, t_k, \hat{\phi})) &< -\upsilon_3\left(\frac{\delta(\epsilon^*)}{2}\right)\frac{\delta^2(\epsilon^*)\frac{\upsilon_2(b_0)}{\upsilon_3(\frac{\delta(\epsilon^*)}{2})\frac{\delta^2(\epsilon^*)}{2\varrho}}}{2\varrho} + \upsilon_2(b_0) \\ &= 0 \end{aligned}$$

which is a contradiction. This proves the asymptotic stability. The proof is completed.

Without loss of generality, assume $\hat{G}$ is in the Frobenius normal form. Namely

$$\hat{G} = \begin{pmatrix} \bar{G}_1 & \bar{G}_{12} & \cdots & \bar{G}_{1m} \\ 0 & \bar{G}_2 & \cdots & \bar{G}_{2m} \\ \vdots & \vdots & \ddots & \vdots \\ 0 & 0 & \cdots & \bar{G}_m \end{pmatrix}, \tag{3.15}$$

where $\bar{G}_1 \in \mathbb{R}^{p_1 \times p_1}, \bar{G}_2 \in \mathbb{R}^{p_2 \times p_2}, \ldots, \bar{G}_m \in \mathbb{R}^{p_m \times p_m}$ are square irreducible matrices. Then, system (3.9) can be rewritten as

$$\begin{aligned} \frac{\partial \hat{e}_i(x,t)}{\partial t} &= (I_{p_i} \otimes D)\Delta\hat{e}_i(x,t) - (I_{p_i} \otimes A)\hat{e}_i(x,t) + (I_{p_i} \otimes B)(\hat{f}_i(z(x,t)) \\ &\quad - \hat{f}_i(w^*(x))) + c\sum_{j=i+1}^{m}(\bar{G}_{ij} \otimes \Gamma)\hat{e}_j(x,t) \\ &\quad + c(\bar{G}_i \otimes \Gamma)\hat{e}_i(x,t), \quad i = 1, 2, \ldots, m, \end{aligned} \tag{3.16}$$

where

$$\begin{aligned}
\hat{e}_1(x,t) &= (e_1^T(x,t), e_2^T(x,t), \ldots, e_{p_1}^T(x,t))^T, \\
\hat{e}_2(x,t) &= (e_{p_1+1}^T(x,t), e_{p_1+2}^T(x,t), \ldots, e_{p_1+p_2}^T(x,t))^T, \\
&\vdots \\
\hat{e}_m(x,t) &= (e_{N-p_m+1}^T(x,t), e_{N-p_m+2}^T(x,t), \ldots, e_N^T(x,t))^T, \\
\hat{f}_i(w^*(x)) &= \underbrace{(f^T(w^*(x)), f^T(w^*(x)), \ldots, f^T(w^*(x)))^T}_{p_i}, \\
\hat{f}_i(z(x,t)) &= (f^T(z_{p_1+\cdots+p_{i-1}+1}(x,t)), f^T(z_{p_1+\cdots+p_{i-1}+2}(x,t)), \\
&\quad \ldots, f^T(z_{p_1+\cdots+p_{i-1}+p_i}(x,t)))^T.
\end{aligned}$$

Theorem 3.6 *If there exist positive definite diagonal matrices $\mathcal{X}_i = \mathrm{diag}(\mathcal{X}_{i1}\, \mathcal{X}_{i2}, \ldots, \mathcal{X}_{ip_i}) \in \mathbb{R}^{p_i \times p_i}, i = 1, 2, \ldots, m$, and positive constants $\alpha_i, i = 1, 2, \ldots, m$, such that*

$$(\alpha\mathcal{X}) \otimes W + c(\alpha\mathcal{X}G + G^T\mathcal{X}\alpha - 2\alpha\mathcal{X}K) \otimes \Gamma < 0, \tag{3.17}$$

where $\alpha = \mathrm{diag}(\alpha_1 I_{p_1}, \alpha_2 I_{p_2}, \ldots, \alpha_m I_{p_m})$, $\mathcal{X} = \mathrm{diag}(\mathcal{X}_1, \mathcal{X}_2, \ldots, \mathcal{X}_m)$, $W = -2\tilde{D} - 2A + BB^T + \Theta$, then the pinning controlled network (3.6) *is synchronized.*

Proof Construct a Lyapunov functional for the system (3.16) as follows:

$$V_1(t) = \sum_{i=1}^{m} \alpha_i \int_\Omega \hat{e}_i^T(x,t)(\mathcal{X}_i \otimes I_n)\hat{e}_i(x,t)dx. \tag{3.18}$$

Calculating the time derivative of $V_1(t)$ along the trajectory of system (3.16), we can get

$$\begin{aligned}
\dot{V}_1(t) &= 2\sum_{i=1}^{m} \alpha_i \int_\Omega \hat{e}_i^T(x,t)(\mathcal{X}_i \otimes I_n)\Big[(I_{p_i} \otimes D)\Delta\hat{e}_i(x,t) - (I_{p_i} \otimes A)\hat{e}_i(x,t) \\
&\quad + (I_{p_i} \otimes B)(\hat{f}_i(z(x,t)) - \hat{f}_i(w^*(x))) + c\sum_{j=i+1}^{m} (\bar{G}_{ij} \otimes \Gamma)\hat{e}_j(x,t) \\
&\quad + c(\bar{G}_i \otimes \Gamma)\hat{e}_i(x,t)\Big]dx \\
&= 2\sum_{i=1}^{m} \alpha_i \int_\Omega \hat{e}_i^T(x,t)\Big[(\mathcal{X}_i \otimes D)\Delta\hat{e}_i(x,t) - (\mathcal{X}_i \otimes A)\hat{e}_i(x,t) \\
&\quad + c\sum_{j=i+1}^{m} \left((\mathcal{X}_i\bar{G}_{ij}) \otimes \Gamma\right)\hat{e}_j(x,t) + c\left((\mathcal{X}_i\bar{G}_i) \otimes \Gamma\right)\hat{e}_i(x,t)
\end{aligned}$$

$$+ (\mathcal{X}_i \otimes B)(\hat{f}_i(z(x,t)) - \hat{f}_i(w^*(x)))\Big]dx. \tag{3.19}$$

From Green's formula and the Dirichlet boundary condition, we have

$$\int_{\Omega} e_{is}(x,t)\Delta e_{is}(x,t)dx = -\sum_{k=1}^{q}\int_{\Omega}\left(\frac{\partial e_{is}(x,t)}{\partial x_k}\right)^2 dx, \tag{3.20}$$

where $i = 1, 2, \ldots, N$, $s = 1, 2, \ldots, n$.

According to Lemma 1.22, we can obtain

$$\begin{aligned}\int_{\Omega} e_i^T(x,t)D\Delta e_i(x,t)dx &= -\sum_{k=1}^{q}\sum_{s=1}^{n}\int_{\Omega} d_s\left(\frac{\partial e_{is}(x,t)}{\partial x_k}\right)^2 dx \\ &\leqslant -\sum_{k=1}^{q}\frac{1}{l_k^2}\sum_{s=1}^{n}\int_{\Omega} d_s\,(e_{is}(x,t))^2 dx \\ &= -\sum_{k=1}^{q}\frac{1}{l_k^2}\int_{\Omega} e_i^T(x,t)De_i(x,t)dx,\end{aligned} \tag{3.21}$$

where $i = 1, 2, \ldots, N$. Obviously, from (3.21), we can get

$$\begin{aligned}&\int_{\Omega}\hat{e}_i^T(x,t)(\mathcal{X}_i\otimes D)\Delta\hat{e}_i(x,t)dx \\ &= \sum_{j=1}^{p_i}\mathcal{X}_{ij}\int_{\Omega} e^T_{p_1+\cdots+p_{i-1}+j}(x,t)D\Delta e_{p_1+\cdots+p_{i-1}+j}(x,t)dx \\ &\leqslant -\sum_{j=1}^{p_i}\mathcal{X}_{ij}\int_{\Omega} e^T_{p_1+\cdots+p_{i-1}+j}(x,t)\tilde{D}e_{p_1+\cdots+p_{i-1}+j}(x,t)dx \\ &= -\int_{\Omega}\hat{e}_i^T(x,t)(\mathcal{X}_i\otimes\tilde{D})\hat{e}_i(x,t)dx\end{aligned}$$

where $i = 1, 2, \ldots, m$.

In view of Lemma 1.14, one has

$$\begin{aligned}&2e_i^T(x,t)B(f(z_i(x,t)) - f(w^*(x)) \\ &\leqslant e_i^T(x,t)BB^Te_i(x,t) + e_i^T(x,t)\Theta e_i(x,t)\end{aligned} \tag{3.22}$$

where $i = 1, 2, \ldots, N$. Then, we can derive from (3.22) that

$$2\hat{e}_i^T(x,t)(\mathcal{X}_i\otimes B)(\hat{f}_i(z(x,t)) - \hat{f}_i(w^*(x))) \leqslant \hat{e}_i^T(x,t)\left[\mathcal{X}_i\otimes(BB^T+\Theta)\right]\hat{e}_i(x,t),$$

where $i = 1, 2, \ldots, m$. Therefore,

$$\begin{aligned}
\dot{V}_1(t) \leqslant & \sum_{i=1}^{m} \alpha_i \int_{\Omega} \hat{e}_i^T(x,t)\Big\{ -2(\mathcal{X}_i \otimes \tilde{D})\hat{e}_i(x,t) + \big[\mathcal{X}_i \otimes (BB^T + \Theta)\big]\hat{e}_i(x,t) \\
& - 2(\mathcal{X}_i \otimes A)\hat{e}_i(x,t) + c\sum_{j=i+1}^{m} 2\left((\mathcal{X}_i \bar{G}_{ij}) \otimes \Gamma\right)\hat{e}_j(x,t) \\
& + 2c\left((\mathcal{X}_i \bar{G}_i) \otimes \Gamma\right)\hat{e}_i(x,t)\Big\}dx \\
= & \int_{\Omega} e^T(x,t)\Big[\hat{\mathcal{X}} \otimes (-2\tilde{D} - 2A + BB^T + \Theta) \\
& + c(\hat{\mathcal{X}}\hat{G} + \hat{G}^T\hat{\mathcal{X}}) \otimes \Gamma\Big]e(x,t)dx, \qquad (3.23)
\end{aligned}$$

where $\hat{\mathcal{X}}_i = \alpha_i \mathcal{X}_i$, $\hat{\mathcal{X}} = \text{diag}(\hat{\mathcal{X}}_1, \hat{\mathcal{X}}_2, \ldots, \hat{\mathcal{X}}_m)$.

From Theorem 3.5 and (3.17), we can conclude that system (3.9) is asymptotically stable. Therefore, the pinning controlled network (3.6) is synchronized. The proof is completed.

Let

$$\mathcal{H}^i = \begin{pmatrix} \mathcal{H}_{11}^i & \mathcal{H}_{12}^i & \cdots & \mathcal{H}_{1i}^i \\ (\mathcal{H}_{12}^i)^T & \mathcal{H}_{22}^i & \cdots & \mathcal{H}_{2i}^i \\ \vdots & \vdots & \ddots & \vdots \\ (\mathcal{H}_{1i}^i)^T & (\mathcal{H}_{2i}^i)^T & \cdots & \mathcal{H}_{ii}^i \end{pmatrix},$$

where $\alpha_i, i = 1, 2, \ldots, m$, are positive constants, and

$$\begin{aligned}
\mathcal{H}_{11}^i &= \alpha_1 \mathcal{X}_1(c\bar{G}_1 + \xi I_{p_1}) + \alpha_1(c\bar{G}_1 + \xi I_{p_1})^T \mathcal{X}_1, \\
\mathcal{H}_{12}^i &= \alpha_1 c \mathcal{X}_1 \bar{G}_{12}, \\
\mathcal{H}_{1i}^i &= \alpha_1 c \mathcal{X}_1 \bar{G}_{1i}, \\
\mathcal{H}_{22}^i &= \alpha_2 \mathcal{X}_2(c\bar{G}_2 + \xi I_{p_2}) + \alpha_2(c\bar{G}_2 + \xi I_{p_2})^T \mathcal{X}_2, \\
\mathcal{H}_{2i}^i &= \alpha_2 c \mathcal{X}_2 \bar{G}_{2i}, \\
\mathcal{H}_{ii}^i &= \alpha_i \mathcal{X}_i(c\bar{G}_i + \xi I_{p_i}) + \alpha_i(c\bar{G}_i + \xi I_{p_i})^T \mathcal{X}_i.
\end{aligned}$$

According to Theorem 3.6, we can easily get the following corollary.

Corollary 3.7 *If there exist positive definite diagonal matrices* $\mathcal{X}_i = \text{diag}(\mathcal{X}_{i1}, \mathcal{X}_{i2}, \ldots, \mathcal{X}_{ip_i})$, $i = 1, 2, \ldots, m$, *such that*

$$\mathcal{X}_i(c\bar{G}_i + \xi I_{p_i}) + (c\bar{G}_i + \xi I_{p_i})^T \mathcal{X}_i < 0, i = 1, 2, \ldots, m, \qquad (3.24)$$

then the pinning controlled network (3.6) *is synchronized.*

Proof From the definition of $\mathcal{H}^i$, we have

$$\mathcal{H}^m = \alpha\mathcal{X}(c\hat{G} + \xi I_N) + (c\hat{G} + \xi I_N)^T\mathcal{X}\alpha,$$

where $\alpha = \text{diag}(\alpha_1 I_{p_1}, \alpha_2 I_{p_2}, \ldots, \alpha_m I_{p_m})$, $\mathcal{X} = \text{diag}(\mathcal{X}_1, \mathcal{X}_2, \ldots, \mathcal{X}_m)$.

According to (3.24), we can easily obtain $\mathcal{H}^1 < 0$. For any $i \in \{1, 2, \ldots, m-1\}$, we suppose

$$\mathcal{H}^i < 0.$$

In what follows, we shall prove that

$$\mathcal{H}^{i+1} < 0.$$

According to Lemma 1.15, $\mathcal{H}^{i+1} < 0$ is equivalent to

$$\mathcal{H}^i - \frac{c^2}{\alpha_{i+1}}\mathcal{Q}\mathcal{W}^{-1}\mathcal{Q}^T < 0, \tag{3.25}$$

where $\mathcal{W} = \mathcal{X}_{i+1}(c\bar{G}_{i+1} + \xi I_{p_{i+1}}) + (c\bar{G}_{i+1} + \xi I_{p_{i+1}})^T\mathcal{X}_{i+1}$, $\mathcal{Q}^T = \Big(\alpha_1\bar{G}^T_{1(i+1)}\mathcal{X}_1, \alpha_2\bar{G}^T_{2(i+1)}\mathcal{X}_2, \ldots, \alpha_i\bar{G}^T_{i(i+1)}\mathcal{X}_i\Big)$. Obviously, by choosing α_{i+1} to be sufficiently larger than α_j for $j < i+1$, condition (3.25) can be satisfied. By mathematical induction, it is easy to conclude that

$$\mathcal{H}^m < 0.$$

Therefore, if condition (3.24) is satisfied, then there exists a positive definite diagonal matrix $\alpha = \text{diag}(\alpha_1 I_{p_1}, \alpha_2 I_{p_2}, \ldots, \alpha_m I_{p_m})$ such that

$$\alpha\mathcal{X}(c\hat{G} + \xi I_N) + (c\hat{G} + \xi I_N)^T\mathcal{X}\alpha < 0.$$

On the other hand,

$$\begin{aligned}
&(\alpha\mathcal{X}) \otimes (-2\tilde{D} - 2A + BB^T + \Theta) + c(\alpha\mathcal{X}\hat{G} + \hat{G}^T\mathcal{X}\alpha) \otimes \Gamma \\
&\leqslant (2\xi\alpha\mathcal{X}) \otimes \Gamma + c(\alpha\mathcal{X}\hat{G} + \hat{G}^T\mathcal{X}\alpha) \otimes \Gamma \\
&= \Big[\alpha\mathcal{X}(c\hat{G} + \xi I_N) + (c\hat{G} + \xi I_N)^T\mathcal{X}\alpha\Big] \otimes \Gamma.
\end{aligned}$$

Since Γ is a positive definite matrix, we can obtain

$$(\alpha\mathcal{X}) \otimes W + c(\alpha\mathcal{X}\hat{G} + \hat{G}^T\mathcal{X}\alpha) \otimes \Gamma < 0,$$

where $W = -2\tilde{D} - 2A + BB^T + \Theta$.

Therefore, we can derive from Theorem 3.6 that the pinning controlled network (3.6) is synchronized. The proof is completed.

Remark 3.8 In Theorem 3.6, a sufficient condition ensuring synchronization is established, in which $\mathcal{X}$ is block diagonal matrix. In this case, the study of the whole network synchronization is reduced to the investigation of the strongly connected components in network (3.6) [see Corollary 3.7].

Let $K = \operatorname{diag}(K_1, K_2, \ldots, K_m)$, $K_i \in \mathbb{R}^{p_i \times p_i}$, $i = 1, 2, \ldots, m$. Since $G = \hat{G} + K$, one has

$$G = \begin{pmatrix} \hat{G}_1 & \bar{G}_{12} & \cdots & \bar{G}_{1m} \\ 0 & \hat{G}_2 & \cdots & \bar{G}_{2m} \\ \vdots & \vdots & \ddots & \vdots \\ 0 & 0 & \cdots & \hat{G}_m \end{pmatrix}, \tag{3.26}$$

where $\hat{G}_1 = \bar{G}_1 + K_1 \in \mathbb{R}^{p_1 \times p_1}$, $\hat{G}_2 = \bar{G}_2 + K_2 \in \mathbb{R}^{p_2 \times p_2}, \ldots, \hat{G}_m = \bar{G}_m + K_m \in \mathbb{R}^{p_m \times p_m}$.

Corollary 3.9 *The CRDNNs* (3.1) *is synchronized if the following condition holds:*

$$-\frac{\lambda_m(\tilde{D})}{\lambda_M(\Gamma)} - \frac{\lambda_m(A)}{\lambda_M(\Gamma)} + \frac{\lambda_M(\frac{\Theta}{2})}{\lambda_m(\Gamma)} + \frac{\lambda_M(\frac{BB^T}{2})}{\lambda_m(\Gamma)} < 0. \tag{3.27}$$

Proof Similar to the proof of Corollary 3.7, we can prove that CRDNNs (3.1) is synchronized if there exist positive definite diagonal matrices $\mathcal{X}_i = \operatorname{diag}(\mathcal{X}_{i1}, \mathcal{X}_{i2}, \ldots, \mathcal{X}_{ip_i})$, $i = 1, 2, \ldots, m$, such that

$$\mathcal{X}_i(c\hat{G}_i + \xi I_{p_i}) + (c\hat{G}_i + \xi I_{p_i})^T \mathcal{X}_i < 0, i = 1, 2, \ldots, m. \tag{3.28}$$

For the sake of clarity, we split the rest of the proof into two cases:

Case 1: $p_i = 1, i \in \{1, 2, \ldots, m\}$.
In this case, $\hat{G}_i = G_{ll} \leqslant 0, l = p_1 + \cdots + p_{i-1} + 1$. Since $\xi < 0$, we have

$$\mathcal{X}_i(cG_{ll} + \xi) + (cG_{ll} + \xi)\mathcal{X}_i < 0$$

for any $\mathcal{X}_i > 0$.

Case 2: $p_i \geqslant 2, i \in \{1, 2, \ldots, m\}$.

Obviously, $\hat{G}_i$ is an irreducible matrix. Then, from Lemma 1.18, we can get that (3.28) holds. The proof is completed.

Remark 3.10 When

$$-\frac{\lambda_m(\tilde{D})}{\lambda_M(\Gamma)} - \frac{\lambda_m(A)}{\lambda_M(\Gamma)} + \frac{\lambda_M(\frac{\Theta}{2})}{\lambda_m(\Gamma)} + \frac{\lambda_M(\frac{BB^T}{2})}{\lambda_m(\Gamma)} < 0,$$

CRDNNs (3.1) can synchronize by itself. Therefore, in this chapter, we always assume

$$-\frac{\lambda_m(\tilde{D})}{\lambda_M(\Gamma)}-\frac{\lambda_m(A)}{\lambda_M(\Gamma)}+\frac{\lambda_M(\frac{\Theta}{2})}{\lambda_m(\Gamma)}+\frac{\lambda_M(\frac{BB^T}{2})}{\lambda_m(\Gamma)}\geqslant 0.$$

From Corollary 3.7, we can easily obtain the following conclusion.

Corollary 3.11 *If there exist positive definite diagonal matrices* $\mathcal{X}_i=\mathrm{diag}(\mathcal{X}_{i1},\mathcal{X}_{i2},\ldots,\mathcal{X}_{ip_i})$, $i=1,2,\ldots,m$, *such that*

$$c>\max_{i=1,2,\ldots,m}\left\{\frac{-2\xi\lambda_M(\mathcal{X}_i)}{\lambda_M(\mathcal{X}_i\bar{G}_i+\bar{G}_i^T\mathcal{X}_i)}\right\},\tag{3.29}$$

where $\mathcal{X}_i\bar{G}_i+\bar{G}_i^T\mathcal{X}_i<0$, *then the pinning controlled network* (3.6) *is synchronized.*

Remark 3.12 According to Corollary 3.11, there exists a critical coupling strength c^* for given $\mathcal{X}_i (i=1,2,\ldots,m)$ such that the pinning controlled network (3.6) will synchronize if

$$c>c^*.$$

Therefore, if

$$\mathcal{X}_i\bar{G}_i+\bar{G}_i^T\mathcal{X}_i<0, i=1,2,\ldots,m$$

are satisfied, then the pinning controlled network (3.6) can synchronize as long as the coupling strength c is large enough.

3.2.3 Adaptive Coupling Strength

The coupling strength c given in (3.29) is very conservative, usually much larger than the needed value. Therefore, an adaptive strategy to tune the coupling strength c is designed in this subsection.

Theorem 3.13 *If there exist positive definite diagonal matrices* $\mathcal{X}_i=\mathrm{diag}(\mathcal{X}_{i1},\mathcal{X}_{i2},\ldots,\mathcal{X}_{ip_i})$, $i=1,2,\ldots,m$, *such that*

$$\mathcal{X}_i\bar{G}_i+\bar{G}_i^T\mathcal{X}_i<0,\tag{3.30}$$

then the pinning controlled network (3.6) *is synchronized under the following adaptive law:*

$$\dot{c}(t)=\beta_1\sum_{i=1}^{m}\alpha_i\int_{\Omega}\hat{e}_i^T(x,t)(\mathcal{X}_i\otimes\Gamma)\hat{e}_i(x,t)dx,\tag{3.31}$$

where $0<\beta_1\in\mathbb{R}$, *and* α_i, $i=1,2,\ldots,m$, *are positive constants satisfying* (3.35).

Proof The pinning controlled network (3.6) with adaptive coupling strength can be described by

$$\begin{aligned}\frac{\partial \hat{e}_i(x,t)}{\partial t} &= (I_{p_i} \otimes D)\Delta \hat{e}_i(x,t) - (I_{p_i} \otimes A)\hat{e}_i(x,t) + (I_{p_i} \otimes B)(\hat{f}_i(z(x,t)) \\ &\quad - \hat{f}_i(w^*(x))) + c(t)\sum_{j=i+1}^{m}(\bar{G}_{ij} \otimes \Gamma)\hat{e}_j(x,t) \\ &\quad + c(t)(\bar{G}_i \otimes \Gamma)\hat{e}_i(x,t), i = 1, 2, \ldots, m. \end{aligned} \tag{3.32}$$

Construct a Lyapunov functional for the system (3.32) as follows:

$$V_2(t) = \sum_{i=1}^{m} \alpha_i \int_{\Omega} \hat{e}_i^T(x,t)(\mathcal{X}_i \otimes I_n)\hat{e}_i(x,t)dx + \frac{r_1}{\beta_1}(c(t) - \hat{c})^2, \tag{3.33}$$

where r_1 and $\hat{c}$ are positive constants.

Calculating the time derivative of $V_2(t)$ along the trajectory of system (3.32), we can get

$$\begin{aligned}\dot{V}_2(t) &= 2\sum_{i=1}^{m} \alpha_i \int_{\Omega} \hat{e}_i^T(x,t)(\mathcal{X}_i \otimes I_n)\Big[(I_{p_i} \otimes D)\Delta \hat{e}_i(x,t) - (I_{p_i} \otimes A)\hat{e}_i(x,t) \\ &\quad + (I_{p_i} \otimes B)(\hat{f}_i(z(x,t)) - \hat{f}_i(w^*(x))) + c(t)\sum_{j=i+1}^{m}(\bar{G}_{ij} \otimes \Gamma)\hat{e}_j(x,t) \\ &\quad + c(t)(\bar{G}_i \otimes \Gamma)\hat{e}_i(x,t)\Big]dx - 2r_1\hat{c}\sum_{i=1}^{m}\alpha_i \int_{\Omega} \hat{e}_i^T(x,t)(\mathcal{X}_i \otimes \Gamma)\hat{e}_i(x,t)dx \\ &\quad + 2r_1 c(t)\sum_{i=1}^{m}\alpha_i \int_{\Omega} \hat{e}_i^T(x,t)(\mathcal{X}_i \otimes \Gamma)\hat{e}_i(x,t)dx \\ &\leqslant \int_{\Omega} e^T(x,t)\Big[(\alpha\mathcal{X}) \otimes (-2\tilde{D} - 2A + BB^T + \Theta) + c(t)(\alpha\mathcal{X}\hat{G} + \hat{G}^T\mathcal{X}\alpha \\ &\quad + 2r_1\alpha\mathcal{X}) \otimes \Gamma - 2r_1\hat{c}\,((\alpha\mathcal{X}) \otimes \Gamma)\Big]e(x,t)dx,\end{aligned}$$

where $\alpha = \mathrm{diag}(\alpha_1 I_{p_1}, \alpha_2 I_{p_2}, \ldots, \alpha_m I_{p_m})$, $\mathcal{X} = \mathrm{diag}(\mathcal{X}_1, \mathcal{X}_2, \ldots, \mathcal{X}_m)$, $e(x,t) = (\hat{e}_1^T(x,t), \hat{e}_2^T(x,t), \ldots, \hat{e}_m^T(x,t))^T$.

According to (3.30), there obviously exists a positive constant $\hat{r}_1$ such that

$$\mathcal{X}_i(\bar{G}_i + \hat{r}_1 I_{p_i}) + (\bar{G}_i + \hat{r}_1 I_{p_i})^T\mathcal{X}_i < 0, i = 1, 2, \ldots, m. \tag{3.34}$$

Let $r_1 \leqslant \hat{r}_1$. Then, following similar arguments as in the proof of Corollary 3.7, we can find a positive definite diagonal matrix $\hat{\alpha} = \mathrm{diag}(\hat{\alpha}_1 I_{p_1}, \hat{\alpha}_2 I_{p_2}, \ldots, \hat{\alpha}_m I_{p_m})$ satisfying

$$\hat{\alpha}\mathcal{X}\hat{G} + \hat{G}^T\mathcal{X}\hat{\alpha} + 2r_1\hat{\alpha}\mathcal{X} < 0. \tag{3.35}$$

Let $\alpha = \hat{\alpha}$. Since $c(t) > 0$, we have

$$\begin{aligned}\dot{V}_2(t) \leqslant \int_{\Omega} e^T(x,t)\Big[(\alpha\mathcal{X}) \otimes (-2\tilde{D} - 2A + BB^T + \Theta) \\ - 2r_1\hat{c}\left((\alpha\mathcal{X}) \otimes \Gamma\right)\Big]e(x,t)dx.\end{aligned}$$

By selecting $\hat{c}$ sufficiently large such that

$$-2\tilde{D} - 2A + BB^T + \Theta - 2r_1\hat{c}\Gamma < 0,$$

one obtains

$$\dot{V}_2(t) \leqslant -\varrho\|e(\cdot,t)\|_2^2, \tag{3.36}$$

where $\varrho = \lambda_m\Big((\alpha\mathcal{X}) \otimes (2\tilde{D} + 2A - BB^T - \Theta + 2r_1\hat{c}\Gamma)\Big) > 0$.

Obviously, $V_2(t)$ is non-increasing, and each term of $V_2(t)$ is bounded. Consequently, $c(t)$ is bounded, and $\lim_{t\to+\infty} V_2(t)$ exists and is a non-negative real number. Since $c(t)$ is monotonically increasing (see (3.31)), one can conclude that $c(t)$ asymptotically converges to a finite positive value. Therefore, by the definition of $V_2(t)$, we can derive that

$$\lim_{t\to+\infty}\sum_{i=1}^{m}\alpha_i\int_{\Omega}\hat{e}_i^T(x,t)(\mathcal{X}_i \otimes I_n)\hat{e}_i(x,t)dx$$

exists and is a non-negative real number. In what follows, we shall prove that

$$\lim_{t\to+\infty}\sum_{i=1}^{m}\alpha_i\int_{\Omega}\hat{e}_i^T(x,t)(\mathcal{X}_i \otimes I_n)\hat{e}_i(x,t)dx = 0.$$

If this is not true, we have

$$\lim_{t\to+\infty}\sum_{i=1}^{m}\alpha_i\int_{\Omega}\hat{e}_i^T(x,t)(\mathcal{X}_i \otimes I_n)\hat{e}_i(x,t)dx = \mu > 0.$$

Then, there obviously exists a real number $M > 0$ such that

$$\sum_{i=1}^{m}\alpha_i\int_{\Omega}\hat{e}_i^T(x,t)(\mathcal{X}_i \otimes I_n)\hat{e}_i(x,t)dx > \frac{\mu}{2}$$

for $t \geqslant M$. Therefore,

$$\|e(\cdot,t)\|_2^2 > \frac{\mu}{2\delta}, \quad t \geqslant M, \tag{3.37}$$

where $\delta = \max_{i=1,2,\ldots,m}\{\lambda_M(\alpha_i \mathcal{X}_i)\}$. From (3.36) and (3.37), we can get

$$\dot{V}_2(t) < -\frac{\varrho\mu}{2\delta}, \quad t \geqslant M. \tag{3.38}$$

By integrating (3.38) with respect to t over the time period M to $+\infty$, we can obtain

$$\begin{aligned} -V_2(M) \leqslant V_2(+\infty) - V_2(M) &= \int_M^{+\infty} \dot{V}_2(t)dt \\ &< -\int_M^{+\infty} \frac{\varrho\mu}{2\delta}dt \\ &= -\infty. \end{aligned}$$

This yields a contradiction, and so

$$\lim_{t\to+\infty} \sum_{i=1}^m \alpha_i \int_\Omega \hat{e}_i^T(x,t)(\mathcal{X}_i \otimes I_n)\hat{e}_i(x,t)dx = 0.$$

Then, we can easily obtain

$$\lim_{t\to+\infty} \|e(\cdot,t)\|_2 = 0.$$

Therefore, the pinning controlled network (3.6) is synchronized under the adaptive law (3.31). The proof is completed.

Remark 3.14 Adaptive technique is a very effective method to tune the coupling strength c. Firstly, coupling strength c can be adjusted very quickly by utilizing the adaptive law defined by the synchronization errors [see (3.31)]. Another important advantage of adaptive technique is to adjust coupling strength c for achieving suitable value. Practically, in some realistic systems, complex networks with adaptively adjusted coupling strength may be more reasonable, such as neural networks, biological systems, etc. [128].

3.2.4 Pinning Adaptive Synchronization Criterion

This subsection further investigates the pinning adaptive synchronization of the CRDNNs (3.1). A synchronization criterion is obtained by constructing appropriate Lyapunov functional and using inequality techniques.

Suppose that the vertices $q_1, q_2, \ldots, q_l (1 \leqslant l < N)$ are selected to be pinned. Then, the controlled network can be described by

$$\begin{aligned}
\frac{\partial z_{q_i}(x,t)}{\partial t} &= D\Delta z_{q_i}(x,t) - Az_{q_i}(x,t) + Bf(z_{q_i}(x,t)) + c\sum_{j=1}^{N} G_{q_i j}\Gamma z_j(x,t) \\
&\quad + J - ck_i(t)\Gamma(z_{q_i}(x,t) - w^*(x)),\ i = 1, 2, \ldots, l \\
\frac{\partial z_{q_i}(x,t)}{\partial t} &= D\Delta z_{q_i}(x,t) - Az_{q_i}(x,t) + Bf(z_{q_i}(x,t)) + c\sum_{j=1}^{N} G_{q_i j}\Gamma z_j(x,t) \\
&\quad + J,\ i = l+1, l+2, \ldots, N, \qquad (3.39) \\
\dot{k}_i(t) &= \beta_2 \int_{\Omega} (z_{q_i}(x,t) - w^*(x))^T \Gamma (z_{q_i}(x,t) - w^*(x)) dx, \qquad (3.40)
\end{aligned}$$

where β_2 and $k_i(0)$ are positive constants.

From (3.39), we can get

$$\begin{aligned}
\frac{\partial \hat{e}_i(x,t)}{\partial t} &= (I_{p_i} \otimes D)\Delta \hat{e}_i(x,t) - (I_{p_i} \otimes A)\hat{e}_i(x,t) + (I_{p_i} \otimes B)(\hat{f}_i(z(x,t)) \\
&\quad - \hat{f}_i(w^*(x))) + c\sum_{j=i+1}^{m} (\bar{G}_{ij} \otimes \Gamma)\hat{e}_j(x,t) + c(\hat{G}_i \otimes \Gamma)\hat{e}_i(x,t) \\
&\quad - c(K_i(t) \otimes \Gamma)\hat{e}_i(x,t), \qquad (3.41)
\end{aligned}$$

where $i = 1, 2, \ldots, m$, $\hat{e}_i(x,t)$, $\hat{f}_i(z(x,t))$, $\hat{f}_i(w^*(x))$, $\bar{G}_{ij}$, $\hat{G}_i$ denote the same meanings as those in (3.16) and (3.26), $K_i(t) = \text{diag}(K_{i1}(t), K_{i2}(t), \ldots, K_{ip_i}(t)) \in \mathbb{R}^{p_i \times p_i}$, $K(t) = \text{diag}(K_1(t), K_2(t), \ldots, K_m(t)) = \text{diag}(0, \ldots, 0, \underbrace{k_1(t)}_{q_1}, 0, \ldots, 0, \underbrace{k_2(t)}_{q_2}, 0, \ldots, 0, \underbrace{k_l(t)}_{q_l}, \cdots)$.

Theorem 3.15 *If there exist matrices* $\mathcal{X}_i = \text{diag}(\mathcal{X}_{i1}, \mathcal{X}_{i2}, \ldots, \mathcal{X}_{ip_i}) > 0$, $\hat{\kappa}_i = \text{diag}(\kappa_{i1}, \kappa_{i2}, \ldots, \kappa_{ip_i}) \geqslant 0$, $i = 1, 2, \ldots, m$, *such that*

$$\mathcal{X}_i(c\hat{G}_i - c\hat{\kappa}_i + \xi I_{p_i}) + (c\hat{G}_i - c\hat{\kappa}_i + \xi I_{p_i})^T \mathcal{X}_i < 0, \qquad (3.42)$$

where $\kappa_{ij} = 0 (j = 1, 2, \ldots, p_i)$ *if and only if* $K_{ij}(t) = 0$, *then the controlled network* (3.39) *is synchronized under the adaptive law* (3.40).

Proof Firstly, according to (3.42), there obviously exists a positive definite diagonal matrix $\alpha = \text{diag}(\alpha_1 I_{p_1}, \alpha_2 I_{p_2}, \ldots, \alpha_m I_{p_m}) (\alpha_i \in \mathbb{R}, i = 1, 2, \ldots, m)$ such that

$$\left[\alpha\mathcal{X}(cG - c\hat{\kappa} + \xi I_N) + (cG - c\hat{\kappa} + \xi I_N)^T \mathcal{X}\alpha\right] \otimes \Gamma < 0 \qquad (3.43)$$

where $\mathcal{X} = \text{diag}(\mathcal{X}_1, \mathcal{X}_2, \ldots, \mathcal{X}_m)$, $\hat{\kappa} = \text{diag}(\hat{\kappa}_1, \hat{\kappa}_2, \ldots, \hat{\kappa}_m)$. Consider the following Lyapunov functional:

$$V_3(t) = \sum_{i=1}^{m} \alpha_i \int_{\Omega} \hat{e}_i^T(x,t)(\mathcal{X}_i \otimes I_n)\hat{e}_i(x,t)dx + \sum_{i=1}^{m} \frac{c\alpha_i}{\beta_2}(\hat{K}_i(t) - \kappa_i)^T \mathcal{X}_i(\hat{K}_i(t) - \kappa_i),$$

where $\hat{K}_i(t) = (K_{i1}(t), K_{i2}(t), \ldots, K_{ip_i}(t))^T$, $\kappa_i = (\kappa_{i1}, \kappa_{i2}, \ldots, \kappa_{ip_i})^T$.

Calculating the time derivative of $V_3(t)$ along the trajectory of system (3.41), we can get

$$\begin{aligned}
\dot{V}_3(t) &= 2c\sum_{i=1}^{m}\sum_{j=1}^{p_i} \alpha_i \mathcal{X}_{ij} K_{ij}(t) \int_{\Omega} e^T_{p_1+\cdots+p_{i-1}+j}(x,t)\Gamma e_{p_1+\cdots+p_{i-1}+j}(x,t)dx \\
&- 2c\sum_{i=1}^{m}\sum_{j=1}^{p_i} \alpha_i \mathcal{X}_{ij}\kappa_{ij} \int_{\Omega} e^T_{p_1+\cdots+p_{i-1}+j}(x,t)\Gamma e_{p_1+\cdots+p_{i-1}+j}(x,t)dx \\
&+ 2\sum_{i=1}^{m}\alpha_i \int_{\Omega} \hat{e}_i^T(x,t)(\mathcal{X}_i \otimes I_n)\frac{\partial \hat{e}_i(x,t)}{\partial t}dx \\
&= 2\sum_{i=1}^{m}\alpha_i \int_{\Omega} \hat{e}_i^T(x,t)(\mathcal{X}_i \otimes I_n)\Big[(I_{p_i} \otimes D)\Delta\hat{e}_i(x,t) - (I_{p_i} \otimes A)\hat{e}_i(x,t) \\
&+ (I_{p_i} \otimes B)(\hat{f}_i(z(x,t)) - \hat{f}_i(w^*(x))) + c\sum_{j=i+1}^{m}(\bar{G}_{ij} \otimes \Gamma)\hat{e}_j(x,t) \\
&+ c(\hat{G}_i \otimes \Gamma)\hat{e}_i(x,t) - c(K_i(t) \otimes \Gamma)\hat{e}_i(x,t)\Big]dx \\
&+ 2c\sum_{i=1}^{m}\alpha_i \int_{\Omega} \hat{e}_i^T(x,t)\Big[(\mathcal{X}_i K_i(t) - \mathcal{X}_i\hat{\kappa}_i) \otimes \Gamma\Big]\hat{e}_i(x,t)dx \\
&\leqslant \int_{\Omega} e^T(x,t)\Big[(\alpha\mathcal{X}) \otimes (-2\tilde{D} - 2A + BB^T + \Theta) + c(\alpha\mathcal{X}G + G^T\mathcal{X}\alpha \\
&- 2\alpha\mathcal{X}\hat{\kappa}) \otimes \Gamma\Big]e(x,t)dx \\
&\leqslant \int_{\Omega} e^T(x,t)\Big\{\big[\alpha\mathcal{X}(cG - c\hat{\kappa} + \xi I_N) + (cG - c\hat{\kappa} \\
&+ \xi I_N)^T\mathcal{X}\alpha\big] \otimes \Gamma\Big\}e(x,t)dx,
\end{aligned} \quad (3.44)$$

where $e(x,t) = (e_1^T(x,t), e_2^T(x,t), \ldots, e_N^T(x,t))^T$. Then, following similar arguments as in the proof of Theorem 3.13, we can obtain the desired result immediately.

Remark 3.16 Recently, many authors have studied the pinning synchronization of the traditional complex networks without reaction-diffusion terms [97–101], which are described by ordinary differential equations. The main difficulty for synchronization analysis of the pinning controlled CRDNNs comes from the reaction-diffusion terms, which can not be dealt with by those techniques used in traditional complex networks. By employing Green's formula, Lemma 1.22 and some inequality techniques, several sufficient conditions ensuring the synchronization are established in this chapter.

3.3 Numerical Example

As an application of the above obtained theoretical results, a representative example is given in this section.

Consider the following CRDNNs:

$$\frac{\partial z_i(x,t)}{\partial t} = D\Delta z_i(x,t) - Az_i(x,t) + J + Bf(z_i(x,t)) + c\sum_{j=1}^{6} G_{ij}\Gamma z_j(x,t), \tag{3.45}$$

where $i = 1, 2, \ldots, 6$, $\Omega = \{x \mid -0.5 < x < 0.5\}$, $f_j(\xi) = \frac{|\xi+1|-|\xi-1|}{2}$, $D = \mathrm{diag}(0.15, 0.2, 0.2)$, $J = (0, 0, 0)^T$, $A = \mathrm{diag}(0.2, 0.2, 0.3)$, $\Gamma = \mathrm{diag}(0.7, 0.5, 0.6)$, $c = 2$, the matrices $B = (b_{ij})_{3\times 3}$ and $G = (G_{ij})_{6\times 6}$ are chosen as, respectively

$$B = \begin{pmatrix} 0.2 & 0.3 & 0.2 \\ 0.1 & 0.3 & 0.3 \\ 0.2 & 0.1 & 0.4 \end{pmatrix},$$

$$G = \begin{pmatrix} -0.5 & 0.2 & 0 & 0.3 & 0 & 0 \\ 0 & -0.3 & 0.3 & 0 & 0 & 0 \\ 0.2 & 0 & -0.2 & 0 & 0 & 0 \\ 0 & 0 & 0 & -0.1 & 0.1 & 0 \\ 0 & 0 & 0 & 0 & -0.2 & 0.2 \\ 0 & 0 & 0 & 0.4 & 0 & -0.4 \end{pmatrix}.$$

Obviously, $f_j(\cdot)(j = 1, 2, 3)$ satisfies the Lipschitz condition with $\rho_j = 1$.

Case 1: We choose the vertices 2, 3, 4 and 5 as pinned vertices. Select the parameters as follows: $k_1 = 0.3$, $k_2 = 0.4$, $k_3 = 0.4$, $k_4 = 0.3$.

We can find the following positive definite diagonal matrices $\mathcal{X}_1$ and $\mathcal{X}_2$ satisfying (3.24),

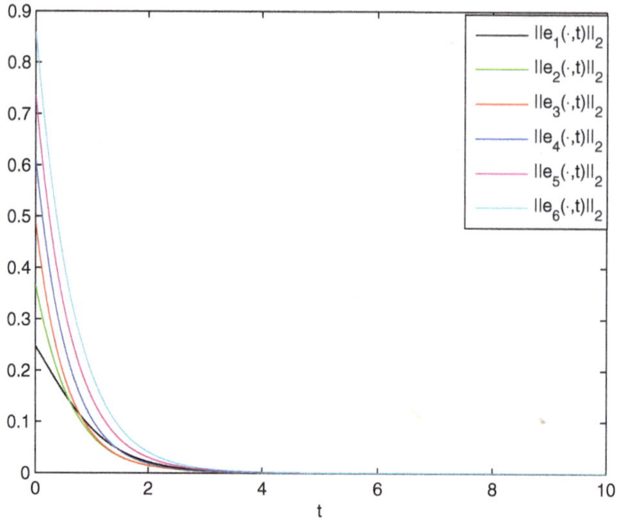

Fig. 3.1 The change processes of $\|e_i(\cdot,t)\|_2, i=1,2,\ldots,6$ ($\|e_1(\cdot,0)\|_2 = 0.2468$, $\|e_2(\cdot,0)\|_2 = 0.3676$, $\|e_3(\cdot,0)\|_2 = 0.4900$, $\|e_4(\cdot,0)\|_2 = 0.6131$, $\|e_5(\cdot,0)\|_2 = 0.7365$ and $\|e_6(\cdot,0)\|_2 = 0.8602$)

$$\mathcal{X}_1 = \text{diag}(0.4, 0.5, 0.3),\ \mathcal{X}_2 = \text{diag}(0.6, 0.3, 0.2).$$

According to Corollary 3.7, the CRDNNs (3.45) under pinning control is synchronized. The simulation results are shown in Fig. 3.1.

On the other hand, it is easy to verify that (3.30) and (3.35) are satisfied if $\mathcal{X}_1 = \text{diag}(0.4, 0.5, 0.3)$, $\mathcal{X}_2 = \text{diag}(0.6, 0.3, 0.2)$, $\hat{\alpha} = \text{diag}(0.5I_3, 3I_3)$ and $r_1 = 0.1$. It follows from Theorem 3.13 that the CRDNNs (3.45) under pinning control and adaptive law (3.31) is synchronized. The simulation results are shown in Figs. 3.2 and 3.3.

Case 2: We can find the following matrices $\mathcal{X}_1$, $\mathcal{X}_2$, $\hat{\kappa}_1$ and $\hat{\kappa}_2$ satisfying (3.42),

$$\mathcal{X}_1 = \text{diag}(0.4, 0.5, 0.3),\ \mathcal{X}_2 = \text{diag}(0.6, 0.3, 0.2),$$
$$\hat{\kappa}_1 = \text{diag}(0, 0.3, 0.4),\quad \hat{\kappa}_2 = \text{diag}(0.4, 0.3, 0).$$

By Theorem 3.15, we know that the CRDNNs (3.45) is synchronized under the pinning adaptive controllers. The simulation results are shown in Figs. 3.4 and 3.5.

From Figs. 3.1, 3.2 and 3.4, we clearly see that $\|e_i(\cdot,t)\|_2, i=1,2,\ldots,6$ are very close to 0 when the time t increases gradually to 4, and this state is maintained along with the increasing of the time. Figs. 3.3 and 3.5 visualize the change processes of $c(t)$ and $k_i(t)(i=1,2,3,4)$ in time interval [0, 10]. The numerical results clearly show that $c(t)$ and $k_i(t)(i=1,2,3,4)$ asymptotically converge to some positive real numbers.

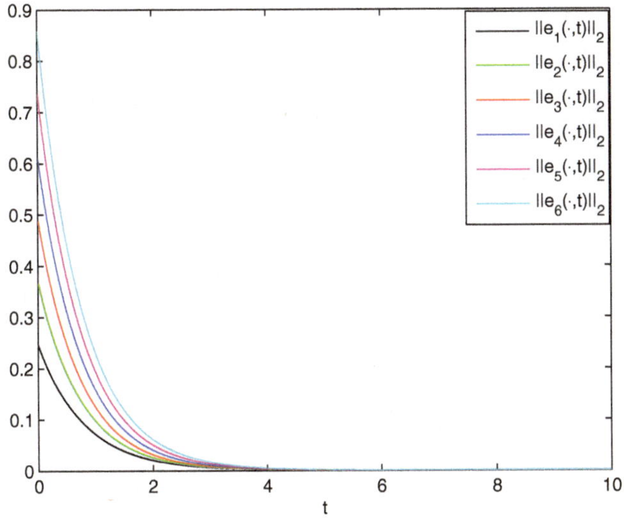

Fig. 3.2 The change processes of $\|e_i(\cdot, t)\|_2, i = 1, 2, \ldots, 6$ ($\|e_1(\cdot, 0)\|_2 = 0.2468$, $\|e_2(\cdot, 0)\|_2 = 0.3676$, $\|e_3(\cdot, 0)\|_2 = 0.4900$, $\|e_4(\cdot, 0)\|_2 = 0.6131$, $\|e_5(\cdot, 0)\|_2 = 0.7365$ and $\|e_6(\cdot, 0)\|_2 = 0.8602$)

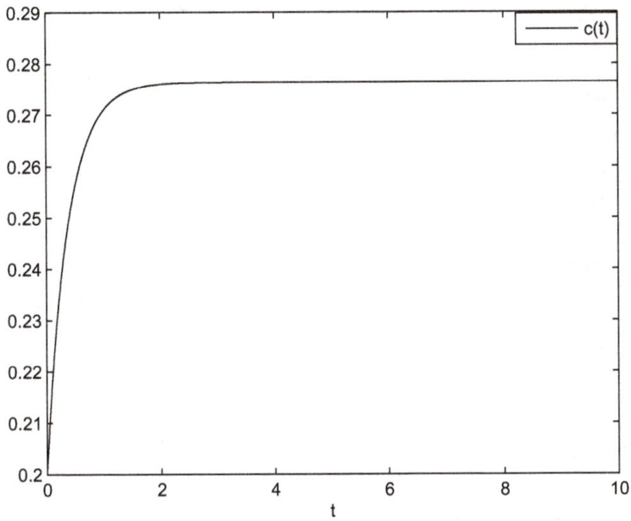

Fig. 3.3 Adaptive coupling strength [$c(0) = 0.2$]

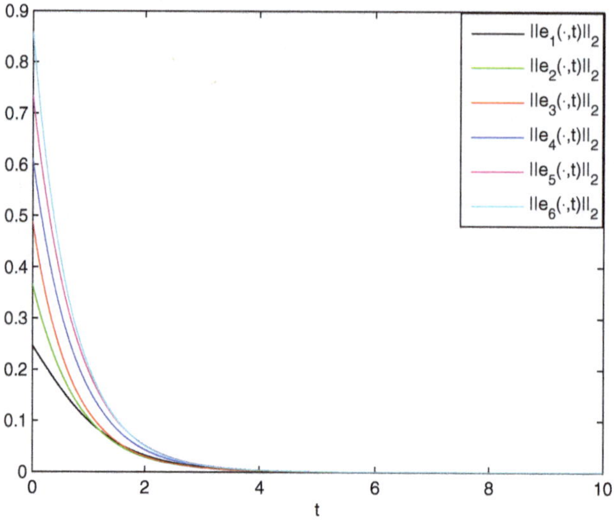

Fig. 3.4 The change processes of $\|e_i(\cdot, t)\|_2, i = 1, 2, \ldots, 6$ ($\|e_1(\cdot, 0)\|_2 = 0.2468$, $\|e_2(\cdot, 0)\|_2 = 0.3676$, $\|e_3(\cdot, 0)\|_2 = 0.4900$, $\|e_4(\cdot, 0)\|_2 = 0.6131$, $\|e_5(\cdot, 0)\|_2 = 0.7365$ and $\|e_6(\cdot, 0)\|_2 = 0.8602$)

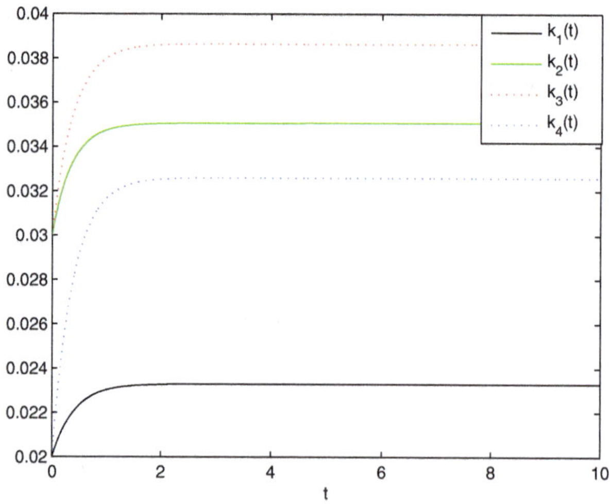

Fig. 3.5 The change processes of $k_i(t), i = 1, 2, 3, 4$ ($k_1(0) = 0.02$, $k_2(0) = 0.03$, $k_3(0) = 0.03$ and $k_4(0) = 0.02$)

Remark 3.17 To the best of our knowledge, very few researchers have investigated the synchronization of CRDNNs by using the pinning control method [97, 101]. In this chapter, the study of the whole network synchronization is reduced to the investigation of the strongly connected components in network. Obviously, our results are totally different from the results in the references [97, 101]. Moreover, in [97, 101], the coupling strength is fixed. In this chapter, an adaptive strategy to tune the coupling strength is designed (see Fig. 3.3).

3.4 Conclusions

In this chapter, pinning synchronization of a class of linearly coupled RDNNs with the Dirichlet boundary conditions has been investigated. Several sufficient conditions ensuring synchronization have been obtained by utilizing the Lyapunov functional method combined with the inequality techniques. Furthermore, some adaptive laws to tune the coupling strength and pinning feedback gains have been developed, and several criteria for synchronization have been established by using the designed adaptive laws. Finally, illustrative simulations have been provided to verify the correctness and effectiveness of the obtained results.

Chapter 4
Impulsive Control for the Synchronization of CRDNNs

4.1 Introduction

Up to date, many researchers have devoted much effort to synchronization problem for CNNs because of its wide applications in different fields. For instance, the linear coupled cellular neural networks have been triumphantly applied to a secure communication system [146] and the electronic circuits [147]. In [76], the authors presented an architecture of CNNs to memorize and reproduce complex oscillatory patterns as synchronization states. Moreover, the research about synchronization of CNNs is a significant step to comprehend brain science [148]. Therefore, studying synchronization problem about coupled neural networks has both practical and theoretical significance [80, 81, 149–157]. Yang et al. [150] studied the finite-time synchronization for an array of coupled neural networks with discontinuous activation functions and mixed delays. A few sufficient conditions were gained to ensure finite-time synchronization of the networks by designing suitable controller. In [151], the authors investigated the exponential synchronization problem for coupled fuzzy neural networks with mixed time-delays and disturbances by utilizing some stochastic analysis methods. Song et al. [153] analyzed the pinning synchronization of coupled delayed neural networks with both constant and delayed couplings.

Nevertheless, in a lot of research works on coupled neural networks [80, 81, 149–157], the diffusion effects are not yet taken into account. Hence, it is extremely important to investigate the synchronization in coupled neural networks with reaction-diffusion terms [95, 97–100, 103]. Liu et al. [97] analyzed the μ-synchronization and pinning control problems for CRDNNs with unbounded time-delays and Dirichlet boundary conditions. In [95], a sufficient condition ensuring synchronization was derived by utilizing the correlation between output strict passivity and synchronization. Wang et al. [98] considered the adaptive synchronization for an array of CRDNNs with time delays.

Impulsive control technique has been widely applied to realize stabilization and synchronization for nonlinear unstable dynamical systems and chaotic systems. The major idea about impulsive control is to change the states of continuous dynamic

J.-L. Wang et al., *Analysis and Control of Coupled Neural Networks with Reaction-Diffusion Terms*, DOI 10.1007/978-981-10-4907-1_4

systems via discontinuous control inputs at certain time instants. From the control point of view, impulsive control method is effective and robust in the research of stability analysis, since it needs only small control gains. Moreover, utilizing the impulsive control method is very advantageous in practical applications due to reduced control cost [116]. More recently, a great deal of attention has been given to impulsive control problem of neural networks with reaction-diffusion terms [30, 116, 158]. Hu et al. [30] gave a few sufficient conditions dependent on the diffusion coefficients to guarantee the global exponential stability and synchronization of delayed RDNNs using the impulsive control strategy. In [116], the authors considered the global exponential stability for Cohen-Grossberg neural networks with reaction-diffusion terms and Dirichlet boundary conditions via the impulsive control method. Yang et al. [158] discussed the problem of stochastic synchronization for RDNNs with mixed delays under impulsive controller. Several sufficient criteria were established by means of the impulsive differential inequality and the properties of random variables. Obviously, it is also beneficial to apply the impulsive control method to study synchronization problem for CRDNNs. Unfortunately, there are very few results concerning the impulsive control for synchronization of CRDNNs. In [101], the authors studied global exponential synchronization for CRDNNs with time-varying delays by adding impulsive controller to a small fraction of nodes.

The main contributions of this chapter are as follows. First, one delay-independent global exponential synchronization condition is derived by using impulsive delay differential inequality. Second, two delay-dependent global exponential synchronization criteria are established by means of suitable Lyapunov functionals and in terms of several linear matrix inequalities. Third, the estimate for exponential convergence rate is provided, which relies on time delay, system parameters and impulsive interval.

4.2 Main Results

In the present chapter, we investigate a CRDNNs with time-varying delay consisting of N identical RDNNs (2.4). The considered mathematical model of the network is characterized by

$$\begin{aligned}\frac{\partial z_i(x,t)}{\partial t} &= D\Delta z_i(x,t) - Az_i(x,t) + J + Bf(z_i(x,t))\\ &\quad + c\sum_{j=1}^{N} G_{ij}\Gamma z_j(x,t-\tau(t)),\end{aligned} \tag{4.1}$$

where $i = 1, 2, \ldots, N$, N denotes the number of nodes in the network; $\tau(t)$ represents the time-varying delay with $0 \leqslant \tau(t) \leqslant \tau$; $z_i(x,t) = (z_{i1}(x,t), z_{i2}(x,t), \ldots, z_{in}(x,t))^T \in \mathbb{R}^n$ denotes the state vector of node i; c is a positive real number, which means the overall coupling strength; $0 < \Gamma = (\gamma_{ij})_{n\times n}$

represents the individual coupling between two nodes; $G = (G_{ij})_{N\times N}$ stands for the topological structure of network and coupling strength between nodes, where G_{ij} is described as follows: if there exists a connection from the node i to the node j $(i \neq j)$, then the coupling $G_{ij} \neq 0$; otherwise, $G_{ij} = 0$ $(i \neq j)$; and the diagonal elements of matrix G satisfy the following condition:

$$G_{ii} = -\sum_{\substack{j=1\\ j\neq i}}^{N} G_{ij},\ i = 1, 2, \ldots, N. \tag{4.2}$$

The initial value and boundary value conditions of network (4.1) are given by

$$z_i(x,t) = \Phi_i(x,t) \in \mathbb{R}^n,\ (x,t) \in \Omega \times [-\tau, 0], \tag{4.3}$$

$$z_i(x,t) = 0,\ (x,t) \in \partial\Omega \times [-\tau, +\infty), \tag{4.4}$$

where $\Phi_i(x,t)(i = 1, 2, \ldots, N)$ is a continuous and bounded function on $\Omega \times [-\tau, 0]$.

Suppose that a vector $w^*(x) = (w_1^*(x), w_2^*(x), \ldots, w_n^*(x))^T \in \mathbb{R}^n$ is an equilibrium solution of the reaction-diffusion neural network (2.4) (or (2.1)), then it satisfies (2.3) and

$$D\Delta w^*(x) - Aw^*(x) + J + Bf(w^*(x)) = 0.$$

Definition 4.1 The network (4.1) is said to achieve global exponential synchronization if there exist constants $\lambda > 0$ and $M \geqslant 1$ such that

$$\|z(\cdot,t,\Phi) - \hat{w}^*(\cdot)\|_2 \leqslant M\|\Phi(\cdot,t) - \hat{w}^*(\cdot)\|_\tau e^{-\lambda t}$$

for all $t \geqslant 0$, where $\Phi(\cdot,t) = (\Phi_1^T(\cdot,t), \Phi_2^T(\cdot,t), \ldots, \Phi_N^T(\cdot,t))^T$, $\hat{w}^*(\cdot) = ((w^*(\cdot))^T, (w^*(\cdot))^T, \ldots, (w^*(\cdot))^T)^T \in \mathbb{R}^{nN}$.

We design the following impulsive control laws to guarantee that network (4.1) is globally exponentially synchronized to desired equilibrium solution:

$$U_i(x,t_k) = c_i^k\big(z_i(x,t_k^-) - w^*(x)\big),\ \ i = 1, 2, \ldots, N, \tag{4.5}$$

where c_i^k denotes a known constant, $k \in \mathbb{N}$.

From (4.1) and (4.5), we can get

$$\begin{cases} \frac{\partial z_i(x,t)}{\partial t} = D\Delta z_i(x,t) - Az_i(x,t) + J + Bf(z_i(x,t)) \\ \qquad\qquad +c\sum_{j=1}^{N} G_{ij}\Gamma z_j(x,t-\tau(t)),\ \ t \neq t_k, \\ I_k(z_i) \ = c_i^k\big(z_i(x,t_k^-) - w^*(x)\big),\ \ t = t_k, \end{cases} \tag{4.6}$$

where $i = 1, 2, \ldots, N$; at time instants t_k, jumps in the state variable $z_i(x,t)$ are characterized by $I_k(z_i) = z_i(x,t_k) - z_i(x,t_k^-)$.

Define $e_i(x,t) = (e_{i1}(x,t), e_{i2}(x,t), \ldots, e_{in}(x,t))^T = z_i(x,t) - w^*(x)$. The dynamics of the error vector $e_i(x,t)$ is given as follows:

$$\begin{cases} \frac{\partial e_i(x,t)}{\partial t} &= D\Delta e_i(x,t) - Ae_i(x,t) + Bf(z_i(x,t)) - Bf(w^*(x)) \\ &\quad +c\sum_{j=1}^{N} G_{ij}\Gamma e_j(x,t-\tau(t)), \;\; t \neq t_k, \\ e_i(x,t_k) = (1+c_i^k)e_i(x,t_k^-), \;\; t = t_k, \end{cases} \tag{4.7}$$

where $i = 1, 2, \ldots, N$.

Theorem 4.2 *If there exist matrices* $0 < P \in \mathbb{R}^{n\times n}, 0 < K = \text{diag}(\alpha_1, \alpha_2, \ldots, \alpha_N) \in \mathbb{R}^{N\times N}$*, real numbers* $\rho > 0, \gamma > 0, \beta_1 > 0$ *and* β *such that*

$$PD + DP \geqslant 0, \tag{4.8}$$

$$\begin{pmatrix} W & \sqrt{K}\otimes(PB) & \sqrt{c}\big((KG)\otimes(P\Gamma)\big) \\ \sqrt{K}\otimes(B^TP) & -I_N\otimes I_n & 0 \\ \sqrt{c}\big((G^TK)\otimes(\Gamma P)\big) & 0 & -\beta_1(K\otimes P) \end{pmatrix} \leqslant 0, \tag{4.9}$$

$$(1+c_i^k)^2 \leqslant \gamma, \; i = 1, 2, \ldots, N, \;\; k \in \mathbb{N}, \tag{4.10}$$

$$\beta + \frac{\ln\gamma}{\rho} + \gamma^{\text{sgn}(\ln\gamma)}c\beta_1 < 0, \tag{4.11}$$

where $W = K\otimes\big[-\sum_{k=1}^{q}\frac{1}{l_k^2}(PD+DP) - PA - AP + \Theta\big] - \beta(K\otimes P)$, $\Theta =$ diag$(\rho_1^2, \rho_2^2, \ldots, \rho_n^2)$, $\sqrt{K} =$diag$(\sqrt{\alpha_1}, \sqrt{\alpha_2}, \ldots, \sqrt{\alpha_N})$, $\rho \geqslant T_{\max}$ *if* $\gamma < 1$*, otherwise,* $\rho \leqslant T_{\min}$*; then the system* (4.7) *achieves global exponential synchronization with decay rate* $\lambda/2$*,* $\lambda > 0$ *is a unique solution of* $\lambda + \beta + \frac{\ln\gamma}{\rho} + \gamma^{\text{sgn}(\ln\gamma)}c\beta_1 e^{\lambda\tau} = 0$.

Proof Construct the following Lyapunov functional:

$$V_1(t) = \sum_{i=1}^{N}\alpha_i\int_\Omega e_i^T(x,t)Pe_i(x,t)dx.$$

The upper right derivative along the trajectory of system (4.7) is calculated as

$$\begin{aligned} D^+V_1(t) =& 2\sum_{i=1}^{N}\alpha_i\int_\Omega e_i^T(x,t)P\frac{\partial e_i(x,t)}{\partial t}dx \\ =& \sum_{i=1}^{N}\alpha_i\int_\Omega\Big[e_i^T(x,t)PD\Delta e_i(x,t) + (\Delta e_i(x,t))^TDPe_i(x,t) \\ & + e_i^T(x,t)(-PA-AP)e_i(x,t) + 2e_i^T(x,t)PB\big(f(z_i(x,t)) \\ & - f(w^*(x))\big) + 2c\sum_{j=1}^{N}G_{ij}e_i^T(x,t)P\Gamma e_j(x,t-\tau(t))\Big]dx. \end{aligned} \tag{4.12}$$

By using the boundary condition and Green's formula, it follows that

$$\int_{\Omega} e_{il}(x,t)\Delta e_{ij}(x,t)dx = -\sum_{k=1}^{q}\int_{\Omega}\frac{\partial e_{il}(x,t)}{\partial x_k}\frac{\partial e_{ij}(x,t)}{\partial x_k}dx,$$

where $l, j \in \{1, 2, \ldots, n\}, i = 1, 2, \ldots, N$.

Letting $P = (p_{jl})_{n\times n}$, one has

$$\begin{aligned}
&\sum_{i=1}^{N}\int_{\Omega} e_i^T(x,t)PD\Delta e_i(x,t)dx \\
&= \sum_{i=1}^{N}\sum_{j=1}^{n}\sum_{l=1}^{n} p_{jl}d_l\int_{\Omega} e_{ij}(x,t)\Delta e_{il}(x,t)dx \\
&= -\sum_{k=1}^{q}\sum_{i=1}^{N}\sum_{j=1}^{n}\sum_{l=1}^{n} p_{jl}d_l\int_{\Omega}\frac{\partial e_{ij}(x,t)}{\partial x_k}\frac{\partial e_{il}(x,t)}{\partial x_k}dx \\
&= -\sum_{k=1}^{q}\sum_{i=1}^{N}\int_{\Omega}\left(\frac{\partial e_i(x,t)}{\partial x_k}\right)^T PD\frac{\partial e_i(x,t)}{\partial x_k}dx,
\end{aligned}$$

$$\begin{aligned}
&\sum_{i=1}^{N}\int_{\Omega}(\Delta e_i(x,t))^T DPe_i(x,t)dx \\
&= \sum_{i=1}^{N}\sum_{j=1}^{n}\sum_{l=1}^{n} d_j p_{jl}\int_{\Omega} e_{il}(x,t)\Delta e_{ij}(x,t)dx \\
&= -\sum_{k=1}^{q}\sum_{i=1}^{N}\int_{\Omega}\left(\frac{\partial e_i(x,t)}{\partial x_k}\right)^T DP\frac{\partial e_i(x,t)}{\partial x_k}dx.
\end{aligned}$$

Then, we can get

$$\begin{aligned}
&\int_{\Omega}\left[e_i^T(x,t)PD\Delta e_i(x,t) + (\Delta e_i(x,t))^T DPe_i(x,t)\right]dx \\
&= -\sum_{k=1}^{q}\int_{\Omega}\left(\frac{\partial e_i(x,t)}{\partial x_k}\right)^T (PD+DP)\frac{\partial e_i(x,t)}{\partial x_k}dx.
\end{aligned}$$

It's obvious that there exists a real matrix $\Psi \in \mathbb{R}^{n\times n}$ such that

$$PD + DP = \Psi^T\Psi,$$

then

$$
\begin{aligned}
&\left(\frac{\partial e_i(x,t)}{\partial x_k}\right)^T (PD+DP)\frac{\partial e_i(x,t)}{\partial x_k}\\
=&\left(\frac{\partial e_i(x,t)}{\partial x_k}\right)^T \Psi^T\Psi\frac{\partial e_i(x,t)}{\partial x_k}\\
=&\left(\frac{\partial(\Psi e_i(x,t))}{\partial x_k}\right)^T \frac{\partial(\Psi e_i(x,t))}{\partial x_k}.
\end{aligned}
$$

Let $\vartheta_i(x,t)=\Psi e_i(x,t)$, for $(x,t)\in\partial\Omega\times[-\tau,+\infty)$ from condition (4.4), we get $\vartheta_i(x,t)=\Psi e_i(x,t)=0$. Then, we can obtain

$$
\begin{aligned}
&\sum_{k=1}^{q}\int_{\Omega}\left(\frac{\partial\vartheta_i(x,t)}{\partial x_k}\right)^T\frac{\partial\vartheta_i(x,t)}{\partial x_k}dx\\
\geqslant&\sum_{k=1}^{q}\frac{1}{l_k^2}\int_{\Omega}\vartheta_i^T(x,t)\vartheta_i(x,t)dx\\
=&\sum_{k=1}^{q}\frac{1}{l_k^2}\int_{\Omega}e_i^T(x,t)(PD+DP)e_i(x,t)dx.
\end{aligned}
\tag{4.13}
$$

Obviously,

$$
\begin{aligned}
&2\sum_{i=1}^{N}e_i^T(x,t)PB\big(f(z_i(x,t))-f(w^*(x))\big)\\
\leqslant&\sum_{i=1}^{N}e_i^T(x,t)PBB^TPe_i(x,t)+\sum_{i=1}^{N}e_i^T(x,t)\Theta e_i(x,t)\\
=&\sum_{i=1}^{N}e_i^T(x,t)(PBB^TP+\Theta)e_i(x,t).
\end{aligned}
\tag{4.14}
$$

It follows from (4.12) to (4.14) that

$$
\begin{aligned}
D^+V_1(t)\leqslant&\sum_{i=1}^{N}\alpha_i\int_{\Omega}e_i^T(x,t)\Big[-\sum_{k=1}^{q}\frac{1}{l_k^2}(PD+DP)-PA-AP+PBB^TP\\
&+\Theta\Big]e_i(x,t)dx+2c\sum_{i=1}^{N}\sum_{j=1}^{N}\alpha_iG_{ij}\int_{\Omega}e_i^T(x,t)P\Gamma e_j(x,t-\tau(t))dx\\
=&2c\int_{\Omega}e^T(x,t)\big((KG)\otimes(P\Gamma)\big)e(x,t-\tau(t))dx
\end{aligned}
$$

$$+\int_{\Omega} e^T(x,t)\Big\{K\otimes\Big[-\sum_{k=1}^{q}\frac{1}{l_k^2}(PD+DP)-PA-AP+PBB^TP$$
$$+\Theta\Big]\Big\}e(x,t)dx. \tag{4.15}$$

Moreover, we can easily derive

$$\begin{aligned}
&2e^T(x,t)\big((KG)\otimes(P\Gamma)\big)e(x,t-\tau(t))\\
&\leqslant \beta_1^{-1}e^T(x,t)\big((KG)\otimes(P\Gamma)\big)(K^{-1}\otimes P^{-1})\big((G^TK)\otimes(\Gamma P)\big)e(x,t)\\
&\quad+\beta_1 e^T(x,t-\tau(t))(K\otimes P)e(x,t-\tau(t)).
\end{aligned} \tag{4.16}$$

By (4.9), (4.15) and (4.16), we can verify that

$$\begin{aligned}
D^+V_1(t)&\leqslant \int_{\Omega} e^T(x,t)\Big\{K\otimes\Big[-\sum_{k=1}^{q}\frac{1}{l_k^2}(PD+DP)-PA-AP+PBB^TP\\
&\quad+\Theta\Big]+c\beta_1\int_{\Omega} e^T(x,t-\tau(t))(K\otimes P)e(x,t-\tau(t))dx\\
&\quad+c\beta_1^{-1}\big((KG)\otimes(P\Gamma)\big)(K^{-1}\otimes P^{-1})\big((G^TK)\otimes(\Gamma P)\big)\Big\}e(x,t)dx\\
&\leqslant c\beta_1\int_{\Omega} e^T(x,t-\tau(t))(K\otimes P)e(x,t-\tau(t))dx\\
&\quad+\beta\int_{\Omega} e^T(x,t)(K\otimes P)e(x,t)dx\\
&=\beta V_1(t)+c\beta_1 V_1(t-\tau(t)).
\end{aligned} \tag{4.17}$$

Furthermore

$$\begin{aligned}
V_1(t_k)&=\sum_{i=1}^{N}\alpha_i(1+c_i^k)^2\int_{\Omega} e_i^T(x,t_k^-)Pe_i(x,t_k^-)dx\\
&\leqslant \gamma\sum_{i=1}^{N}\alpha_i\int_{\Omega} e_i^T(x,t_k^-)Pe_i(x,t_k^-)dx\\
&=\gamma V_1(t_k^-).
\end{aligned}$$

By Lemma 1.20 and (4.11), we obtain

$$V_1(t)\leqslant \alpha^+\lambda_M(P)\gamma^{\mathrm{sgn}(\ln\gamma)}e^{-\lambda t}\|\Phi(\cdot,0)-\hat{w}^*(\cdot)\|_\tau^2,\quad t\geqslant 0, \tag{4.18}$$

where $\alpha^+ = \max_{i=1,2,\ldots,N}\{\alpha_i\}$, $\lambda>0$ is a unique solution of

$$\lambda + \beta + \frac{\ln \gamma}{\rho} + \gamma^{\mathrm{sgn}(\ln \gamma)} c\beta_1 e^{\lambda \tau} = 0,$$

in which $\rho \geqslant T_{\max}$ if $\gamma < 1$, otherwise, $\rho \leqslant T_{\min}$.

On the basis of the definition of $V_1(t)$, we can get

$$\alpha^- \lambda_m(P)\|z(\cdot, t, \Phi) - \hat{w}^*(\cdot)\|_2^2 \leqslant V_1(t), \tag{4.19}$$

where $\alpha^- = \min_{i=1,2,\ldots,N}\{\alpha_i\}$.

By (4.18) and (4.19), we have

$$\begin{aligned} &\|z(\cdot, t, \Phi) - \hat{w}^*(\cdot)\|_2 \\ &\leqslant \sqrt{\frac{\alpha^+ \lambda_M(P)\gamma^{\mathrm{sgn}(\ln \gamma)}}{\alpha^- \lambda_m(P)}} \|\Phi(\cdot, 0) - \hat{w}^*(\cdot)\|_\tau e^{-\lambda t/2}, \quad t \geqslant 0. \end{aligned} \tag{4.20}$$

The proof is completed. □

Theorem 4.3 *Let $\dot{\tau}(t) \leqslant \sigma < 1$. If there exist matrices $0 < P \in \mathbb{R}^{n\times n}, 0 < Q \in \mathbb{R}^{n\times n}, 0 < K = \mathrm{diag}(\alpha_1, \alpha_2, \ldots, \alpha_N) \in \mathbb{R}^{N\times N}$ and positive constants β_2, β_3 such that*

$$\text{(I) } PD + DP \geqslant 0, \tag{4.21}$$

$$\text{(II) } \begin{pmatrix} \Xi_1 & \sqrt{K} \otimes (PB) & \Xi_2 \\ \sqrt{K} \otimes (B^T P) & -I_N \otimes I_n & 0 \\ \Xi_2{}^T & 0 & -I_N \otimes Q \end{pmatrix} \leqslant 0, \tag{4.22}$$

where $\Xi_1 = K \otimes \big[-\sum_{k=1}^{q} \frac{1}{l_k^2}(PD+DP) - PA - AP + \Theta\big] + \frac{I_N \otimes Q}{1-\sigma} - \beta_2(K \otimes P)$, $\Xi_2 = c\big((KG) \otimes (P\Gamma)\big)$, $\Theta = \mathrm{diag}(\rho_1^2, \rho_2^2, \ldots, \rho_n^2)$, $\sqrt{K} = \mathrm{diag}(\sqrt{\alpha_1}, \sqrt{\alpha_2}, \ldots, \sqrt{\alpha_N})$;

(III) there exists a positive integer $m \geqslant 1$ such that $t_{k-m} \leqslant t_k - \tau < t_{k-m+1}$ for any $k \geqslant m, k \in \mathbb{N}$, and the discrete system

$$a(k+1) = A_k(m)a(k), \quad k \geqslant m,$$

is globally exponentially stable with decay rate $\eta > 0$, where $A_k(m)$ is

$$\begin{pmatrix} 0 & 1 & 0 & \cdots & 0 & 0 \\ 0 & 0 & 1 & \cdots & 0 & 0 \\ \vdots & \vdots & \vdots & \cdots & \vdots & \vdots \\ 0 & 0 & 0 & \cdots & 0 & 1 \\ \chi_{k-m} & \chi_{k-m+1} & \chi_{k-m+2} & \cdots & \chi_{k-2} & \tilde{\chi}_{k-1} \end{pmatrix},$$

in which $\beta_3 = \frac{\lambda_M(Q)}{\alpha^-(1-\sigma)\lambda_m(P)}$, $\epsilon_k = \max_{i=1,2,\ldots,N}\{(1+c_i^k)^2\}$, $\tilde{\chi}_{k-1} = \epsilon_k e^{\beta_2 T_{k-1}} + \chi_{k-1}$, $\chi_{k-l} = \beta_3 T_{k-l} e^{\beta_2 T_{k-l}}$, $l = 1, 2, \ldots, m$; *then system* (4.7) *achieves global exponential synchronization with decay rate* $\eta/(2T_{\max})$.

Proof Let us consider the Lyapunov functional $\overline{V}(t) = V_1(t) + V_2(t)$ as

$$V_1(t) = \int_\Omega e^T(x,t)(K \otimes P)e(x,t)dx,$$
$$V_2(t) = \frac{1}{1-\sigma}\int_{t-\tau(t)}^{t}\int_\Omega e^T(x,s)(I_N \otimes Q)e(x,s)dxds.$$

By the proof of Theorem 4.2, let us know

$$\begin{aligned} D^+V_1(t) \leqslant{} & 2c\int_\Omega e^T(x,t)\big((KG)\otimes(P\Gamma)\big)e(x,t-\tau(t))dx \\ & + \int_\Omega e^T(x,t)\Big\{K \otimes \Big[-\sum_{k=1}^{q}\frac{1}{l_k^2}(PD+DP) \\ & - PA - AP + PBB^TP + \Theta\Big]\Big\}e(x,t)dx. \end{aligned} \tag{4.23}$$

Furthermore, we can easily derive

$$\begin{aligned} & 2e^T(x,t)\big((KG)\otimes(P\Gamma)\big)e(x,t-\tau(t)) \\ \leqslant{} & ce^T(x,t)\big((KG)\otimes(P\Gamma)\big)(I_N \otimes Q^{-1})\big((G^TK)\otimes(\Gamma P)\big)e(x,t) \\ & + \frac{1}{c}e^T(x,t-\tau(t))(I_N\otimes Q)e(x,t-\tau(t)). \end{aligned} \tag{4.24}$$

Therefore,

$$\begin{aligned} D^+V_1(t) \leqslant{} & \int_\Omega e^T(x,t)\Big\{K \otimes \Big[-\sum_{k=1}^{q}\frac{1}{l_k^2}(PD+DP) - PA \\ & - AP + PBB^TP + \Theta\Big] + c^2\big((KG)\otimes(P\Gamma)\big) \\ & \times (I_N \otimes Q^{-1})\big((G^TK)\otimes(\Gamma P)\big)\Big\}e(x,t)dx \\ & + \int_\Omega e^T(x,t-\tau(t))(I_N\otimes Q)e(x,t-\tau(t))dx. \end{aligned} \tag{4.25}$$

The upper right derivative $D^+V_2(t)$ along the solution of system (4.7) gives

$$\begin{aligned} D^+V_2(t) &= -\frac{1-\dot{\tau}(t)}{1-\sigma}\int_\Omega e^T(x,t-\tau(t))(I_N\otimes Q)e(x,t-\tau(t))dx \\ &\quad + \frac{1}{1-\sigma}\int_\Omega e^T(x,t)(I_N\otimes Q)e(x,t)dx \\ &\leqslant -\int_\Omega e^T(x,t-\tau(t))(I_N\otimes Q)e(x,t-\tau(t))dx \\ &\quad + \frac{1}{1-\sigma}\int_\Omega e^T(x,t)(I_N\otimes Q)e(x,t)dx. \end{aligned} \tag{4.26}$$

By (4.22), (4.25) and (4.26), we can get

$$\begin{aligned} D^+\overline{V}(t) &= D^+V_1(t) + D^+V_2(t) \\ &\leqslant \int_\Omega e^T(x,t)\Big\{K\otimes\Big[-\sum_{k=1}^{q}\frac{1}{l_k^2}(PD+DP) - PA - AP + PBB^TP \\ &\quad + \Theta\Big] + c^2\big((KG)\otimes(P\Gamma)\big)(I_N\otimes Q^{-1})\big((G^TK)\otimes(\Gamma P)\big) \\ &\quad + \frac{I_N\otimes Q}{1-\sigma}\Big\}e(x,t)dx \\ &\leqslant \beta_2\int_\Omega e^T(x,t)(K\otimes P)e(x,t)dx \\ &= \beta_2 V_1(t) \\ &\leqslant \beta_2\overline{V}(t). \end{aligned} \tag{4.27}$$

From (4.27), we have

$$\overline{V}(t) \leqslant \overline{V}(t_{k-1})e^{\beta_2(t-t_{k-1})}, \quad t\in[t_{k-1},t_k). \tag{4.28}$$

Besides, for $t = t_k$, one has

$$\begin{aligned} V_1(t_k) &= \int_\Omega e^T(x,t_k)(K\otimes P)e(x,t_k)dx \\ &= \sum_{i=1}^{N}\alpha_i\int_\Omega e_i^T(x,t_k)Pe_i(x,t_k)dx \\ &= \sum_{i=1}^{N}\alpha_i(1+c_i^k)^2\int_\Omega e_i^T(x,t_k^-)Pe_i(x,t_k^-)dx \\ &\leqslant \epsilon_k V_1(t_k^-). \end{aligned} \tag{4.29}$$

By (4.28) and the definition of $V_2(t)$, for $t = t_k, k \geqslant m \geqslant 1$, we obtain

$$
\begin{aligned}
V_2(t_k) &= \frac{1}{1-\sigma}\int_{t_k-\tau(t_k)}^{t_k}\int_{\Omega} e^T(x,s)(I_N \otimes Q)e(x,s)dxds \\
&\leqslant \frac{1}{1-\sigma}\sum_{i=1}^{N}\int_{t_{k-m}}^{t_k}\int_{\Omega} e_i^T(x,s)Qe_i(x,s)dxds \\
&\leqslant \frac{\lambda_M(Q)}{\alpha^-(1-\sigma)\lambda_m(P)}\sum_{i=1}^{N}\alpha_i\int_{t_{k-m}}^{t_k}\int_{\Omega} e_i^T(x,s)Pe_i(x,s)dxds \\
&\leqslant \frac{\lambda_M(Q)}{\alpha^-(1-\sigma)\lambda_m(P)}\sum_{l=1}^{m}\int_{t_{k-l}}^{t_{k-l+1}} V_1(s)ds \\
&\leqslant \frac{\lambda_M(Q)}{\alpha^-(1-\sigma)\lambda_m(P)}\sum_{l=1}^{m} T_{k-l}e^{\beta_2 T_{k-l}}\overline{V}(t_{k-l}) \\
&= \beta_3\sum_{l=1}^{m} T_{k-l}e^{\beta_2 T_{k-l}}\overline{V}(t_{k-l}),
\end{aligned} \tag{4.30}
$$

where $\alpha^- = \min_{i=1,2,\ldots,N}\{\alpha_i\}$.

Thus, it follows from (4.29) and (4.30) that

$$
\begin{aligned}
\overline{V}(t_k) &\leqslant (\epsilon_k + \beta_3 T_{k-1})e^{\beta_2 T_{k-1}}\overline{V}(t_{k-1}) + \beta_3\sum_{l=2}^{m} T_{k-l}e^{\beta_2 T_{k-l}}\overline{V}(t_{k-l}) \\
&= \tilde{\chi}_{k-1}\overline{V}(t_{k-1}) + \sum_{l=1}^{m-1}\chi_{k-l-1}\overline{V}(t_{k-l-1}).
\end{aligned} \tag{4.31}
$$

Let

$$
\begin{cases}
v_1(k) = \overline{V}(t_k), \\
v_2(k) = \overline{V}(t_{k+1}), \\
\vdots \\
v_m(k) = \overline{V}(t_{k+m-1}),
\end{cases}
$$

and $W(k) = (v_1(k), v_2(k), \ldots, v_m(k))^T$. By use of (4.31), it follows that

$$
W(k+1-m) \leqslant A_k(m)W(k-m).
$$

And consider the comparison system as

$$
\begin{cases}
a(k+1) = A_k(m)a(k), \quad k \geqslant m, \\
a(m) = W(0).
\end{cases} \tag{4.32}
$$

In view of the comparison principle [111], we derive that

$$W(k-m) \leqslant a(k).$$

The discrete system (4.32) is globally exponentially stable with decay rate $\eta > 0$, so there exists a constant $\varrho \geqslant 1$ such that the following inequality holds:

$$\sqrt{\sum_{j=0}^{m-1} \overline{V}^2(t_{j+k-m})} \leqslant \varrho e^{-\eta(k-m)} \sqrt{\sum_{j=0}^{m-1} \overline{V}^2(t_j)}, \quad k \geqslant m.$$

Hence, we have

$$\overline{V}(t_k) \leqslant \varrho e^{-\eta(k-m+1)} \sqrt{\sum_{j=0}^{m-1} \overline{V}^2(t_j)}, \quad k \geqslant m-1. \tag{4.33}$$

For $k = 1, 2, \ldots, m-1$, it follows that

$$\begin{aligned}
V_2(t_k) &= \frac{1}{1-\sigma} \int_{t_k-\tau(t_k)}^{t_k} \int_{\Omega} e^T(x,s)(I_N \otimes Q)e(x,s)dxds \\
&\leqslant \sum_{i=1}^{N} \frac{1}{1-\sigma} \int_{t_0-\tau}^{t_0} \int_{\Omega} e_i^T(x,s)Qe_i(x,s)dxds \\
&\quad + \sum_{i=1}^{N} \frac{1}{1-\sigma} \int_{t_0}^{t_k} \int_{\Omega} e_i^T(x,s)Qe_i(x,s)dxds \\
&\leqslant \frac{\tau\lambda_M(Q)}{1-\sigma} \|\Phi(\cdot,0) - \hat{w}^*(\cdot)\|_\tau^2 + \frac{\lambda_M(Q)}{\alpha^-(1-\sigma)\lambda_m(P)} \sum_{l=0}^{k-1} \int_{t_l}^{t_{l+1}} V_1(s)ds \\
&\leqslant \frac{\tau\lambda_M(Q)}{1-\sigma} \|\Phi(\cdot,0) - \hat{w}^*(\cdot)\|_\tau^2 + \beta_3 \sum_{l=0}^{k-1} T_l e^{\beta_2 T_l} \overline{V}(t_l),
\end{aligned} \tag{4.34}$$

$$V_1(t_k) \leqslant \epsilon_k e^{\beta_2 T_{k-1}} \overline{V}(t_{k-1}). \tag{4.35}$$

Moreover, by the definitions of $V_1(t)$ and $V_2(t)$, we obtain

$$\begin{aligned}
V_1(0) &\leqslant \alpha^+ \lambda_M(P) \|\Phi(\cdot,0) - \hat{w}^*(\cdot)\|_\tau^2, \\
V_2(0) &\leqslant \frac{\tau\lambda_M(Q)}{1-\sigma} \|\Phi(\cdot,0) - \hat{w}^*(\cdot)\|_\tau^2.
\end{aligned} \tag{4.36}$$

Therefore, from (4.34)–(4.36), there's a positive constant ρ, which is relied on $\beta_3, \beta_2, \alpha^+, \lambda_M(P), \tau, \lambda_M(Q), \sigma, \epsilon_k, T_l, k = 1, 2, \ldots, m-1, l = 0, 1, \ldots, m-2$, such that

$$\sum_{j=0}^{m-1} \overline{V}^2(t_j) \leqslant \rho^2 \|\Phi(\cdot,0) - \hat{w}^*(\cdot)\|_\tau^4. \tag{4.37}$$

Thus, by (4.32), (4.36) and definition of $\overline{V}(t)$, one has

$$\begin{aligned} &\|z(\cdot,t,\Phi) - \hat{w}^*(\cdot)\|_2^2 \\ \leqslant\ & \frac{\overline{V}(t)}{\alpha^-\lambda_m(P)} \\ \leqslant\ & \frac{\rho\varrho}{\alpha^-\lambda_m(P)} e^{-\eta(t/T_{\max}-m)} e^{\beta_2 T_{\max}} \|\Phi(\cdot,0) - \hat{w}^*(\cdot)\|_\tau^2, \quad t \geqslant 0. \end{aligned} \tag{4.38}$$

Namely,

$$\|z(\cdot,t,\Phi) - \hat{w}^*(\cdot)\|_2 \leqslant M\|\Phi(\cdot,0) - \hat{w}^*(\cdot)\|_\tau e^{-\eta t/(2T_{\max})}, \quad t \geqslant 0, \tag{4.39}$$

where $M = \sqrt{(\rho\varrho/\alpha^-\lambda_m(P))e^{\beta_2 T_{\max}+\eta m}}$. The proof is completed. □

When $\tau \leqslant T_{\min}$, it is easy to derive the following theorem.

Theorem 4.4 *Let $\dot{\tau}(t) \leqslant \sigma < 1$. Suppose $\tau \leqslant t_k - t_{k-1}$ for any $k \in \mathbb{N}$, and that there exist matrices $0 < P \in \mathbb{R}^{n\times n}, 0 < Q \in \mathbb{R}^{n\times n}, 0 < K = \mathrm{diag}(\alpha_1, \alpha_2, \ldots, \alpha_N) \in \mathbb{R}^{N\times N}$ and positive constants β_2, β_3 such that*

$$\text{(I) } PD + DP \geqslant 0, \tag{4.40}$$

$$\text{(II) } \begin{pmatrix} \Xi_1 & \sqrt{K}\otimes(PB) & \Xi_2 \\ \sqrt{K}\otimes(B^T P) & -I_N\otimes I_n & 0 \\ \Xi_2^{\ T} & 0 & -I_N\otimes Q \end{pmatrix} \leqslant 0, \tag{4.41}$$

where $\Xi_1 = K\otimes\left[-\sum_{k=1}^{q}\frac{1}{l_k^2}(PD+DP) - PA - AP + \Theta\right] + \frac{I_N\otimes Q}{1-\sigma} - \beta_2(K\otimes P)$, $\Xi_2 = c\big((KG)\otimes(P\Gamma)\big)$, $\Theta = \mathrm{diag}(\rho_1^2, \rho_2^2, \ldots, \rho_n^2)$, $\sqrt{K} = \mathrm{diag}(\sqrt{\alpha_1}, \sqrt{\alpha_2}, \ldots, \sqrt{\alpha_N}\,)$;

(III) for any $k \in \mathbb{N}$

$$max_{i=1,2,\ldots,N}\{(1+c_i^k)^2\} \leqslant \gamma_k < 1, \tag{4.42}$$

(IV) there is a positive constant ϑ such that

$$\ln(\gamma_k + \beta_3\tau) + \beta_2 T_{k-1} \leqslant -\vartheta, \tag{4.43}$$

in which $\beta_3 = \frac{\lambda_M(Q)}{\alpha^-(1-\sigma)\lambda_m(P)}$, then system (4.7) achieves global exponential synchronization with decay rate $\vartheta/(2T_{\max})$.

Proof Let us define the same Lyapunov functional $\overline{V}(t) = V_1(t) + V_2(t)$ as in Theorem 4.3

$$V_1(t) = \int_\Omega e^T(x,t)(K \otimes P)e(x,t)dx,$$
$$V_2(t) = \frac{1}{1-\sigma}\int_{t-\tau(t)}^{t}\int_\Omega e^T(x,s)(I_N \otimes Q)e(x,s)dxds.$$

According to the demonstration of Theorem 4.3, we can know that (4.23)–(4.29) hold. For $t = t_k$, one derives from the definition of $V_2(t)$ that

$$V_2(t_k) \leqslant \beta_3 \int_{t_k-\tau}^{t_k} V_1(s)ds \leqslant \beta_3\tau e^{\beta_2 T_{k-1}}\overline{V}(t_{k-1}). \tag{4.44}$$

Therefore, for $t = t_k$, by (4.28), (4.29) and (4.44), we can get

$$\begin{aligned}\overline{V}(t_k) &\leqslant \gamma_k e^{\beta_2 T_{k-1}}\overline{V}(t_{k-1}) + \beta_3\tau e^{\beta_2 T_{k-1}}\overline{V}(t_{k-1})\\ &= e^{\ln(\gamma_k+\beta_3\tau)+\beta_2 T_{k-1}}\overline{V}(t_{k-1})\\ &\leqslant e^{-\vartheta}\overline{V}(t_{k-1}).\end{aligned} \tag{4.45}$$

From (4.45), for any $k \in \mathbb{N}$, then one obtains

$$\overline{V}(t_k) \leqslant e^{-\vartheta}\overline{V}(t_{k-1}) \leqslant e^{-k\vartheta}\overline{V}(0). \tag{4.46}$$

Hence, for $t \geqslant 0$, it follows from (4.36) and (4.46) that

$$\begin{aligned}&\|z(\cdot,t,\Phi) - \hat{w}^*(\cdot)\|_2^2\\ &\leqslant \frac{\alpha^+\lambda_M(P)+\mu}{\alpha^-\lambda_m(P)} e^{-\vartheta(t/T_{\max}-1)} e^{\beta_2 T_{\max}} \|\Phi(\cdot,0) - \hat{w}^*(\cdot)\|_\tau^2,\end{aligned} \tag{4.47}$$

where $\mu = \frac{\tau\lambda_M(Q)}{1-\sigma}$.

Namely

$$\|z(\cdot,t,\Phi) - \hat{w}^*(\cdot)\|_2 \leqslant M\|\Phi(\cdot,0) - \hat{w}^*(\cdot)\|_\tau e^{-\vartheta t/(2T_{\max})}, \quad t \geqslant 0, \tag{4.48}$$

where $M = \sqrt{\big((\alpha^+\lambda_M(P)+\mu)/(\alpha^-\lambda_m(P))\big)e^{\beta_2 T_{\max}+\vartheta}}$. This completes the proof. □

Remark 4.5 It's worth noting that Theorems 4.2–4.4 not only can be applied to judge global exponential synchronization of CRDNNs (4.7), but also can estimate exponential convergence rate for CRDNNs (4.7). In addition, according to the condition (IV) of Theorem 4.4, we all know that time delay plays a key part in the global exponential synchronization of CRDNNs (4.7).

Remark 4.6 The major difficulty for synchronization investigation of impulsive controlled CRDNNs comes from the reaction-diffusion terms, which can not be disposed by those methods used in traditional complex networks [159–165]. By means of Lyapunov functionals, Green's formula, impulsive delay differential inequality and comparison principle, several sufficient delay-independent and delay-dependent global exponential synchronization conditions are established in the present chapter, which are relied on the reaction-diffusion terms.

Remark 4.7 As far as we know, very few authors have investigated the synchronization of CRDNNs via impulsive control [101]. Yang, Cao and Yang [101] proposed a coupled delayed RDNNs, and analyzed the synchronization of such network model by designing pinning-impulsive controllers. In this chapter, we consider a CRDNNs model with time-varying coupling delay. By utilizing Lyapunov functional method, impulsive delay differential inequality and comparison principle, a few delay-independent and delay-dependent global exponential synchronization criteria are established.

Remark 4.8 Based on the impulsive delay differential inequality, one delay-independent global exponential synchronization condition is gained in Theorem 4.2. Generally, the delay-dependent synchronization criterion is less conservative than the delay-independent one. Thus, one delay-dependent global exponential synchronization condition is obtained by employing the comparison principle in Theorem 4.3.

4.3 Numerical Examples

In this section, we provide two representative examples to show the effectiveness of the above-obtained theoretical results.

Example 4.9 Consider the following CRDNNs consisting of five identical nodes with time-varying delay, in which each node is a 3-dimensional reaction-diffusion neural network:

$$\frac{\partial w_i(x,t)}{\partial t} = d_i \frac{\partial^2 w_i(x,t)}{\partial x^2} - a_i w_i(x,t) + J_i + \sum_{j=1}^{3} b_{ij} f_j(w_j(x,t)), \qquad (4.49)$$

where $i = 1, 2, 3$, $\Omega = \{x \mid -0.5 < x < 0.5\}$, $d_1 = 0.4$, $d_2 = 0.5$, $d_3 = 0.6$, $a_1 = 0.5$, $a_2 = 0.4$, $a_3 = 0.2$, $J_1 = J_2 = J_3 = 0$, $f_j(\xi) = \frac{|\xi+1|-|\xi-1|}{4}$. Taking the matrix $B = (b_{ij})_{3\times 3}$ as follows:

$$B = \begin{pmatrix} 0.6 & 0.4 & 0.5 \\ 0.4 & 0.3 & 0.6 \\ 0.3 & 0.5 & 0.4 \end{pmatrix}.$$

It is obvious that $f_j(\cdot)(j = 1, 2, 3)$ satisfies the Lipschitz condition with $\rho_j = 0.5$, and $w^*(x) = (0, 0, 0)^T \in \mathbb{R}^3$ is an equilibrium solution of the network (4.49).

Setting $c = 0.6$, $c_i^k = -0.5$, $\Gamma = \text{diag}(0.4, 0.6, 0.5)$ and $\tau(t) = 0.3 - 0.3e^{-t}$. Taking the matrix G as follows:

$$G = \begin{pmatrix} -0.3 & 0.3 & 0 & 0 & 0 \\ 0 & -0.5 & 0 & 0.5 & 0 \\ 0 & 0 & -0.7 & 0 & 0.7 \\ 0.1 & 0 & 0.4 & -0.5 & 0 \\ 0 & 0.6 & 0 & 0 & -0.6 \end{pmatrix}.$$

Take $\beta = 0.5$, $\beta_1 = 0.2$ and $K = \text{diag}(0.3, 0.5, 0.6, 0.5, 0.4)$. We can verify that conditions (4.8), (4.9), (4.10) and (4.11) are satisfied if $\rho = 0.1$, $\gamma = 0.25$ and the following positive definite matrix:

$$P = \begin{pmatrix} 0.7853 & -0.0796 & -0.0624 \\ -0.0796 & 0.7247 & -0.0460 \\ -0.0624 & -0.0460 & 0.6895 \end{pmatrix}.$$

According to Theorem 4.2, the network (4.7) with the above parameters achieves global exponential synchronization in the sense of Definition 4.1. Figure 4.1 shows the simulation results.

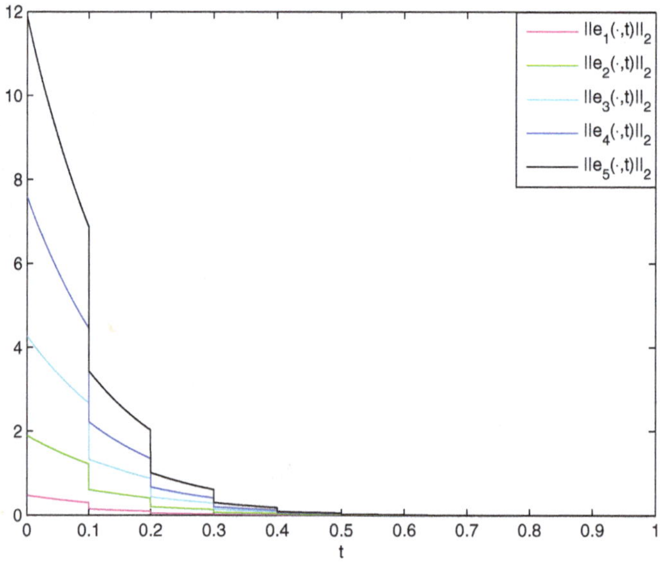

Fig. 4.1 The change processes of $\|e_i(\cdot, t)\|_2, i = 1, 2, \ldots, 5$

Example 4.10 Consider the following CRDNNs consisting of five identical nodes with time-varying delay, in which each node is a 3-dimensional reaction-diffusion neural network:

$$\frac{\partial w_i(x,t)}{\partial t} = d_i \frac{\partial^2 w_i(x,t)}{\partial x^2} - a_i w_i(x,t) + J_i + \sum_{j=1}^{3} b_{ij} f_j(w_j(x,t)), \tag{4.50}$$

where $i = 1, 2, 3,\ f_j(\xi) = \frac{|\xi+1|-|\xi-1|}{4}$, $\Omega = \{x \mid -0.5 < x < 0.5\}, a_1 = 0.4, a_2 = 0.6, a_3 = 0.5, d_1 = 0.4, d_2 = 0.2, d_3 = 0.3, J_1 = J_2 = J_3 = 0$. Taking $B = (b_{ij})_{3\times3}$ as follows:

$$B = \begin{pmatrix} 0.6 & 0.2 & 0.3 \\ 0.7 & 0.3 & 0.5 \\ 0.5 & 0.4 & 0.4 \end{pmatrix}.$$

Obviously, $f_j(\cdot)$ satisfies the Lipschitz condition with $\rho_j = 0.5$ $(j = 1, 2, 3)$, and $w^*(x) = (0, 0, 0)^T \in \mathbb{R}^3$ is an equilibrium point for the network (4.50).

Setting $c_i^k = -0.5, c = 0.2, \tau(t) = 0.1 - 0.1e^{-t}, \tau = 0.1$ and $\Gamma = \mathrm{diag}(0.6, 0.5, 0.2)$. Choosing the matrix G as follows:

$$G = \begin{pmatrix} -0.4 & 0.4 & 0 & 0 & 0 \\ 0 & -0.6 & 0 & 0.6 & 0 \\ 0 & 0 & -0.8 & 0 & 0.8 \\ 0.5 & 0 & 0.2 & -0.7 & 0 \\ 0 & 0.3 & 0 & 0 & -0.3 \end{pmatrix}.$$

Take $\beta_2 = 0.4$, $K = \mathrm{diag}(0.6, 0.5, 0.4, 0.5, 0.7)$ and $T_{k-1} = 0.1$. We can easily verify that conditions (4.40), (4.41), (4.42) and (4.43) are satisfied if $\beta_3 = 3.4937$, $\vartheta = 0.2359$, $\gamma_k = 0.25$ and the following positive definite matrices:

$$P = \begin{pmatrix} 0.8068 & -0.1068 & -0.0661 \\ -0.1068 & 0.9356 & -0.1448 \\ -0.0661 & -0.1448 & 0.8965 \end{pmatrix},\ Q = \begin{pmatrix} 0.7537 & -0.1477 & -0.0998 \\ -0.1477 & 0.5745 & -0.1788 \\ -0.0998 & -0.1788 & 0.7057 \end{pmatrix}.$$

By use of Theorem 4.4, we can see that the network (4.7) realizes global exponential synchronization as depicted in Fig. 4.2.

As seen in Figs. 4.1 and 4.2, from the change processes of $\|e_i(\cdot,t)\|_2, i = 1, 2, \ldots, 5$, we can clearly find $\|e_i(\cdot,t)\|_2, i = 1, 2, \ldots, 5$, are very close to 0 when the time t increases gradually to 0.7s, and this state keeps up along with the increasing of the time. These results demonstrate that $\|e_i(\cdot,t)\|_2, i = 1, 2, \ldots, 5$, asymptotically converges to 0. Hence, CRDNNs (4.7) achieves global exponential synchronization.

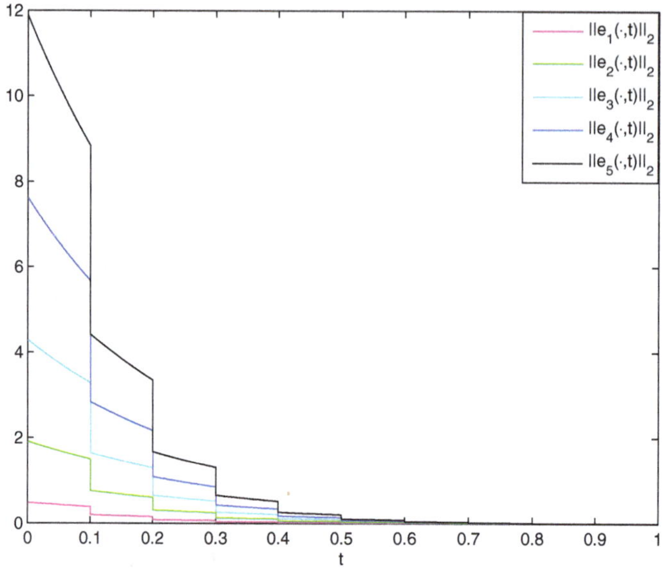

Fig. 4.2 The change processes of $\|e_i(\cdot, t)\|_2, i = 1, 2, \ldots, 5$

4.4 Conclusions

The global exponential synchronization of CNNs with reaction-diffusion terms via impulsive control has been investigated in this chapter, and several synchronization criteria have been established by construction of appropriate Lyapunov functionals and utilization of a few inequality techniques. It is worth noting that the global exponential synchronization conditions are relied on the reaction-diffusion terms. Furthermore, two numerical examples have been provided to verify the validity of the main results. Our future direction is to study the synchronization for CRDNNs with impulsive effects and adaptive coupling.

Chapter 5
Novel Adaptive Strategies for Synchronization of CRDNNs

5.1 Introduction

In most existing works on the complex networks, they always assume that the node state is only dependent on the time. But, in reality, the node state is not only dependent on the time, but also intensively dependent on space variable in many circumstances. As a well-known example of complex networks, food webs attract increasing attention of researchers from different fields in recent years. A food web can be characterized by a model of complex network, in which a node represents a species. To our knowledge, species are usually inhomogeneously distributed in a bounded habitat and the different population densities of predators and preys may cause different population movements, thus it is important and interesting to investigate their spatial density in order to better protect and control their population. In such a case, the state variable of node will represent the spatial density of the species. Obviously, the spatial density of the species is seriously dependent on the time and space. More recently, spatially and temporally complex dynamical network models have been used to describe the dynamic changes of spatial density of species in a bounded spatial habitat by investigators [118, 166, 167]. Moreover, the diffusion phenomena could not be ignored in neural networks and electric circuits once electrons transport in a nonuniform electromagnetic field [28, 30, 33]. Many RDNNs can result in a complex network by mutually coupling [97]. However, very few researchers have investigated the synchronization in an array of linearly coupled neural networks with reaction-diffusion terms [97, 98]. Wang et al. [98] discussed the adaptive synchronization problem for linearly coupled neural networks with reaction-diffusion terms and time delays. Based on the LaSalle invariant principle of functional differential equations and the adaptive feedback control technique, some sufficient conditions for adaptive synchronization were obtained. In the modeling of chemical reactions, it is also essential to consider the diffusion effects. Practically, there are many reaction-diffusion phenomena in nature and discipline fields. Therefore, it is interesting to further study the synchronization of CRDNNs with spatially and temporally varying state variables.

J.-L. Wang et al., *Analysis and Control of Coupled Neural Networks with Reaction-Diffusion Terms*, DOI 10.1007/978-981-10-4907-1_5

On the other hand, recent years have witnessed a rapid growing interest in adaptive control [168–170], which is an important control technique and has been widely used to synchronize complex dynamical networks. Adaptive control has been developed as an effective method to design controller feedback gains for implementing synchronization in complex dynamical networks [67, 68, 120, 134–137, 171, 172]. For instance, Zhou et al. [67] investigated the locally and globally adaptive synchronization of an uncertain complex dynamical network. In [136], Chai et al. studied the synchronization problem of fractional-order dynamical complex networks. Some sufficient conditions ensuring local and global synchronization were derived by using adaptive pinning control. Besides, adaptive control method can also be utilized to tune the strengths of the interconnections among network nodes (see [41, 138–140]). In [138], the global asymptotic stability of two local adaptive strategies, named respectively edge-based and vertex-based adaptation laws, were discussed. In [41], an adaptive strategy to tune the coupling weights of the network was designed based on local information of node dynamics. Moreover, the case that a small fraction of coupling weights can be adjusted was also taken into account. Obviously, it is also beneficial to apply the local adaptive strategies (vertex-based adaptive strategy and edge-based adaptive strategy) to study the synchronization problem of complex dynamical networks with spatially and temporally varying state variables. However, the synchronization of spatially and temporally complex dynamical networks with adaptive coupling has not yet been investigated. Therefore, it is important and interesting to study the synchronization of spatially and temporally complex dynamical networks using the local adaptive strategies.

The main contributions of this chapter are as follows. First, we propose two kinds of linearly coupled neural networks with reaction-diffusion terms, which generalize some existing complex dynamical network models to some extent. Second, we respectively investigate the synchronization problem of these two kinds of complex network models using edge-based adaptive strategy, and some novel adaptive strategies to tune all (or a small fraction of) the coupling strengths are designed. Third, practically, many adaptive control schemes [67, 68, 98, 120, 135–137] were proposed to guarantee the synchronization of complex networks in recent years. However, these adaptive control strategies are based on a special solution of an isolate node of the networks, which may be difficult to obtain in some engineering applications. In this chapter, we remove this restriction.

5.2 Adaptive Synchronization in CRDNNs

In this section, we propose a CRDNNs consisting of N linearly and diffusively coupled identical nodes, in which each node is an n-dimensional reaction-diffusion neural network. Two edge-based adaptive strategies to tune the coupling strengths among network nodes are designed. By utilizing the designed adaptive laws, two sufficient conditions are obtained to guarantee the synchronization of the proposed network model.

5.2.1 Network Model

Consider a weighted CRDNNs consisting of N such RDNNs Sect. (2.4) with linear couplings, which is characterized by

$$\frac{\partial z_i(x,t)}{\partial t} = D\Delta z_i(x,t) - Az_i(x,t) + J + Bf(z_i(x,t)) + c\sum_{j=1}^{N} G_{ij}(t)\Gamma z_j(x,t),\ i = 1, 2, \ldots, N \tag{5.1}$$

where $z_i(x,t) = (z_{i1}(x,t), z_{i2}(x,t), \ldots, z_{in}(x,t))^T \in \mathbb{R}^n$ is the state vector of node i; c is a positive real number, which represents the overall coupling strength; $\Gamma = \mathrm{diag}(\gamma_1, \gamma_2, \ldots, \gamma_n) \in \mathbb{R}^{n\times n}$ is a positive definite inner coupling matrix. The weighted time-varying adjacency matrix $G(t) = (G_{ij}(t))_{N\times N}$ is defined as follows: If there is an edge between node i and node j at time t, then the element $G_{ij}(t) = G_{ji}(t) > 0$; otherwise, $G_{ij}(t) = G_{ji}(t) = 0 (i \neq j)$, and the diagonal elements of matrix $G(t)$ are defined by

$$G_{ii}(t) = -\sum_{\substack{j=1\\ j\neq i}}^{N} G_{ij}(t),\quad i = 1, 2, \ldots, N.$$

The topological structure of CRDNNs is fixed in this chapter. That is, if there is no connection between nodes i and $j (i \neq j)$, then $G_{ij}(t) = G_{ji}(t) = 0$ for all t. The neighbor set of node i is denoted by $\mathcal{N}_i = \{j \in V : (i, j) \in \mathcal{E}\}$.

In this chapter, we always assume CRDNNs (5.1) is connected. The initial value and boundary value conditions associated with network (5.1) are given in the form

$$z_i(x, 0) = \Phi_i(x) \in \mathbb{R}^n, x \in \Omega, \tag{5.2}$$

$$z_i(x,t) = 0, (x,t) \in \partial\Omega \times [0, +\infty) \tag{5.3}$$

where $\Phi_i(x) (i = 1, 2, \ldots, N)$ is bounded and continuous on Ω.

Remark 5.1 In recent years, synchronization in an array of linearly coupled neural networks without reaction-diffusion has been extensively studied by the researchers, see ([85, 88, 173–175] and relevant references therein). However, strictly speaking, diffusion effects can not be avoided in the neural networks when electrons are moving in asymmetric electromagnetic fields, thus we must consider the diffusion effects in neural networks. As a natural extension of the existing network models, we propose a new CRDNNs consisting of N linearly coupled identical RDNNs.

Definition 5.2 The CRDNNs (5.1) is said to achieve synchronization if

$$\lim_{t\to+\infty} \|z_i(\cdot,t) - \frac{1}{N}\sum_{j=1}^{N} z_j(\cdot,t)\|_2 = 0, \text{ for all } i = 1, 2, ..., N.$$

5.2.2 *Adaptive Strategy for Synchronization of CRDNNs*

Let $\bar{z}(x,t) = \frac{1}{N}\sum_{j=1}^{N} z_j(x,t)$. Then, we have

$$\frac{\partial \bar{z}(x,t)}{\partial t} = D\Delta\bar{z}(x,t) - A\bar{z}(x,t) + J + \frac{1}{N}\sum_{j=1}^{N} Bf(z_j(x,t)). \tag{5.4}$$

Defining $e_i(x,t) = z_i(x,t) - \bar{z}(x,t)$, then the dynamics of the error vector $e_i(x,t)$ is governed by the following equation:

$$\begin{aligned}\frac{\partial e_i(x,t)}{\partial t} &= D\Delta e_i(x,t) - Ae_i(x,t) - \frac{1}{N}\sum_{j=1}^{N} Bf(z_j(x,t))\\ &\quad + Bf(z_i(x,t)) + c\sum_{j=1}^{N} G_{ij}(t)\Gamma e_j(x,t)\end{aligned} \tag{5.5}$$

where $i = 1, 2, \ldots, N$.

Theorem 5.3 *The CRDNNs (5.1) is synchronized under the following adaptive law:*

$$\begin{aligned}\dot{G}_{ij}(t) &= \beta_{ij}\int_{\Omega}(z_i(x,t) - z_j(x,t))^T\Gamma(z_i(x,t) - z_j(x,t))dx,\\ G_{ij}(0) &= G_{ji}(0) > 0,\end{aligned} \tag{5.6}$$

$(i,j) \in \mathcal{E}$, *where* $\beta_{ij} = \beta_{ji}$ *are positive constants.*

Proof Construct a Lyapunov functional for the system (5.5) as follows:

$$V_1(t) = \frac{1}{2}\sum_{i=1}^{N}\int_{\Omega} e_i^T(x,t)e_i(x,t)dx + \sum_{i=1}^{N}\sum_{j\in\mathcal{N}_i}\frac{c}{4\beta_{ij}}(G_{ij}(t) - p_{ij})^2, \tag{5.7}$$

where $p_{ij} = p_{ji}(i \neq j, i, j = 1, 2, \ldots, N)$ are non-negative constants, and $p_{ij} = 0(i \neq j)$ if and only if $G_{ij}(t) = 0$.

Calculating the time derivative of $V_1(t)$ along the trajectory of system (5.5), we can get

$$\begin{aligned}
\dot{V}_1(t) &= \sum_{i=1}^{N}\int_{\Omega} e_i^T(x,t)\frac{\partial e_i(x,t)}{\partial t}dx + \sum_{i=1}^{N}\sum_{j\in\mathcal{N}_i}\frac{c}{2\beta_{ij}}(G_{ij}(t)-p_{ij})\dot{G}_{ij}(t) \\
&= \sum_{i=1}^{N}\int_{\Omega} e_i^T(x,t)\Big[D\Delta e_i(x,t) - Ae_i(x,t) - \frac{1}{N}\sum_{j=1}^{N}Bf(z_j(x,t)) \\
&\quad + Bf(z_i(x,t)) + c\sum_{j=1}^{N}G_{ij}(t)\Gamma e_j(x,t)\Big]dx + \frac{c}{2}\sum_{i=1}^{N}\sum_{j\in\mathcal{N}_i}(G_{ij}(t) \\
&\quad - p_{ij})\int_{\Omega}(z_i(x,t)-z_j(x,t))^T\Gamma(z_i(x,t)-z_j(x,t))dx \\
&= \sum_{i=1}^{N}\int_{\Omega} e_i^T(x,t)\Big[D\Delta e_i(x,t) - Ae_i(x,t) - \frac{1}{N}\sum_{j=1}^{N}Bf(z_j(x,t)) \\
&\quad + Bf(\bar{z}(x,t)) - Bf(\bar{z}(x,t)) + c\sum_{j=1}^{N}G_{ij}(t)\Gamma e_j(x,t) \\
&\quad + Bf(z_i(x,t))\Big]dx + \frac{c}{2}\sum_{i=1}^{N}\sum_{j\in\mathcal{N}_i}(G_{ij}(t)-p_{ij})\int_{\Omega}(e_i(x,t) \\
&\quad - e_j(x,t))^T\Gamma(e_i(x,t)-e_j(x,t))dx.
\end{aligned} \tag{5.8}$$

From Green's formula and the Dirichlet boundary condition, we then have

$$\int_{\Omega} e_i^T(x,t)D\Delta e_i(x,t)dx = -\sum_{k=1}^{q}\int_{\Omega}\left(\frac{\partial e_i(x,t)}{\partial x_k}\right)^T D\frac{\partial e_i(x,t)}{\partial x_k}dx. \tag{5.9}$$

According to Lemma 1.22, we can obtain

$$\begin{aligned}
\int_{\Omega} e_i^T(x,t)D\Delta e_i(x,t)dx &\leqslant -\lambda_m(D)\sum_{k=1}^{q}\int_{\Omega}\left(\frac{\partial e_i(x,t)}{\partial x_k}\right)^T\frac{\partial e_i(x,t)}{\partial x_k}dx \\
&\leqslant \sum_{k=1}^{q}\frac{-\lambda_m(D)}{l_k^2}\int_{\Omega} e_i^T(x,t)e_i(x,t)dx.
\end{aligned} \tag{5.10}$$

Since $\sum_{i=1}^{N} e_i^T(x,t) = 0$, we have

$$\sum_{i=1}^{N} e_i^T(x,t)\Big[Bf(\bar{z}(x,t)) - \frac{1}{N}\sum_{j=1}^{N}Bf(z_j(x,t))\Big] = 0. \tag{5.11}$$

Define the matrix $P=(p_{ij})_{N\times N}$, where $p_{ii}=-\sum_{j=1, j\neq i}^{N} p_{ij},\ i=1,2,\ldots,N$. From Lemma 1.17, we can easily derive

$$
\begin{aligned}
&\sum_{i=1}^{N}\sum_{j\in\mathcal{N}_i}(G_{ij}(t)-p_{ij})(e_i(x,t)-e_j(x,t))^T\Gamma(e_i(x,t)-e_j(x,t))\\
&=-2\sum_{i=1}^{N}\sum_{j=1}^{N}(G_{ij}(t)-p_{ij})e_i^T(x,t)\Gamma e_j(x,t). \qquad (5.12)
\end{aligned}
$$

Let $\Theta=\mathrm{diag}(\rho_1^2,\rho_2^2,\ldots,\rho_n^2)$. Obviously

$$
\begin{aligned}
&e_i^T(x,t)B(f(z_i(x,t))-f(\bar{z}(x,t)))\\
&\leqslant \frac{1}{2}e_i^T(x,t)BB^Te_i(x,t)+\frac{1}{2}e_i^T(x,t)\Theta e_i(x,t). \qquad (5.13)
\end{aligned}
$$

It follows from (5.10)–(5.13) that

$$
\begin{aligned}
\dot{V}_1(t)\leqslant&\sum_{i=1}^{N}\int_{\Omega}e_i^T(x,t)\left(\sum_{k=1}^{q}\frac{-\lambda_m(D)}{l_k^2}I_n-A+\frac{\Theta}{2}+\frac{BB^T}{2}\right)e_i(x,t)dx\\
&+c\sum_{i=1}^{N}\sum_{j=1}^{N}p_{ij}\int_{\Omega}e_i^T(x,t)\Gamma e_j(x,t)dx\\
=&\int_{\Omega}e^T(x,t)\left[\sum_{k=1}^{q}\frac{-\lambda_m(D)}{l_k^2}(I_N\otimes I_n)+I_N\otimes\frac{\Theta}{2}+I_N\otimes\frac{BB^T}{2}\right.\\
&\left.-I_N\otimes A+c(P\otimes\Gamma)\right]e(x,t)dx \qquad (5.14)
\end{aligned}
$$

where $e(x,t)=(e_1^T(x,t),e_2^T(x,t),\ldots,e_N^T(x,t))^T$.

According to Lemma 1.17, there obviously exists a unitary matrix $\Psi=(\psi_1,\psi_2,\ldots,\psi_N)\in\mathbb{R}^{N\times N}$ such that

$$
\Psi^TP\Psi=\Lambda
$$

with $\Lambda=\mathrm{diag}(\lambda_1,\lambda_2,\ldots,\lambda_N)$. $\lambda_i, i=1,2,\ldots,N$, are the eigenvalues of P and $0=\lambda_1>\lambda_2\geqslant\lambda_3\geqslant\cdots\geqslant\lambda_N$. Let $y(x,t)=(y_1^T(x,t),y_2^T(x,t),\ldots,y_N^T(x,t))^T=(\Psi^T\otimes I_n)e(x,t)$. Since $\psi_1=\frac{1}{\sqrt{N}}(1,1,\ldots,1)^T$, one has

$$
y_1(x,t)=(\psi_1^T\otimes I_n)e(x,t)=0.
$$

Then, we can derive from (5.14) that

$$\begin{aligned}\dot{V}_1(t) &\leqslant \int_\Omega e^T(x,t)\Big[\sum_{k=1}^{q}\frac{-\lambda_m(D)}{l_k^2}(I_N\otimes I_n)+I_N\otimes\frac{\Theta}{2}+I_N\otimes\frac{BB^T}{2}\\&-I_N\otimes A+c(\Psi\otimes I_n)(\Lambda\otimes\Gamma)(\Psi^T\otimes I_n)\Big]e(x,t)dx\\&=\int_\Omega e^T(x,t)\Big[\sum_{k=1}^{q}\frac{-\lambda_m(D)}{l_k^2}(I_N\otimes I_n)+I_N\otimes\frac{\Theta}{2}+I_N\otimes\frac{BB^T}{2}\\&-I_N\otimes A\Big]e(x,t)dx+\int_\Omega cy^T(x,t)(\Lambda\otimes\Gamma)y(x,t)dx.\end{aligned}\tag{5.15}$$

Since Γ is a positive definite diagonal matrix, we have

$$y^T(x,t)(\Lambda\otimes\Gamma)y(x,t)\leqslant\lambda_2 y^T(x,t)(I_N\otimes\Gamma)y(x,t).$$

Therefore,

$$\begin{aligned}\dot{V}_1(t) &\leqslant \int_\Omega e^T(x,t)\Big[\sum_{k=1}^{q}\frac{-\lambda_m(D)}{l_k^2}(I_N\otimes I_n)+I_N\otimes\frac{\Theta}{2}+I_N\otimes\frac{BB^T}{2}\\&-I_N\otimes A\Big]e(x,t)dx+\int_\Omega c\lambda_2 y^T(x,t)(I_N\otimes\Gamma)y(x,t)dx\\&=\int_\Omega e^T(x,t)\Big[\sum_{k=1}^{q}\frac{-\lambda_m(D)}{l_k^2}(I_N\otimes I_n)+I_N\otimes\frac{\Theta}{2}+I_N\otimes\frac{BB^T}{2}\\&-I_N\otimes A\Big]e(x,t)dx\\&+\int_\Omega c\lambda_2 e^T(x,t)(\Psi\otimes I_n)(I_N\otimes\Gamma)(\Psi^T\otimes I_n)e(x,t)dx\\&=\int_\Omega e^T(x,t)\Big[\sum_{k=1}^{q}\frac{-\lambda_m(D)}{l_k^2}(I_N\otimes I_n)+I_N\otimes\frac{\Theta}{2}+I_N\otimes\frac{BB^T}{2}\\&-I_N\otimes A+c\lambda_2(I_N\otimes\Gamma)\Big]e(x,t)dx.\end{aligned}$$

By selecting p_{ij} sufficiently large such that

$$c\lambda_2\gamma+\frac{\lambda_M(BB^T)}{2}-\lambda_m(A)+\frac{\theta}{2}\leqslant 0,\tag{5.16}$$

one obtains

$$I_N\otimes\frac{\Theta}{2}-I_N\otimes A+I_N\otimes\frac{BB^T}{2}+c\lambda_2(I_N\otimes\Gamma)\leqslant 0$$

where $\gamma = \min_{i=1,2,\ldots,n}\{\gamma_i\}$, $\theta = \max_{i=1,2,\ldots,n}\{\rho_i^2\}$. Therefore,

$$\dot{V}_1(t) \leqslant -\sum_{k=1}^{q} \frac{\lambda_m(D)}{l_k^2} \|e(\cdot, t)\|_2^2. \tag{5.17}$$

Obviously, $V_1(t)$ is non-increasing, and each term of $V_1(t)$ is bounded. Consequently, all the coupling strengths $G_{ij}(t)((i, j) \in \mathcal{E})$ are bounded, and $\lim_{t\to+\infty} V_1(t)$ exists and is a non-negative real number. Since $G_{ij}(t)$ is monotonically increasing (see (5.6)), one can conclude that $G_{ij}(t)((i, j) \in \mathcal{E})$ asymptotically converges to a finite positive value. Therefore, by the definition of $V_1(t)$, we can derive that $\lim_{t\to+\infty} \|e(\cdot, t)\|_2^2$ exists and is a non-negative real number. In what follows, we shall prove that

$$\lim_{t\to+\infty} \|e(\cdot, t)\|_2^2 = 0.$$

If this is not true, we have

$$\lim_{t\to+\infty} \|e(\cdot, t)\|_2^2 = \mu(\mu > 0).$$

Then, there obviously exists a real number $M > 0$ such that

$$\|e(\cdot, t)\|_2^2 > \frac{\mu}{2}$$

for $t \geqslant M$. From (5.17), we can get

$$\dot{V}_1(t) < -\sum_{k=1}^{q} \frac{\lambda_m(D)\mu}{2l_k^2}, \quad t \geqslant M. \tag{5.18}$$

By integrating (5.18) with respect to t over the time period M to $+\infty$, we can obtain

$$\begin{aligned} -V_1(M) \leqslant V_1(+\infty) - V_1(M) &= \int_M^{+\infty} \dot{V}_1(t)dt \\ &< -\int_M^{+\infty} \sum_{k=1}^{q} \frac{\lambda_m(D)\mu}{2l_k^2} dt \\ &= -\infty. \end{aligned}$$

This yields a contradiction, and so

$$\lim_{t\to+\infty} \|e(\cdot, t)\|_2 = 0.$$

Therefore, the CRDNNs (5.1) achieves synchronization under the adaptive law (5.6). The proof is completed.

Remark 5.4 Recently, in order to guarantee the synchronization of complex dynamical networks, some adaptive strategies to tune the coupling weights of the networks have been proposed based on local information of node dynamics [41, 138–140]. However, in these existing works, they always assume that the node state is only dependent on the time. But, in reality, the node state is not only dependent on the time, but also intensively dependent on space variable in many circumstances. To the best of our knowledge, this is the first to consider the synchronization of spatially and temporally complex dynamical networks with adaptive coupling. It is very important in theories and applications and also is a very challenging problem.

5.2.3 Adaptive Pinning Strategy for Synchronization of CRDNNs

In Theorem 5.3, all the coupling strengths are updated according to the adaptive law (5.6). Here, consider the case where only a small fraction of the coupling strengths is adapted.

Suppose that $\hat{\mathcal{E}}$ is a subset of $\mathcal{E}$, and CRDNNs (5.1) is connected through the undirected edges $\hat{\mathcal{E}}$.

For the convenience, we denote

$$H_{ij} = \begin{cases} G_{ij}(0), & \text{if } (i,j) \in \mathcal{E} - \hat{\mathcal{E}}, \\ -\sum_{\substack{j=1 \\ j \neq i}}^{N} H_{ij}, & \text{if } i = j, \\ 0, & \text{otherwise.} \end{cases}$$

Theorem 5.5 *The CRDNNs (5.1) is synchronized under the following adaptive law:*

$$\begin{aligned} &\dot{G}_{ij}(t) = \beta_{ij} \int_{\Omega} (z_i(x,t) - z_j(x,t))^T \Gamma (z_i(x,t) - z_j(x,t)) dx, \text{ if } (i,j) \in \hat{\mathcal{E}}, \\ &\dot{G}_{ij}(t) = 0, \text{ if } (i,j) \in \mathcal{E} - \hat{\mathcal{E}} \end{aligned} \tag{5.19}$$

where $\beta_{ij} = \beta_{ji}$ are positive constants.

Proof Define the following Lyapunov functional for the system (5.5)

$$V_2(t) = \frac{1}{2} \sum_{i=1}^{N} \int_{\Omega} e_i^T(x,t) e_i(x,t) dx + \sum_{i=1}^{N} \sum_{(i,j) \in \hat{\mathcal{E}}} \frac{c}{4\beta_{ij}} (G_{ij}(t) - p_{ij})^2, \tag{5.20}$$

where p_{ij} is defined as follows: if $(i,j) \in \hat{\mathcal{E}}$, then $p_{ij} = p_{ji} > 0$; otherwise, $p_{ij} = 0 (i \neq j)$. Now we define

$$p_{ii} = -\sum_{\substack{j=1 \\ j \neq i}}^{N} p_{ij}, i = 1, 2, \ldots, N$$

$$P = (p_{ij})_{N \times N}.$$

In the following, we calculate the time derivative $\dot{V}_2(t)$ along the solution of system (5.5)

$$\begin{aligned}
\dot{V}_2(t) &= \sum_{i=1}^{N} \int_{\Omega} e_i^T(x,t) \frac{\partial e_i(x,t)}{\partial t} dx + \sum_{i=1}^{N} \sum_{(i,j) \in \hat{\mathcal{E}}} \frac{c}{2\beta_{ij}} (G_{ij}(t) - p_{ij}) \dot{G}_{ij}(t) \\
&= \sum_{i=1}^{N} \int_{\Omega} e_i^T(x,t) \Big[D \Delta e_i(x,t) - A e_i(x,t) - \frac{1}{N} \sum_{j=1}^{N} B f(z_j(x,t)) \\
&\quad + B f(\bar{z}(x,t)) - B f(\bar{z}(x,t)) + c \sum_{j=1}^{N} G_{ij}(t) \Gamma e_j(x,t) \\
&\quad + B f(z_i(x,t)) \Big] dx + \frac{c}{2} \sum_{i=1}^{N} \sum_{(i,j) \in \hat{\mathcal{E}}} (G_{ij}(t) - p_{ij}) \int_{\Omega} (e_i(x,t) \\
&\quad - e_j(x,t))^T \Gamma (e_i(x,t) - e_j(x,t)) dx \\
&\leqslant \sum_{i=1}^{N} \int_{\Omega} e_i^T(x,t) \Big(\sum_{k=1}^{q} \frac{-\lambda_m(D)}{l_k^2} I_n - A + \frac{\Theta}{2} + \frac{B B^T}{2} \Big) e_i(x,t) dx \\
&\quad + c \sum_{i=1}^{N} \sum_{j=1}^{N} H_{ij} \int_{\Omega} e_i^T(x,t) \Gamma e_j(x,t) dx \\
&\quad + c \sum_{i=1}^{N} \sum_{j=1}^{N} p_{ij} \int_{\Omega} e_i^T(x,t) \Gamma e_j(x,t) dx \\
&= \int_{\Omega} e^T(x,t) \Big[\sum_{k=1}^{q} \frac{-\lambda_m(D)}{l_k^2} (I_N \otimes I_n) + I_N \otimes \frac{\Theta}{2} - I_N \otimes A + I_N \otimes \frac{B B^T}{2} \\
&\quad + c(H \otimes \Gamma) + c(P \otimes \Gamma) \Big] e(x,t) dx,
\end{aligned} \tag{5.21}$$

where $e(x,t) = (e_1^T(x,t), e_2^T(x,t), \ldots, e_N^T(x,t))^T$, $H = (H_{ij})_{N \times N}$, $\Theta = \text{diag}(\rho_1^2, \rho_2^2, \ldots, \rho_n^2)$.

Then, following similar arguments as in the proof of Theorem 5.3, we can obtain the desired result immediately.

Remark 5.6 In the past few years, some authors [41, 138–140] have studied the adaptive synchronization of the traditional complex networks, which are described by ordinary differential equations. In Theorems 5.3 and 5.5, two sufficient conditions are obtained to guarantee the synchronization of the CRDNNs (5.1) by utilizing the Lyapunov functional method combined with the inequality techniques. The main difficulty for synchronization analysis of system (5.5) comes from the reaction-diffusion terms, which can not be dealt with by those techniques used in traditional complex networks [41, 138–140].

5.3 Adaptive Synchronization in CRDNNs with Time-Varying Coupling Delay

In this section, a more general CRDNNs model is introduced. With the help of Lyapunov functionals and some inequality techniques, several sufficient conditions are established to ensure that the proposed network model is synchronized under the designed adaptive laws.

5.3.1 Network Model

In this section, a more general network is analyzed,

$$\begin{aligned}\frac{\partial z_i(x,t)}{\partial t} &= D\Delta z_i(x,t) - Az_i(x,t) + J + Bf(z_i(x,t)) + c\sum_{j=1}^{N} G_{ij}(t)\Gamma z_j(x,t)\\ &\quad + \hat{c}\sum_{j=1}^{N} \hat{G}_{ij}\hat{\Gamma} z_j(x,t-\tau(t)) \end{aligned} \tag{5.22}$$

where $i = 1, 2, \ldots, N$, $\tau(t)$ is the time-varying delay with $0 \leqslant \tau(t) \leqslant \tau$.

x, Ω, $z_i(x,t)$, Δ, c, D, A, J, B, Γ, $f(\cdot)$, and $G(t) = (G_{ij}(t))_{N\times N}$ denote the same meanings as those in network (5.1). $\hat{\Gamma} = \mathrm{diag}(\hat{\gamma}_1, \hat{\gamma}_2, \ldots, \hat{\gamma}_n) \in \mathbb{R}^{n\times n}$ is a positive definite matrix, which describes the coupling relation between two nodes for delayed configuration; $\hat{c}$ is a positive real number, which represents the overall coupling strength for delayed configuration; the constant matrix $\hat{G} = (\hat{G}_{ij})_{N\times N}$ represents the topological structure of network and coupling strength between nodes for delayed configuration, where $\hat{G}_{ij}$ is defined as follows: if there exists a connection between node i and node j, then $\hat{G}_{ij} = \hat{G}_{ji} > 0$; otherwise, $\hat{G}_{ij} = \hat{G}_{ji} = 0 (i \neq j)$, and the diagonal elements of matrix $\hat{G}$ are defined by

$$\hat{G}_{ii} = -\sum_{\substack{j=1\\ j\neq i}}^{N} \hat{G}_{ij}, \quad i = 1, 2, \ldots, N.$$

Similarly, we also assume CRDNNs (5.22) is connected. The initial value and boundary value conditions associated with network (5.22) are given in form

$$z_i(x,t) = \Phi_i(x,t) \in \mathbb{R}^n,\ (x,t) \in \Omega \times [-\tau, 0], \tag{5.23}$$
$$z_i(x,t) = 0,\ (x,t) \in \partial\Omega \times [-\tau, +\infty) \tag{5.24}$$

where $i = 1, 2, \ldots, N$, $\Phi_i(x,t)$ is bounded and continuous on $\Omega \times [-\tau, 0]$.

Remark 5.7 It should be noted that network (5.22) represents a more general complex dynamical network than that considered in the literature in the following sense:

(1) The time delay is considered in order to describe more accurately the evolutionary processes of the nodes.

(2) The case that the coupling configurations are related to the current states and the delayed states is also taken into account.

(3) In this network model, the state of node is dependent on the time and space, which can better reflect the dynamical behavior of real systems in some circumstances.

Definition 5.8 The CRDNNs (5.22) is said to achieve synchronization if

$$\lim_{t\to+\infty} \|z_i(\cdot,t) - \frac{1}{N}\sum_{j=1}^{N} z_j(\cdot,t)\|_2 = 0, \text{ for all } i = 1, 2, \ldots, N.$$

5.3.2 Adaptive Strategy for Synchronization of CRDNNs

Let $\bar{z}(x,t) = \frac{1}{N}\sum_{j=1}^{N} z_j(x,t)$ and $e_i(x,t) = z_i(x,t) - \bar{z}(x,t)$. Then, the dynamics of the error vector $e_i(x,t)$ is governed by the following equation:

$$\frac{\partial e_i(x,t)}{\partial t} = D\Delta e_i(x,t) - Ae_i(x,t) - \frac{1}{N}\sum_{j=1}^{N} Bf(z_j(x,t)) + Bf(z_i(x,t)) + c\sum_{j=1}^{N} G_{ij}(t)\Gamma e_j(x,t) + \hat{c}\sum_{j=1}^{N} \hat{G}_{ij}\hat{\Gamma} e_j(x, t-\tau(t)), \tag{5.25}$$

where $i = 1, 2, \ldots, N$.

Theorem 5.9 *Let $\dot{\tau}(t) \leqslant \sigma < 1$. The CRDNNs (5.22) is synchronized under the following adaptive law:*

$$\dot{G}_{ij}(t) = \beta_{ij}\int_{\Omega} (z_i(x,t) - z_j(x,t))^T \Gamma (z_i(x,t) - z_j(x,t))dx,$$
$$G_{ij}(0) = G_{ji}(0) > 0, \tag{5.26}$$

$(i, j) \in \mathcal{E}$, where $\beta_{ij} = \beta_{ji}$ are positive constants.

Proof Construct a Lyapunov functional for the system (5.25) as follows:

$$V_3(t) = \frac{1}{2}\sum_{i=1}^{N}\int_{\Omega} e_i^T(x,t)e_i(x,t)dx + \sum_{i=1}^{N}\sum_{j\in\mathcal{N}_i}\frac{c}{4\beta_{ij}}(G_{ij}(t)-p_{ij})^2$$
$$+\frac{\hat{c}}{2}\sum_{i=1}^{N}\sum_{j=1}^{N}\frac{|\hat{G}_{ij}|}{1-\sigma}\int_{t-\tau(t)}^{t}\int_{\Omega} e_j^T(x,s)\hat{\Gamma}e_j(x,s)dxds$$

where $p_{ij} = p_{ji}(i \neq j, i, j = 1, 2, \ldots, N)$ are non-negative constants, and $p_{ij} = 0(i \neq j)$ if and only if $G_{ij}(t) = 0$.

Calculating the time derivative of $V_3(t)$ along the trajectory of system (5.25), we can get

$$\dot{V}_3(t) \leqslant \sum_{i=1}^{N}\int_{\Omega} e_i^T(x,t)\frac{\partial e_i(x,t)}{\partial t}dx + \sum_{i=1}^{N}\sum_{j\in\mathcal{N}_i}\frac{c}{2\beta_{ij}}(G_{ij}(t)-p_{ij})\dot{G}_{ij}(t)$$
$$-\frac{\hat{c}}{2}\sum_{i=1}^{N}\sum_{j=1}^{N}|\hat{G}_{ij}|\int_{\Omega} e_j^T(x,t-\tau(t))\hat{\Gamma}e_j(x,t-\tau(t))dx$$
$$+\frac{\hat{c}}{2}\sum_{i=1}^{N}\sum_{j=1}^{N}\frac{|\hat{G}_{ij}|}{1-\sigma}\int_{\Omega} e_j^T(x,t)\hat{\Gamma}e_j(x,t)dx$$
$$=\sum_{i=1}^{N}\int_{\Omega} e_i^T(x,t)\Big[D\Delta e_i(x,t) - Ae_i(x,t) + \hat{c}\sum_{j=1}^{N}\hat{G}_{ij}\hat{\Gamma}e_j(x,t-\tau(t))$$
$$-\frac{1}{N}\sum_{j=1}^{N}Bf(z_j(x,t)) + Bf(z_i(x,t)) + c\sum_{j=1}^{N}G_{ij}(t)\Gamma e_j(x,t)\Big]dx$$
$$+\frac{c}{2}\sum_{i=1}^{N}\sum_{j\in\mathcal{N}_i}(G_{ij}(t)-p_{ij})\int_{\Omega}(z_i(x,t)-z_j(x,t))^T\Gamma(z_i(x,t)$$
$$-z_j(x,t))dx + \frac{\hat{c}}{2}\sum_{i=1}^{N}\sum_{j=1}^{N}\frac{|\hat{G}_{ij}|}{1-\sigma}\int_{\Omega} e_j^T(x,t)\hat{\Gamma}e_j(x,t)dx$$
$$-\frac{\hat{c}}{2}\sum_{i=1}^{N}\sum_{j=1}^{N}|\hat{G}_{ij}|\int_{\Omega} e_j^T(x,t-\tau(t))\hat{\Gamma}e_j(x,t-\tau(t))dx$$
$$\leqslant\sum_{i=1}^{N}\int_{\Omega} e_i^T(x,t)\Big[D\Delta e_i(x,t) - Ae_i(x,t) + Bf(\bar{z}(x,t)) - Bf(\bar{z}(x,t))$$
$$-\frac{1}{N}\sum_{j=1}^{N}Bf(z_j(x,t)) + Bf(z_i(x,t)) + c\sum_{j=1}^{N}G_{ij}(t)\Gamma e_j(x,t)\Big]dx$$

$$
\begin{aligned}
&+\frac{c}{2}\sum_{i=1}^{N}\sum_{j\in\mathcal{N}_i}(G_{ij}(t)-p_{ij})\int_{\Omega}(e_i(x,t)-e_j(x,t))^T\Gamma(e_i(x,t)\\
&-e_j(x,t))dx+\frac{\hat{c}}{2}\sum_{i=1}^{N}\sum_{j=1}^{N}\frac{|\hat{G}_{ij}|}{1-\sigma}\int_{\Omega}e_j^T(x,t)\hat{\Gamma}e_j(x,t)dx\\
&+\frac{\hat{c}}{2}\sum_{i=1}^{N}\sum_{j=1}^{N}|\hat{G}_{ij}|\int_{\Omega}e_i^T(x,t)\hat{\Gamma}e_i(x,t)dx\\
\leqslant&\sum_{i=1}^{N}\int_{\Omega}e_i^T(x,t)\Big[\sum_{k=1}^{q}\frac{-\lambda_m(D)}{l_k^2}I_n+\frac{\Theta}{2}+\frac{BB^T}{2}+\frac{\hat{c}|\hat{G}_{ii}|(2-\sigma)}{1-\sigma}\hat{\Gamma}\\
&-A\Big]e_i(x,t)dx+c\sum_{i=1}^{N}\sum_{j=1}^{N}p_{ij}\int_{\Omega}e_i^T(x,t)\Gamma e_j(x,t)dx,
\end{aligned}
$$

where $\Theta=\mathrm{diag}(\rho_1^2,\rho_2^2,\ldots,\rho_n^2)$.

Define the matrix $P=(p_{ij})_{N\times N}$, where $p_{ii}=-\sum_{\substack{j=1\\ j\neq i}}^{N}p_{ij}$, $i=1,2,\ldots,N$. Then, we can obtain

$$
\begin{aligned}
\dot{V}_3(t)\leqslant&\int_{\Omega}e^T(x,t)\Big[\sum_{k=1}^{q}\frac{-\lambda_m(D)}{l_k^2}(I_N\otimes I_n)+I_N\otimes\frac{\Theta}{2}-I_N\otimes A\\
&+I_N\otimes\frac{BB^T}{2}+\hat{c}(\bar{G}\otimes\hat{\Gamma})+c(P\otimes\Gamma)\Big]e(x,t)dx
\end{aligned}
$$

where $\bar{G}=\mathrm{diag}(\frac{|\hat{G}_{11}|(2-\sigma)}{1-\sigma},\frac{|\hat{G}_{22}|(2-\sigma)}{1-\sigma},\ldots,\frac{|\hat{G}_{NN}|(2-\sigma)}{1-\sigma})$, $e(x,t)=(e_1^T(x,t),e_2^T(x,t),\ldots,e_N^T(x,t))^T$.

By a minor modification of the proof of Theorem 5.3, we can easily derive

$$
\dot{V}_3(t)\leqslant-\sum_{k=1}^{q}\frac{\lambda_m(D)}{l_k^2}\|e(\cdot,t)\|_2^2.
$$

Then, we have

$$
\|e(\cdot,t)\|_2^2\leqslant-\frac{\dot{V}_3(t)}{\sum_{k=1}^{q}\frac{\lambda_m(D)}{l_k^2}}.
$$

Since $V_3(t)$ is monotonically decreasing and bounded, we can conclude that $V_3(t)$ asymptotically converges to a finite non-negative value.

Thus, $\lim_{t\to+\infty}\int_0^t\|e(\cdot,s)\|_2^2ds$ exists and is finite. Since $0\leqslant\tau(t)\leqslant\tau$, we can get

$$
\lim_{t\to+\infty}\int_{t-\tau(t)}^{t}\int_{\Omega}e_j^T(x,s)\hat{\Gamma}e_j(x,s)dxds=0\ (j=1,2,\ldots,N).
$$

Then, similar to the proof of Theorem 5.3, we can obtain

$$\lim_{t\to+\infty} \|e(\cdot,t)\|_2 = 0.$$

Therefore, the CRDNNs (5.22) achieves synchronization under the adaptive law (5.26). The proof is completed.

5.3.3 *Adaptive Pinning Strategy for Synchronization of CRDNNs*

In Theorem 5.9, all the coupling strengths are adjusted according to the adaptive law (5.26). In this subsection, consider the case where only a small fraction of the coupling strengths is adapted.

Similarly, suppose that $\hat{\mathcal{E}}$ is a subset of $\mathcal{E}$, and CRDNNs (5.22) is connected through the undirected edges $\hat{\mathcal{E}}$.

For the convenience, we denote

$$H_{ij} = \begin{cases} G_{ij}(0), & \text{if } (i,j)\in\mathcal{E}-\hat{\mathcal{E}}, \\ -\sum_{\substack{j=1\\ j\neq i}}^{N} H_{ij}, & \text{if } i=j, \\ 0, & \text{otherwise.} \end{cases}$$

Theorem 5.10 *Let $\dot{\tau}(t)\leqslant\sigma<1$. The CRDNNs (5.22) is synchronized under the following adaptive law:*

$$\begin{aligned} &\dot{G}_{ij}(t) = \beta_{ij}\int_{\Omega}(z_i(x,t)-z_j(x,t))^T\Gamma(z_i(x,t)-z_j(x,t))dx, \text{ if } (i,j)\in\hat{\mathcal{E}}, \\ &\dot{G}_{ij}(t) = 0, \text{ if } (i,j)\in\mathcal{E}-\hat{\mathcal{E}} \end{aligned} \tag{5.27}$$

where $\beta_{ij}=\beta_{ji}$ are positive constants.

Proof Define the following Lyapunov functional for the system (5.25)

$$\begin{aligned} V_4(t) = {}& \frac{1}{2}\sum_{i=1}^{N}\int_{\Omega} e_i^T(x,t)e_i(x,t)dx + \sum_{i=1}^{N}\sum_{(i,j)\in\hat{\mathcal{E}}}\frac{c}{4\beta_{ij}}(G_{ij}(t)-p_{ij})^2 \\ & + \frac{\hat{c}}{2}\sum_{i=1}^{N}\sum_{j=1}^{N}\frac{|\hat{G}_{ij}|}{1-\sigma}\int_{t-\tau(t)}^{t}\int_{\Omega} e_j^T(x,s)\hat{\Gamma}e_j(x,s)dxds, \end{aligned} \tag{5.28}$$

where p_{ij} is defined as follows: if $(i,j)\in\hat{\mathcal{E}}$, then $p_{ij}=p_{ji}>0$; otherwise, $p_{ij}=0 (i\neq j)$. Now we define

$$p_{ii} = -\sum_{\substack{j=1\\ j\neq i}}^{N} p_{ij}, i = 1, 2, \ldots, N$$

$$P = (p_{ij})_{N\times N}.$$

In the following, we calculate the time derivative $\dot{V}_4(t)$ along the solution of system (5.25)

$$\begin{aligned}
\dot{V}_4(t) \leqslant & \sum_{i=1}^{N} \int_{\Omega} e_i^T(x,t)\frac{\partial e_i(x,t)}{\partial t}dx + \sum_{i=1}^{N}\sum_{(i,j)\in\hat{\mathcal{E}}} \frac{c}{2\beta_{ij}}(G_{ij}(t) - p_{ij})\dot{G}_{ij}(t) \\
& - \frac{\hat{c}}{2}\sum_{i=1}^{N}\sum_{j=1}^{N} |\hat{G}_{ij}| \int_{\Omega} e_j^T(x,t-\tau(t))\hat{\Gamma} e_j(x,t-\tau(t))dx \\
& + \frac{\hat{c}}{2}\sum_{i=1}^{N}\sum_{j=1}^{N} \frac{|\hat{G}_{ij}|}{1-\sigma} \int_{\Omega} e_j^T(x,t)\hat{\Gamma} e_j(x,t)dx \\
= & \sum_{i=1}^{N} \int_{\Omega} e_i^T(x,t)\Big[D\Delta e_i(x,t) - Ae_i(x,t) + Bf(\bar{z}(x,t)) - Bf(\bar{z}(x,t)) \\
& + Bf(z_i(x,t)) - \frac{1}{N}\sum_{j=1}^{N} Bf(z_j(x,t)) + c\sum_{j=1}^{N} G_{ij}(t)\Gamma e_j(x,t) \\
& + \hat{c}\sum_{j=1}^{N} \hat{G}_{ij}\hat{\Gamma} e_j(x,t-\tau(t))\Big]dx + \frac{c}{2}\sum_{i=1}^{N}\sum_{(i,j)\in\hat{\mathcal{E}}} (G_{ij}(t) \\
& - p_{ij}) \int_{\Omega} (e_i(x,t) - e_j(x,t))^T \Gamma (e_i(x,t) - e_j(x,t))dx \\
& - \frac{\hat{c}}{2}\sum_{i=1}^{N}\sum_{j=1}^{N} |\hat{G}_{ij}| \int_{\Omega} e_j^T(x,t-\tau(t))\hat{\Gamma} e_j(x,t-\tau(t))dx \\
& + \frac{\hat{c}}{2}\sum_{i=1}^{N}\sum_{j=1}^{N} \frac{|\hat{G}_{ij}|}{1-\sigma} \int_{\Omega} e_j^T(x,t)\hat{\Gamma} e_j(x,t)dx \\
\leqslant & \sum_{i=1}^{N} \int_{\Omega} e_i^T(x,t)\Big[D\Delta e_i(x,t) - Ae_i(x,t) - \frac{1}{N}\sum_{j=1}^{N} Bf(z_j(x,t)) \\
& + Bf(\bar{z}(x,t)) - Bf(\bar{z}(x,t)) + Bf(z_i(x,t)) + c\sum_{j=1}^{N} G_{ij}(t)\Gamma e_j(x,t)\Big]dx \\
& + \frac{c}{2}\sum_{i=1}^{N}\sum_{(i,j)\in\hat{\mathcal{E}}} (G_{ij}(t) - p_{ij}) \int_{\Omega} (e_i(x,t) - e_j(x,t))^T \Gamma (e_i(x,t)
\end{aligned}$$

$$- e_j(x,t))dx + \frac{\hat{c}}{2}\sum_{i=1}^{N}\sum_{j=1}^{N}\frac{|\hat{G}_{ij}|}{1-\sigma}\int_{\Omega} e_j^T(x,t)\hat{\Gamma}e_j(x,t)dx$$

$$+ \frac{\hat{c}}{2}\sum_{i=1}^{N}\sum_{j=1}^{N}|\hat{G}_{ij}|\int_{\Omega} e_i^T(x,t)\hat{\Gamma}e_i(x,t)dx$$

$$\leqslant \sum_{i=1}^{N}\int_{\Omega} e_i^T(x,t)\Big[\sum_{k=1}^{q}\frac{-\lambda_m(D)}{l_k^2}I_n + \frac{\Theta}{2} + \frac{BB^T}{2} + \frac{\hat{c}|\hat{G}_{ii}|(2-\sigma)}{1-\sigma}\hat{\Gamma}$$

$$- A\Big]e_i(x,t)dx + c\sum_{i=1}^{N}\sum_{j=1}^{N}H_{ij}\int_{\Omega} e_i^T(x,t)\Gamma e_j(x,t)dx$$

$$+ c\sum_{i=1}^{N}\sum_{j=1}^{N}p_{ij}\int_{\Omega} e_i^T(x,t)\Gamma e_j(x,t)dx$$

$$= \int_{\Omega} e^T(x,t)[\sum_{k=1}^{q}\frac{-\lambda_m(D)}{l_k^2}(I_N\otimes I_n) + I_N\otimes\frac{\Theta}{2} - I_N\otimes A + I_N\otimes\frac{BB^T}{2}$$

$$+ \hat{c}(\bar{G}\otimes\hat{\Gamma}) + c(H\otimes\Gamma) + c(P\otimes\Gamma)]e(x,t)dx$$

where $\bar{G} = \mathrm{diag}(\frac{|\hat{G}_{11}|(2-\sigma)}{1-\sigma}, \frac{|\hat{G}_{22}|(2-\sigma)}{1-\sigma}, \ldots, \frac{|\hat{G}_{NN}|(2-\sigma)}{1-\sigma})$, $e(x,t) = (e_1^T(x,t), e_2^T(x,t), \ldots, e_N^T(x,t))^T$, $\Theta = \mathrm{diag}(\rho_1^2, \rho_2^2, \ldots, \rho_n^2)$, $H = (H_{ij})_{N\times N}$.

In what follows, by the similar proof of Theorem 5.9, we can obtain the conclusion. Here we omit its proof to avoid the repetition.

5.4 Numerical Examples

In this section, we give two examples and their simulation to show the effectiveness of the above obtained theoretical results.

Example 5.11 Consider a CRDNNs consisting of 5 identical nodes with diffusive coupling, in which each node is a 3-dimensional reaction-diffusion neural network described by

$$\frac{\partial w_i(x,t)}{\partial t} = d_i\frac{\partial^2 w_i(x,t)}{\partial x^2} - a_i w_i(x,t) + J_i + \sum_{j=1}^{3} b_{ij}f_j(w_j(x,t))$$

where $i = 1, 2, 3$, $\Omega = \{x \mid -0.5 < x < 0.5\}$, $f_j(\xi) = \frac{|\xi+1|-|\xi-1|}{2}$, $d_1 = 0.15$, $d_2 = 0.1$, $d_3 = 0.2$, $a_1 = 0.25$, $a_2 = 0.2$, $a_3 = 0.3$, $J_1 = J_2 = J_3 = 0$, and the matrix $B = (b_{ij})_{3\times 3}$ is chosen as

$$B = \begin{pmatrix} 1 & 0.4 & 0.3 \\ 0.5 & 1 & 0.3 \\ 0.6 & 0.5 & 1 \end{pmatrix}.$$

Obviously, $f_j(\cdot)(j = 1, 2, 3)$ satisfies the Lipschitz condition with $\rho_j = 1$.

We take $c = 0.2$ and $\Gamma = \text{diag}(0.5, 0.4, 0.6)$. The matrix $G(0)$ is chosen as

$$G(0) = \begin{pmatrix} -1.2 & 0.4 & 0.3 & 0.2 & 0.3 \\ 0.4 & -0.9 & 0.5 & 0 & 0 \\ 0.3 & 0.5 & -0.8 & 0 & 0 \\ 0.2 & 0 & 0 & -0.6 & 0.4 \\ 0.3 & 0 & 0 & 0.4 & -0.7 \end{pmatrix}.$$

Case 1: Choose $\beta_{12} = \beta_{21} = 0.2, \beta_{13} = \beta_{31} = 0.3, \beta_{14} = \beta_{41} = 0.4, \beta_{15} = \beta_{51} = 0.2, \beta_{23} = \beta_{32} = 0.2, \beta_{45} = \beta_{54} = 0.3$. According to Theorem 5.3, the CRDNNs achieves synchronization. The simulation results are shown in Figs. 5.1 and 5.2.

Case 2: We select a portion of undirected links in the network. Choose $\beta_{12} = \beta_{21} = 0.2, \beta_{13} = \beta_{31} = 0.3, \beta_{15} = \beta_{51} = 0.2, \beta_{45} = \beta_{54} = 0.3$. It follows from Theorem 5.5 that the CRDNNs achieves synchronization. The simulation results are shown in Figs. 5.3 and 5.4.

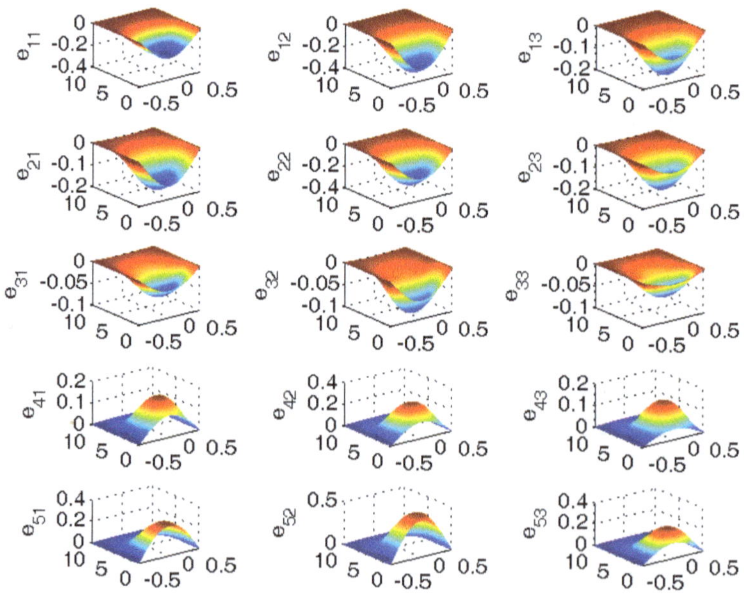

Fig. 5.1 The change processes of $e_{ij}(x, t), i = 1, 2, \ldots, 5, j = 1, 2, 3$

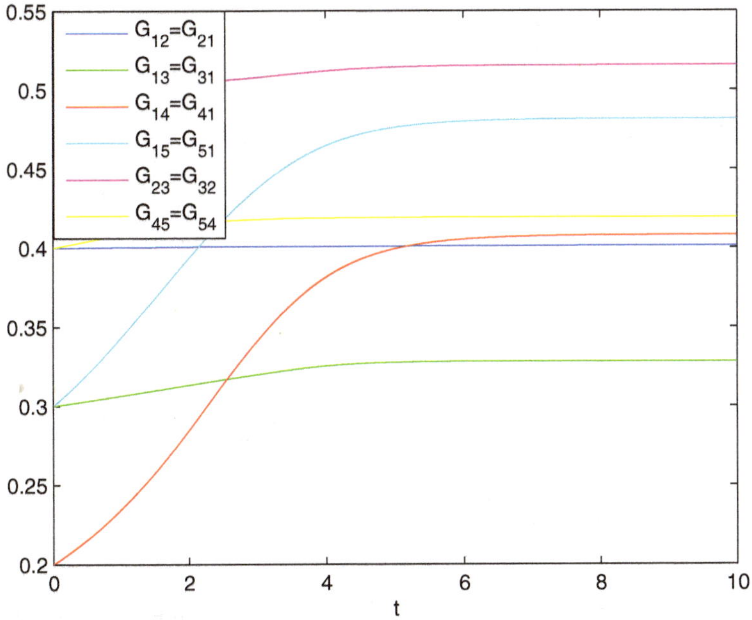

Fig. 5.2 Adaptive coupling strengths

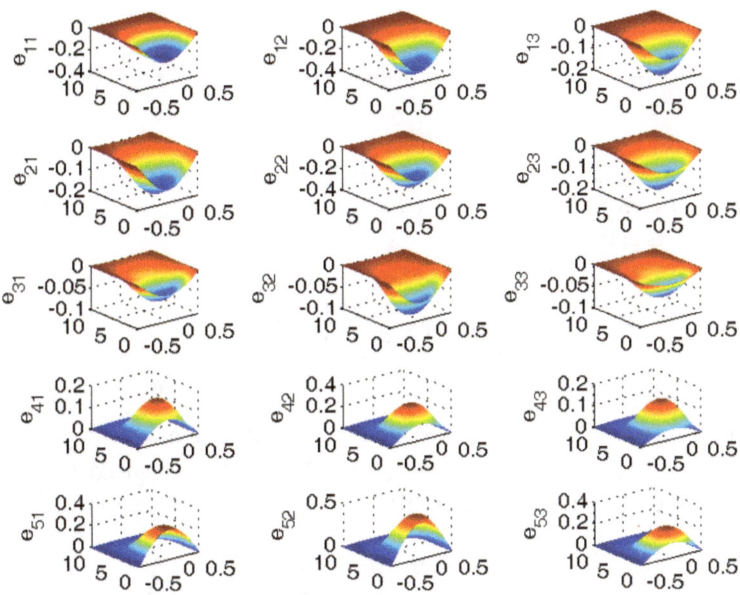

Fig. 5.3 The change processes of $e_{ij}(x,t)$, $i=1,2,\ldots,5$, $j=1,2,3$

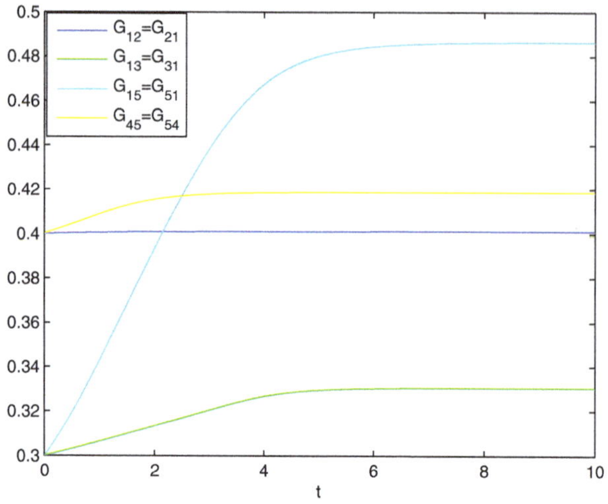

Fig. 5.4 Adaptive coupling strengths

Example 5.12 Consider a CRDNNs consisting of 5 identical nodes with diffusive and delay coupling, in which each node is a 3-dimensional reaction-diffusion neural network described by

$$\frac{\partial w_i(x,t)}{\partial t} = d_i \frac{\partial^2 w_i(x,t)}{\partial x^2} - a_i w_i(x,t) + J_i + \sum_{j=1}^{3} b_{ij} f_j(w_j(x,t))$$

where $i = 1, 2, 3$, $\Omega = \{x \mid -0.5 < x < 0.5\}$, $f_j(\xi) = \frac{|\xi+1|-|\xi-1|}{4}$, $d_1 = 0.1$, $d_2 = 0.15$, $d_3 = 0.1$, $a_1 = 0.5$, $a_2 = 0.3$, $a_3 = 0.2$, $J_1 = J_2 = J_3 = 0$, and the matrix $B = (b_{ij})_{3\times3}$ is chosen as

$$B = \begin{pmatrix} 1 & 0.5 & 0.4 \\ 0.3 & 1 & 0.8 \\ 0.4 & 0.7 & 1 \end{pmatrix}.$$

Obviously, $f_j(\cdot)(j = 1, 2, 3)$ satisfies the Lipschitz condition with $\rho_j = \frac{1}{2}$.

We take $c = 0.3$, $\hat{c} = 0.4$, $\Gamma = \mathrm{diag}(0.4, 0.3, 0.5)$, $\hat{\Gamma} = \mathrm{diag}(0.6, 0.5, 0.8)$ and $\tau(t) = 0.2 - 0.2e^{-t}$. The matrices $G(0)$ and $\hat{G}$ are chosen as, respectively,

$$G(0) = \begin{pmatrix} -0.9 & 0 & 0.3 & 0.4 & 0.2 \\ 0 & -0.5 & 0.5 & 0 & 0 \\ 0.3 & 0.5 & -0.9 & 0.1 & 0 \\ 0.4 & 0 & 0.1 & -0.7 & 0.2 \\ 0.2 & 0 & 0 & 0.2 & -0.4 \end{pmatrix},$$

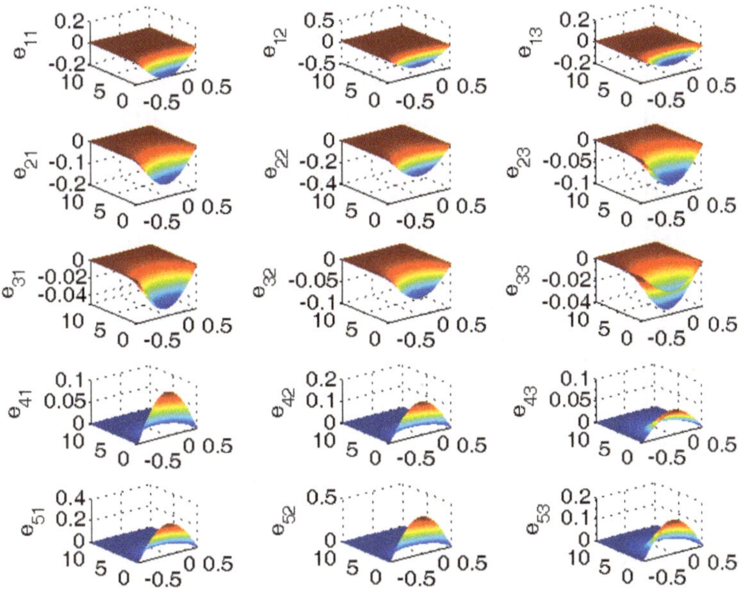

Fig. 5.5 The change processes of $e_{ij}(x,t), i = 1, 2, \ldots, 5, j = 1, 2, 3$

$$\hat{G} = \begin{pmatrix} -1.1 & 0 & 0.4 & 0.3 & 0.4 \\ 0 & -0.6 & 0.6 & 0 & 0 \\ 0.4 & 0.6 & -1.2 & 0.2 & 0 \\ 0.3 & 0 & 0.2 & -0.8 & 0.3 \\ 0.4 & 0 & 0 & 0.3 & -0.7 \end{pmatrix}.$$

Case 1: Choose $\beta_{13} = \beta_{31} = 0.4, \beta_{14} = \beta_{41} = 0.5, \beta_{15} = \beta_{51} = 0.3, \beta_{23} = \beta_{32} = 0.2, \beta_{34} = \beta_{43} = 0.4, \beta_{45} = \beta_{54} = 0.3$. According to Theorem 5.9, the CRDNNs achieves synchronization. The simulation results are shown in Figs. 5.5 and 5.6.

Case 2: We select a portion of undirected links in the network. Choose $\beta_{15} = \beta_{51} = 0.3, \beta_{23} = \beta_{32} = 0.2, \beta_{34} = \beta_{43} = 0.4, \beta_{45} = \beta_{54} = 0.3$. It follows from Theorem 4.2 that the CRDNNs achieves synchronization. The simulation results are shown in Figs. 5.7 and 5.8.

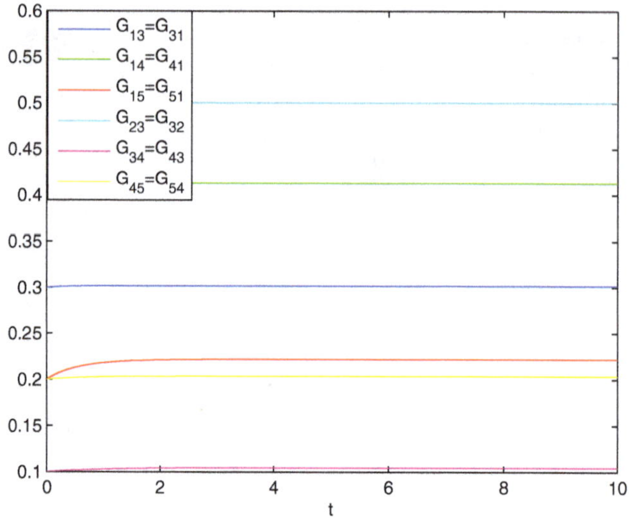

Fig. 5.6 Adaptive coupling strengths

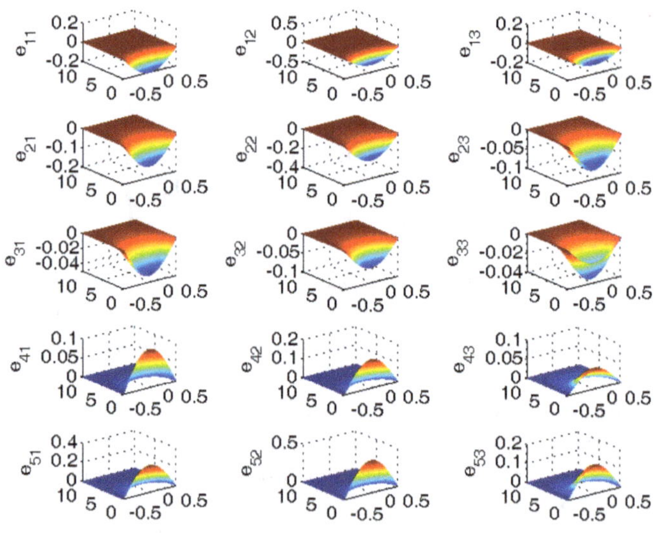

Fig. 5.7 The change processes of $e_{ij}(x, t)$, $i = 1, 2, \ldots, 5$, $j = 1, 2, 3$

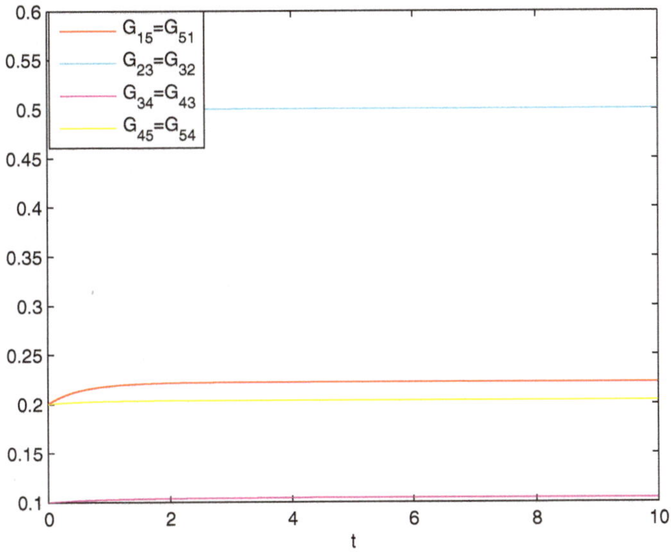

Fig. 5.8 Adaptive coupling strengths

5.5 Conclusions

Two spatially and temporally complex dynamical network models have been introduced, which generalize some existing complex network models to some extent. The only difference between them is whether delay coupling terms are incorporated in the mathematical model of network. Several novel edge-based adaptive strategies to tune the coupling strengths among network nodes have been proposed. By constructing suitable Lyapunov functionals and utilizing the designed adaptive laws, some sufficient conditions have been established to ensure that the proposed network models are synchronized. Illustrative simulations have been provided to verify the correctness and effectiveness of the obtained results.

Chapter 6
Synchronization and Adaptive Control of CRDNNs with Hybrid Coupling

6.1 Introduction

To our knowledge, in most existing works on spatially and temporally complex dynamical networks, only state coupling was considered. Practically, in some cases, the more realistic network models must additionally include the spatial diffusion coupling terms [122]. In [122], Wu and Chen considered the asymptotical synchronization and $\mathcal{H}_\infty$ synchronization of N-coupled linear partial differential systems via spatial diffusion coupling. Therefore, it is interesting to further study the synchronization in an array of CRDNNs with spatial diffusion coupling and state coupling. Moreover, external disturbances widely exist in complex networks, which may lead the network to an unanticipated state and even destroy the synchronization. Hence, the $\mathcal{H}_\infty$ synchronization problem is also investigated for an array of CRDNNs.

In many circumstances, complex dynamical networks cannot synchronize by themselves, thus some control strategies should be adopted to achieve synchronization. Adaptive control, as an important control technique, in the past ten years, has been widely used to synchronize complex dynamical networks. On the one hand, adaptive control has been developed as an effective method to design controller feedback gains [67, 68, 120, 134–137]. On the other hand, adaptive control method can also be utilized to tune the strengths of the interconnections among network nodes [41, 138–140]. In [41], Yu et al. considered distributed adaptive control of synchronization in complex networks. Some adaptive strategies to tune the coupling weights of the network were designed based on local information of node dynamics. In these existing works [41, 138–140], the node state is only dependent on the time. Obviously, it is also beneficial to apply the local adaptive strategies (vertex-based adaptive strategy and edge-based adaptive strategy) to study the synchronization problem of spatially and temporally complex dynamical networks.

The main contributions of this chapter are as follows. First, we propose a general model of an array of N linearly coupled RDNNs with hybrid coupling, which generalizes some existing complex network models to some extent. Second, we respectively investigate the synchronization and $\mathcal{H}_\infty$ synchronization of the

J.-L. Wang et al., *Analysis and Control of Coupled Neural Networks with Reaction-Diffusion Terms*, DOI 10.1007/978-981-10-4907-1_6

proposed network model, and some sufficient conditions ensuring the synchronization and $\mathcal{H}_\infty$ synchronization are established. Third, the adaptive synchronization problem and adaptive $\mathcal{H}_\infty$ synchronization problem are also studied using edge-based adaptive strategy, and several sufficient conditions are obtained to guarantee the synchronization and $\mathcal{H}_\infty$ synchronization.

6.2 Synchronization in CRDNNs with Hybrid Coupling

6.2.1 Network Model

In this chapter, we consider a CRDNNs consisting of N such RDNNs (2.4) with hybrid coupling. The mathematical model of the network can be described as follows:

$$\begin{aligned}\frac{\partial z_i(x,t)}{\partial t} &= D\Delta z_i(x,t) - Az_i(x,t) + Bf(z_i(x,t)) + J + c\sum_{j=1}^{N} G_{ij}\varGamma\Delta z_j(x,t)\\ &\quad + \hat{c}\sum_{j=1}^{N}\hat{G}_{ij}\hat{\varGamma} z_j(x,t) \end{aligned} \tag{6.1}$$

where $i = 1, 2, \ldots, N$, $z_i(x,t) = (z_{i1}(x,t), z_{i2}(x,t), \ldots, z_{in}(x,t))^T \in \mathbb{R}^n$ is the state vector of node i; c and $\hat{c}$ are positive real numbers, which represent the overall coupling strength; $\varGamma = \text{diag}(\gamma_1, \gamma_2, \ldots, \gamma_n) \in \mathbb{R}^{n\times n}$ and $\hat{\varGamma} = \text{diag}(\hat{\gamma}_1, \hat{\gamma}_2, \ldots, \hat{\gamma}_n) \in \mathbb{R}^{n\times n}$ are positive definite inner coupling matrices; $G = (G_{ij})_{N\times N}$ and $\hat{G} = (\hat{G}_{ij})_{N\times N}$ represent the topological structure of network and coupling strength between nodes, where G_{ij} ($\hat{G}_{ij}$ can be defined similarly) is defined as follows: if there exists a connection between node i and node j, then $G_{ij} = G_{ji} > 0$; otherwise, $G_{ij} = G_{ji} = 0 (i \neq j)$, and the diagonal elements of matrix G are defined by $G_{ii} = -\sum_{\substack{j=1\\ j\neq i}}^{N} G_{ij}, i = 1, 2, \ldots, N$.

The initial value and boundary value conditions associated with network (6.1) are given in the form

$$z_i(x,0) = \varPhi_i(x) \in \mathbb{R}^n, \quad x \in \varOmega, \tag{6.2}$$

$$z_i(x,t) = 0, \quad (x,t) \in \partial\varOmega \times [0, +\infty), \tag{6.3}$$

where $\varPhi_i(x)$ is bounded and continuous on $\varOmega$.

Remark 6.1 In reaction-diffusion networks, different diffusion of node may cause different changes of other nodes. For example, as is well known, different diffusion of species may cause different movements of other species in food webs [123, 124]. Another example occurs in the control design of reaction-diffusion networks. If the spatial diffusion coupling is considered in controller, better control effect may be obtained. Therefore, it is interesting to consider the spatial diffusion coupling in CRDNNs. This chapter proposes a CRDNNs model consisting of N identical RDNNs with hybrid coupling, which generalizes some existing complex network models to some extent [80–90, 97, 98].

6.2.2 Synchronization of CRDNNs

Definition 6.2 The CRDNNs (6.1) is said to achieve synchronization if

$$\lim_{t\to+\infty} \|z_i(\cdot,t) - \frac{1}{N}\sum_{j=1}^{N} z_j(\cdot,t)\|_2 = 0, \text{ for all } i = 1, 2, \ldots, N.$$

Let $\bar{z}(x,t) = \frac{1}{N}\sum_{j=1}^{N} z_j(x,t)$. Then, we have

$$\frac{\partial \bar{z}(x,t)}{\partial t} = D\Delta\bar{z}(x,t) - A\bar{z}(x,t) + J + \frac{1}{N}\sum_{j=1}^{N} Bf(z_j(x,t)). \tag{6.4}$$

Define $e_i(x,t) = (e_{i1}(x,t), e_{i2}(x,t), \ldots, e_{in}(x,t))^T = z_i(x,t) - \bar{z}(x,t)$, then the dynamics of the error vector $e_i(x,t)$ is governed by the following equation:

$$\begin{aligned}\frac{\partial e_i(x,t)}{\partial t} &= D\Delta e_i(x,t) - Ae_i(x,t) - \frac{1}{N}\sum_{j=1}^{N} Bf(z_j(x,t)) + Bf(z_i(x,t)) \\ &\quad + c\sum_{j=1}^{N} G_{ij}\Gamma\Delta e_j(x,t) + \hat{c}\sum_{j=1}^{N} \hat{G}_{ij}\hat{\Gamma} e_j(x,t)\end{aligned} \tag{6.5}$$

where $i = 1, 2, \ldots, N$.

By using Kronecker product, we can rewrite network (6.5) in a compact form as follows:

$$\begin{aligned}\frac{\partial e(x,t)}{\partial t} &= (I_N \otimes D)\Delta e(x,t) - (I_N \otimes A)e(x,t) - \mathbf{1}^N \otimes \left(\frac{1}{N}\sum_{j=1}^{N} Bf(z_j(x,t))\right) \\ &\quad + (I_N \otimes B)\hat{f}(z(x,t)) + c(G \otimes \Gamma)\Delta e(x,t) + \hat{c}(\hat{G} \otimes \hat{\Gamma})e(x,t)\end{aligned} \tag{6.6}$$

where $e(x,t) = (e_1^T(x,t), e_2^T(x,t), \ldots, e_N^T(x,t))^T$, $\mathbf{1}^N = (1, 1, \ldots, 1)^T \in \mathbb{R}^N$, $\hat{f}(z(x,t)) = (f^T(z_1(x,t)), f^T(z_2(x,t)), \ldots, f^T(z_N(x,t)))^T$.

For the convenience, we denote

$$\begin{aligned}\Theta &= \mathrm{diag}(\rho_1^2, \rho_2^2, \ldots, \rho_n^2), \\ \hat{f}(\bar{z}(x,t)) &= (f^T(\bar{z}(x,t)), f^T(\bar{z}(x,t)), \ldots, f^T(\bar{z}(x,t)))^T.\end{aligned}$$

Theorem 6.3 *If there exists a matrix $P \in \mathbb{R}^{n\times n} > 0$ such that*

$$W \geqslant 0, \tag{6.7}$$

$$-\sum_{k=1}^{q} \frac{1}{l_k^2} W + \hat{W} < 0, \tag{6.8}$$

where $W = I_N \otimes (PD) + I_N \otimes (DP) + cG \otimes (P\Gamma) + cG \otimes (\Gamma P)$, $\hat{W} = -I_N \otimes (PA) - I_N \otimes (AP) + \hat{c}\hat{G} \otimes (P\hat{\Gamma}) + I_N \otimes (PBB^T P + \Theta) + \hat{c}\hat{G} \otimes (\hat{\Gamma} P)$, *then the CRDNNs (6.1) is synchronized in the sense of Definition 6.2.*

Proof First, construct a Lyapunov functional for system (6.6) as follows:

$$V_1(t) = \int_{\Omega} e^T(x,t)(I_N \otimes P)e(x,t)dx. \tag{6.9}$$

According to the definition of $V_1(t)$, we can obtain

$$\lambda_m(P)\|e(\cdot,t)\|_2^2 \leqslant V_1(t) \leqslant \lambda_M(P)\|e(\cdot,t)\|_2^2. \tag{6.10}$$

The derivative of $V_1(t)$ along the trajectory of system (6.6) is given as follows:

$$\begin{aligned}
\dot{V}_1(t) = 2\int_{\Omega} & e^T(x,t)[(I_N \otimes (PD))\Delta e(x,t) - (I_N \otimes (PA))e(x,t) \\
& + (I_N \otimes (PB))\hat{f}(\bar{z}(x,t)) - \mathbf{1}^N \otimes \left(\frac{1}{N}\sum_{j=1}^{N} PBf(z_j(x,t))\right) \\
& + (I_N \otimes (PB))(\hat{f}(z(x,t)) - \hat{f}(\bar{z}(x,t))) + c(G \otimes (P\Gamma))\Delta e(x,t) \\
& + \hat{c}(\hat{G} \otimes (P\hat{\Gamma}))e(x,t)]dx.
\end{aligned}$$

Since $\sum_{i=1}^{N} e_i^T(x,t) = 0$, we can get

$$\begin{aligned}
& e^T(x,t)\left[(I_N \otimes (PB))\hat{f}(\bar{z}(x,t)) - \mathbf{1}^N \otimes \left(\frac{1}{N}\sum_{j=1}^{N} PBf(z_j(x,t))\right)\right] \\
&= \sum_{i=1}^{N} e_i^T(x,t)(PBf(\bar{z}(x,t)) - \frac{1}{N}\sum_{j=1}^{N} PBf(z_j(x,t))) \\
&= 0.
\end{aligned} \tag{6.11}$$

From Green's formula and the boundary condition, we then have

$$\int_{\Omega} e_{il}(x,t)\Delta e_{ij}(x,t)dx = -\sum_{k=1}^{q}\int_{\Omega} \frac{\partial e_{il}(x,t)}{\partial x_k}\frac{\partial e_{ij}(x,t)}{\partial x_k}dx$$

where $l, j = 1, 2, \ldots, n$, $i = 1, 2, \ldots, N$.

Letting $P = (p_{jl})_{n\times n}$, we can obtain

$$\begin{aligned}
&\int_{\Omega} e^T(x,t)(I_N \otimes (PD))\Delta e(x,t)dx \\
&= \sum_{i=1}^{N} \int_{\Omega} e_i^T(x,t) PD\Delta e_i(x,t)dx \\
&= \sum_{i=1}^{N}\sum_{j=1}^{n}\sum_{l=1}^{n} p_{jl}d_l \int_{\Omega} e_{ij}(x,t)\Delta e_{il}(x,t)dx \\
&= -\sum_{k=1}^{q}\sum_{i=1}^{N}\sum_{j=1}^{n}\sum_{l=1}^{n} p_{jl}d_l \int_{\Omega} \frac{\partial e_{ij}(x,t)}{\partial x_k}\frac{\partial e_{il}(x,t)}{\partial x_k}dx \\
&= -\sum_{k=1}^{q} \int_{\Omega} \left(\frac{\partial e(x,t)}{\partial x_k}\right)^T (I_N \otimes (PD))\frac{\partial e(x,t)}{\partial x_k}dx, \\
&\int_{\Omega} (\Delta e(x,t))^T (I_N \otimes (DP))e(x,t)dx \\
&= \sum_{i=1}^{N}\sum_{j=1}^{n}\sum_{l=1}^{n} d_j p_{jl} \int_{\Omega} e_{il}(x,t)\Delta e_{ij}(x,t)dx \\
&= -\sum_{k=1}^{q} \int_{\Omega} \left(\frac{\partial e(x,t)}{\partial x_k}\right)^T (I_N \otimes (DP))\frac{\partial e(x,t)}{\partial x_k}dx, \\
&\int_{\Omega} e^T(x,t)(G \otimes (P\Gamma))\Delta e(x,t)dx \\
&= \sum_{i=1}^{N}\sum_{j=1}^{N}\sum_{l=1}^{n}\sum_{s=1}^{n} G_{ij}p_{ls}\gamma_s \int_{\Omega} e_{il}(x,t)\Delta e_{js}(x,t)dx \\
&= -\sum_{k=1}^{q} \int_{\Omega} \left(\frac{\partial e(x,t)}{\partial x_k}\right)^T (G \otimes (P\Gamma))\frac{\partial e(x,t)}{\partial x_k}dx, \\
&\int_{\Omega} (\Delta e(x,t))^T (G \otimes (\Gamma P))e(x,t)dx \\
&= \sum_{i=1}^{N}\sum_{j=1}^{N}\sum_{l=1}^{n}\sum_{s=1}^{n} G_{ij}\gamma_l p_{ls} \int_{\Omega} \Delta e_{il}(x,t)e_{js}(x,t)dx \\
&= -\sum_{k=1}^{q} \int_{\Omega} \left(\frac{\partial e(x,t)}{\partial x_k}\right)^T (G \otimes (\Gamma P))\frac{\partial e(x,t)}{\partial x_k}dx. \qquad (6.12)
\end{aligned}$$

By (6.12), we can get

$$
\begin{aligned}
&2\int_{\Omega} e^T(x,t)[I_N \otimes (PD) + c(G \otimes (P\Gamma))]\Delta e(x,t)dx \\
&= -\sum_{k=1}^{q}\int_{\Omega}\left(\frac{\partial e(x,t)}{\partial x_k}\right)^T W\frac{\partial e(x,t)}{\partial x_k}dx.
\end{aligned}
$$

On the other hand, there obviously exists a real matrix Q such that $W = Q^TQ$, then

$$
\left(\frac{\partial e(x,t)}{\partial x_k}\right)^T W\frac{\partial e(x,t)}{\partial x_k} = \left(\frac{\partial(Qe(x,t))}{\partial x_k}\right)^T \frac{\partial(Qe(x,t))}{\partial x_k}.
$$

Let $y(x,t) = Qe(x,t)$, for $(x,t) \in \partial\Omega \times [0,+\infty)$ from the boundary condition (6.3), we have $y(x,t) = Qe(x,t) = 0$. In view of Lemma 1.22, one has

$$
\begin{aligned}
&\sum_{k=1}^{q}\int_{\Omega}\left(\frac{\partial y(x,t)}{\partial x_k}\right)^T \frac{\partial y(x,t)}{\partial x_k}dx \\
&\geqslant \sum_{k=1}^{q}\frac{1}{l_k^2}\int_{\Omega} e^T(x,t)We(x,t)dx. \qquad (6.13)
\end{aligned}
$$

Furthermore, we can easily derive

$$
\begin{aligned}
&e^T(x,t)(I_N \otimes (PB))(\hat{f}(z(x,t)) - \hat{f}(\bar{z}(x,t))) \\
&\leqslant \frac{1}{2}e^T(x,t)[I_N \otimes (PBB^TP + \Theta)]e(x,t). \qquad (6.14)
\end{aligned}
$$

Thus, we can derive from (6.10)–(6.14) that

$$
\begin{aligned}
\dot{V}_1(t) &\leqslant \int_{\Omega} e^T(x,t)\left(-\sum_{k=1}^{q}\frac{1}{l_k^2}W + \hat{W}\right)e(x,t)dx \\
&\leqslant -\frac{\lambda_m(\sum_{k=1}^{q}\frac{1}{l_k^2}W - \hat{W})}{\lambda_M(P)}V_1(t). \qquad (6.15)
\end{aligned}
$$

From (6.15), we have

$$
V_1(t) \leqslant V_1(0)e^{-\frac{\lambda_m(\sum_{k=1}^{q}\frac{1}{l_k^2}W-\hat{W})}{\lambda_M(P)}t}. \qquad (6.16)
$$

By (6.10) and (6.16), we can get

$$
\|e(\cdot,t)\|_2 \leqslant \sqrt{\frac{\lambda_M(P)}{\lambda_m(P)}}\|e(\cdot,0)\|_2 e^{-\frac{\lambda_m(\sum_{k=1}^{q}\frac{1}{l_k^2}W-\hat{W})}{2\lambda_M(P)}t}.
$$

Therefore, we can conclude that the CRDNNs (6.1) is synchronized. The proof is completed.

6.2.3 $\mathcal{H}_\infty$ Synchronization of CRDNNs

Due to the external disturbance is ubiquitous, and may lead the network to an unanticipated state and even destroy the synchronization, we must discuss the robustness to resist external disturbance of the complex dynamical networks. Therefore, it is essential to consider the external disturbance in the CRDNNs (6.1), and investigate its $\mathcal{H}_\infty$ synchronization problem.

In this section, we consider the following CRDNNs model:

$$\begin{aligned}\frac{\partial z_i(x,t)}{\partial t} &= D\Delta z_i(x,t) - Az_i(x,t) + Bf(z_i(x,t)) + J + c\sum_{j=1}^{N} G_{ij}\Gamma\Delta z_j(x,t)\\ &\quad + \hat{c}\sum_{j=1}^{N}\hat{G}_{ij}\hat{\Gamma}z_j(x,t) + v_i(x,t)\end{aligned} \tag{6.17}$$

where $i = 1, 2, \ldots, N$, $z_i(x,t)$, $\Omega, D, \Delta, A, B, f(\cdot), c, \hat{c}, \Gamma, \hat{\Gamma}, J, G=(G_{ij})_{N\times N}$, $\hat{G} = (\hat{G}_{ij})_{N\times N}$ denote the same meanings as these in network (6.1), $v_i(x,t) = (v_{i1}(x,t), v_{i2}(x,t), \ldots, v_{in}(x,t))^T \in \mathbb{R}^n$ represents the external disturbance and is square integrable, i.e., $\int_0^{t_f}\int_\Omega v_i^T(x,t)v_i(x,t)dxdt < +\infty$ for any $t_f \geqslant 0$.

The initial value and boundary value conditions associated with network (6.17) are given in the form

$$z_i(x,0) = \Phi_i(x) \in \mathbb{R}^n, \quad x \in \Omega, \tag{6.18}$$

$$z_i(x,t) = 0, \quad (x,t) \in \partial\Omega \times [0, +\infty) \tag{6.19}$$

where $\Phi_i(x)$ is bounded and continuous on Ω.

Let $\bar{z}(x,t) = \frac{1}{N}\sum_{j=1}^{N} z_j(x,t)$, $e_i(x,t) = z_i(x,t) - \bar{z}(x,t)$ and $v(x,t) = (v_1^T(x,t), v_2^T(x,t), \ldots, v_N^T(x,t))^T$. We can obtain

$$\begin{aligned}\frac{\partial e(x,t)}{\partial t} &= (I_N \otimes D)\Delta e(x,t) - (I_N \otimes A)e(x,t) + (I_N \otimes B)\hat{f}(z(x,t))\\ &\quad + c(G\otimes\Gamma)\Delta e(x,t) + \hat{c}(\hat{G}\otimes\hat{\Gamma})e(x,t) + v(x,t)\\ &\quad - \mathbf{1}^N \otimes \left(\frac{1}{N}\sum_{j=1}^{N} Bf(z_j(x,t))\right)\\ &\quad - \mathbf{1}^N \otimes \left(\frac{1}{N}\sum_{j=1}^{N} v_j(x,t)\right)\end{aligned} \tag{6.20}$$

where $\hat{f}(z(x,t))$, $e(x,t)$ and $\mathbf{1}^N$ denote the same meanings as these in system (6.6).

Definition 6.4 The CRDNNs (6.17) is $\mathcal{H}_\infty$ synchronized with the disturbance attenuation level γ if the error vector $e(x,t)$ satisfies:

$$\int_0^{t_f}\int_\Omega e^T(x,t)e(x,t)dxdt \leqslant V(0,e(\cdot,0)) + \gamma^2\int_0^{t_f}\int_\Omega v^T(x,t)v(x,t)dxdt$$

for any $t_f \geqslant 0$ and some non-negative function V.

Theorem 6.5 *If there exist a positive constant γ and a matrix $P \in \mathbb{R}^{n\times n} > 0$ such that*

$$W \geqslant 0 \tag{6.21}$$

$$-\sum_{k=1}^{q}\frac{1}{l_k^2}W + \tilde{W} + \frac{1}{\gamma^2}(I_N \otimes P^2) \leqslant 0 \tag{6.22}$$

where $\Theta = \mathrm{diag}(\rho_1^2, \rho_2^2, \ldots, \rho_n^2)$, $W = I_N \otimes (PD) + I_N \otimes (DP) + cG \otimes (P\Gamma) + cG \otimes (\Gamma P)$, $\tilde{W} = -I_N \otimes (PA) - I_N \otimes (AP) + \hat{c}\hat{G} \otimes (P\hat{\Gamma}) + \hat{c}\hat{G} \otimes (\hat{\Gamma}P) + I_N \otimes (PBB^T P + \Theta + I_n)$, *then the CRDNNs (6.17) is $\mathcal{H}_\infty$ synchronized with the disturbance attenuation level γ.*

Proof Take the same Lyapunov functional $V_1(t)$ as in Theorem 6.3. By (6.21) and (6.22), we can easily derive

$$\begin{aligned}
&\int_0^{t_f}\int_\Omega (e^T(x,t)e(x,t) - \gamma^2 v^T(x,t)v(x,t))dxdt \\
&= \int_0^{t_f}[\int_\Omega (e^T(x,t)e(x,t) - \gamma^2 v^T(x,t)v(x,t))dx + \dot{V}_1(t)]dt + V_1(0) - V_1(t_f) \\
&= V_1(0) - V_1(t_f) + \int_0^{t_f}\int_\Omega e^T(x,t)e(x,t)dxdt - \gamma^2\int_0^{t_f}\int_\Omega v^T(x,t)v(x,t)dxdt \\
&\quad + 2\int_0^{t_f}\int_\Omega e^T(x,t)[(I_N \otimes (PD))\Delta e(x,t) \\
&\quad - (I_N \otimes (PA))e(x,t) + (I_N \otimes P)v(x,t) \\
&\quad - \mathbf{1}^N \otimes \left(\frac{1}{N}\sum_{j=1}^{N} Pv_j(x,t)\right) + \hat{c}(\hat{G} \otimes (P\hat{\Gamma}))e(x,t) \\
&\quad + c(G \otimes (P\Gamma))\Delta e(x,t) + (I_N \otimes (PB))\hat{f}(z(x,t))
\end{aligned}$$

$$
\begin{aligned}
&+ (I_N \otimes (PB))\hat{f}(\bar{z}(x,t)) - (I_N \otimes (PB))\hat{f}(\bar{z}(x,t)) \\
&-\mathbf{1}^N \otimes \left(\frac{1}{N} \sum_{j=1}^{N} PBf(z_j(x,t)) \right) \Bigg] dxdt \\
&\leqslant V_1(0) - V_1(t_f) - \int_0^{t_f} \int_{\Omega} \Big[\gamma v(x,t) \\
&- \frac{1}{\gamma}(I_N \otimes P)e(x,t) \Big]^T \left[\gamma v(x,t) - \frac{1}{\gamma}(I_N \otimes P)e(x,t) \right] dxdt \\
&+ \int_0^{t_f} \int_{\Omega} e^T(x,t) \left[-\sum_{k=1}^{q} \frac{1}{l_k^2} W + \tilde{W} + \frac{1}{\gamma^2}(I_N \otimes P^2) \right] e(x,t) dxdt \\
&\leqslant V_1(0).
\end{aligned}
\tag{6.23}
$$

where $\hat{f}(\bar{z}(x,t)) = (f^T(\bar{z}(x,t)), \ldots, f^T(\bar{z}(x,t)))^T$.

From (6.23), we can get

$$
\begin{aligned}
&\int_0^{t_f} \int_{\Omega} e^T(x,t) e(x,t) dxdt \\
&\leqslant V_1(0) + \gamma^2 \int_0^{t_f} \int_{\Omega} v^T(x,t) v(x,t) dxdt.
\end{aligned}
$$

Therefore, the CRDNNs (6.17) is $\mathcal{H}_\infty$ synchronized with the disturbance attenuation level γ. The proof is completed.

Remark 6.6 To the best of our knowledge, this is the first to consider the $\mathcal{H}_\infty$ synchronization problem of CRDNNs with spatial diffusion coupling and state coupling, which is a very important and challenging problem. Theorem 6.5 gives a sufficient condition to ensure that CRDNNs (6.17) is $\mathcal{H}_\infty$ synchronized in the sense of Definition 6.4. From (6.22), we can easily derive

$$
\begin{aligned}
&-\sum_{k=1}^{q} \frac{1}{l_k^2} W - I_N \otimes (PA) - I_N \otimes (AP) + \hat{c}\hat{G} \otimes (P\hat{\Gamma}) \\
&+ I_N \otimes (PBB^T P + \Theta) + \hat{c}\hat{G} \otimes (\hat{\Gamma} P) \\
&\leqslant -I_N \otimes I_n - \frac{1}{\gamma^2}(I_N \otimes P^2) \\
&< 0.
\end{aligned}
$$

Obviously, Theorem 6.5 can also guarantee that the CRDNNs (6.17) is synchronized in the sense of Definition 6.2 when $v(x,t) \equiv 0$.

6.3 Adaptive Synchronization of CRDNNs with Hybrid Coupling

6.3.1 Network Model

Let $\mathcal{V} = \{1, 2, \ldots, N\}$ and $\mathcal{E} \subset \mathcal{V} \times \mathcal{V}$ respectively denote the set of nodes and the set of undirected links in the networks (6.1) and (6.17).

The control objective here is to design some adaptive strategies to tune the coupling strengths of the network towards reaching synchronization or $\mathcal{H}_\infty$ synchronization. Here, consider the case where only the coupling strengths $\hat{G}_{ij}$ are adapted. In this case, the CRDNNs (6.1) and (6.17) can be expressed respectively by

$$\begin{aligned}\frac{\partial z_i(x,t)}{\partial t} &= D\Delta z_i(x,t) - Az_i(x,t) + Bf(z_i(x,t)) + J + c\sum_{j=1}^{N} G_{ij}\Gamma\Delta z_j(x,t)\\ &\quad + \hat{c}\sum_{j=1}^{N} \hat{G}_{ij}(t)\hat{\Gamma} z_j(x,t),\end{aligned} \tag{6.24}$$

$$\begin{aligned}\frac{\partial z_i(x,t)}{\partial t} &= D\Delta z_i(x,t) - Az_i(x,t) + Bf(z_i(x,t)) + J + c\sum_{j=1}^{N} G_{ij}\Gamma\Delta z_j(x,t)\\ &\quad + \hat{c}\sum_{j=1}^{N} \hat{G}_{ij}(t)\hat{\Gamma} z_j(x,t) + v_i(x,t)\end{aligned} \tag{6.25}$$

where $i = 1, 2, \ldots, N$, $\hat{G}(t) = (\hat{G}_{ij}(t))_{N\times N}$ is a time-varying matrix, where $\hat{G}_{ij}(t)$ is defined as follows: if there is an edge between node i and node j at time t, then the element $\hat{G}_{ij}(t) = \hat{G}_{ji}(t) > 0$; otherwise, $\hat{G}_{ij}(t) = \hat{G}_{ji}(t) = 0 (i \neq j)$, and the diagonal elements of matrix $\hat{G}(t)$ are defined by $\hat{G}_{ii}(t) = -\sum_{\substack{j=1\\ j\neq i}}^{N} \hat{G}_{ij}(t)$, $i = 1, 2, \ldots, N$. In this section, we always assume networks (6.24) and (6.25) are connected, and their topological structures are fixed. That is, if there is no connection between nodes i and $j (i \neq j)$, then $\hat{G}_{ij}(t) = \hat{G}_{ji}(t) = 0$ for all t.

6.3.2 Adaptive Strategies for Synchronization of CRDNNs

Let $\bar{z}(x,t) = \frac{1}{N}\sum_{j=1}^{N} z_j(x,t)$ and $e_i(x,t) = z_i(x,t) - \bar{z}(x,t)$. Then, we can obtain from (6.24) that

$$\frac{\partial e_i(x,t)}{\partial t} = D\Delta e_i(x,t) - Ae_i(x,t) - \frac{1}{N}\sum_{j=1}^{N} Bf(z_j(x,t)) + Bf(z_i(x,t)) + c\sum_{j=1}^{N} G_{ij}\Gamma\Delta e_j(x,t) + \hat{c}\sum_{j=1}^{N} \hat{G}_{ij}(t)\hat{\Gamma} e_j(x,t) \tag{6.26}$$

where $i = 1, 2, \ldots, N$.

Suppose that $\hat{\mathcal{E}}$ is a subset of $\mathcal{E}$, and CRDNNs (6.24) is connected through the undirected edges $\hat{\mathcal{E}}$. For the convenience, we denote

$$H_{ij} = \begin{cases} \hat{G}_{ij}(0), & \text{if } (i,j) \in \mathcal{E} - \hat{\mathcal{E}}, \\ -\sum_{j=1, j\neq i}^{N} H_{ij}, & \text{if } i = j, \\ 0, & \text{otherwise.} \end{cases}$$

Theorem 6.7 *If there exists a matrix $P \in \mathbb{R}^{n\times n} > 0$ such that*

$$W \geqslant 0, \tag{6.27}$$

$$P\hat{\Gamma} + \hat{\Gamma}P > 0, \tag{6.28}$$

where $W = I_N \otimes (PD) + I_N \otimes (DP) + cG \otimes (P\Gamma) + cG \otimes (\Gamma P)$, then the CRDNNs (6.24) is synchronized under the following adaptive law:

$$\begin{aligned} \dot{\hat{G}}_{ij}(t) &= \beta_{ij}\int_{\Omega} (z_i(x,t) - z_j(x,t))^T \frac{P\hat{\Gamma} + \hat{\Gamma}P}{2}(z_i(x,t) \\ &\quad - z_j(x,t))dx, \qquad \text{if } (i,j) \in \hat{\mathcal{E}}, \\ \dot{\hat{G}}_{ij}(t) &= 0, \qquad \text{if } (i,j) \in \mathcal{E} - \hat{\mathcal{E}} \end{aligned} \tag{6.29}$$

where $\beta_{ij} = \beta_{ji}$ are positive constants.

Proof Construct a Lyapunov functional for the system (6.26) as follows:

$$V_2(t) = \frac{1}{2}\sum_{i=1}^{N}\int_{\Omega} e_i^T(x,t)Pe_i(x,t)dx + \sum_{i=1}^{N}\sum_{(i,j)\in\hat{\mathcal{E}}} \frac{\hat{c}}{4\beta_{ij}}(\hat{G}_{ij}(t) - \sigma_{ij})^2 \tag{6.30}$$

where σ_{ij} is defined as follows: if $(i,j) \in \hat{\mathcal{E}}$, then $\sigma_{ij} = \sigma_{ji} > 0$; otherwise, $\sigma_{ij} = 0 (i \neq j)$.

Calculating the time derivative of $V_2(t)$ along the trajectory of system (6.26), we can get

$$
\begin{aligned}
\dot{V}_2(t) &= \sum_{i=1}^{N} \int_{\Omega} e_i^T(x,t) P \frac{\partial e_i(x,t)}{\partial t} dx + \sum_{i=1}^{N} \sum_{(i,j)\in\hat{\mathcal{E}}} \frac{\hat{c}}{2\beta_{ij}} (\hat{G}_{ij}(t) - \sigma_{ij}) \dot{\hat{G}}_{ij}(t) \\
&= \sum_{i=1}^{N} \int_{\Omega} e_i^T(x,t) P \Big[D\Delta e_i(x,t) + Bf(\bar{z}(x,t)) - Bf(\bar{z}(x,t)) - Ae_i(x,t) \\
&\quad - \frac{1}{N} \sum_{j=1}^{N} Bf(z_j(x,t)) + Bf(z_i(x,t)) + \hat{c} \sum_{j=1}^{N} \hat{G}_{ij}(t) \hat{\Gamma} e_j(x,t) \\
&\quad + c \sum_{j=1}^{N} G_{ij} \Gamma \Delta e_j(x,t) \Big] dx + \frac{\hat{c}}{2} \sum_{i=1}^{N} \sum_{(i,j)\in\hat{\mathcal{E}}} (\hat{G}_{ij}(t) - \sigma_{ij}) \int_{\Omega} (e_i(x,t) \\
&\quad - e_j(x,t))^T \frac{P\hat{\Gamma} + \hat{\Gamma} P}{2} (e_i(x,t) - e_j(x,t)) dx.
\end{aligned} \tag{6.31}
$$

Since $\sum_{i=1}^{N} e_i^T(x,t) = 0$, we have

$$
\sum_{i=1}^{N} e_i^T(x,t) P \left[Bf(\bar{z}(x,t)) - \frac{1}{N} \sum_{j=1}^{N} Bf(z_j(x,t)) \right] = 0. \tag{6.32}
$$

Let $\Theta = \mathrm{diag}(\rho_1^2, \rho_2^2, \ldots, \rho_n^2)$. Obviously

$$
\begin{aligned}
& e_i^T(x,t) PB[f(z_i(x,t)) - f(\bar{z}(x,t))] \\
\leqslant\ & \frac{1}{2} e_i^T(x,t) PBB^T P e_i(x,t) + \frac{1}{2} e_i^T(x,t) \Theta e_i(x,t).
\end{aligned} \tag{6.33}
$$

Define the matrix $\Sigma = (\sigma_{ij})_{N\times N}$, where $\sigma_{ii} = -\sum_{\substack{j=1 \\ j\neq i}}^{N} \sigma_{ij}$, $i = 1, 2, \ldots, N$. Then, we can derive from (6.27) and (6.31)–(6.33) that

$$
\begin{aligned}
\dot{V}_2(t) \leqslant\ & \sum_{i=1}^{N} \int_{\Omega} e_i^T(x,t) \left(\frac{\Theta}{2} + \frac{PBB^T P}{2} - \frac{PA + AP}{2} \right) e_i(x,t) dx \\
& + \sum_{i=1}^{N} \int_{\Omega} e_i^T(x,t) (PD\Delta e_i(x,t) + c \sum_{j=1}^{N} G_{ij} P \Gamma \Delta e_j(x,t)) dx \\
& + \hat{c} \sum_{i=1}^{N} \sum_{j=1}^{N} H_{ij} \int_{\Omega} e_i^T(x,t) \frac{P\hat{\Gamma} + \hat{\Gamma} P}{2} e_j(x,t) dx \\
& + \hat{c} \sum_{i=1}^{N} \sum_{j=1}^{N} \sigma_{ij} \int_{\Omega} e_i^T(x,t) \frac{P\hat{\Gamma} + \hat{\Gamma} P}{2} e_j(x,t) dx
\end{aligned}
$$

$$
\begin{aligned}
&= \int_{\Omega} e^T(x,t)\left[I_N \otimes \left(\frac{\Theta}{2} + \frac{PBB^TP}{2} - \frac{PA+AP}{2}\right)\right.\\
&\quad \left. + \hat{c}\left(\Sigma \otimes \frac{P\hat{\Gamma} + \hat{\Gamma}P}{2}\right)\right] e(x,t)dx\\
&\quad + \int_{\Omega} e^T(x,t)(I_N \otimes (PD) + cG \otimes (P\Gamma))\Delta e(x,t)dx\\
&\quad + \hat{c}\int_{\Omega} e^T(x,t)\left(H \otimes \frac{P\hat{\Gamma} + \hat{\Gamma}P}{2}\right)e(x,t)dx\\
&\leqslant \int_{\Omega} e^T(x,t)\left[I_N \otimes \left(\frac{\Theta}{2} + \frac{PBB^TP}{2} - \frac{PA+AP}{2}\right)\right.\\
&\quad \left. + \hat{c}\left(\Sigma \otimes \frac{P\hat{\Gamma} + \hat{\Gamma}P}{2}\right) - \sum_{k=1}^{q} \frac{W}{2l_k^2}\right] e(x,t)dx
\end{aligned}
\tag{6.34}
$$

where $e(x,t) = (e_1^T(x,t), e_2^T(x,t), \ldots, e_N^T(x,t))^T$, $H = (H_{ij})_{N\times N}$.

According to Lemma 1.17, there obviously exists a unitary matrix $\Psi = (\psi_1, \psi_2, \ldots, \psi_N) \in \mathbb{R}^{N\times N}$ such that $\Psi^T\Sigma\Psi = \Lambda$ with $\Lambda = \text{diag}(\lambda_1, \lambda_2, \ldots, \lambda_N)$. $\lambda_i, i = 1, 2, \ldots, N$, are the eigenvalues of Σ and $0 = \lambda_1 > \lambda_2 \geqslant \lambda_3 \geqslant \ldots \geqslant \lambda_N$. Let $\eta(x,t) = (\eta_1^T(x,t), \eta_2^T(x,t), \ldots, \eta_N^T(x,t))^T = (\Psi^T \otimes I_n)e(x,t)$. Since $\psi_1 = \frac{1}{\sqrt{N}}(1, 1, \ldots, 1)^T$, one has $\eta_1(x,t) = (\psi_1^T \otimes I_n)e(x,t) = 0$. Then, we can derive from (6.34) that

$$
\begin{aligned}
\dot{V}_2(t) &\leqslant \int_{\Omega} e^T(x,t)\left[I_N \otimes \left(\frac{\Theta}{2} + \frac{PBB^TP}{2} - \frac{PA+AP}{2}\right) - \sum_{k=1}^{q}\frac{W}{2l_k^2}\right.\\
&\quad \left. + \hat{c}(\Psi \otimes I_n)\left(\Lambda \otimes \frac{P\hat{\Gamma} + \hat{\Gamma}P}{2}\right)(\Psi^T \otimes I_n)\right] e(x,t)dx\\
&= \int_{\Omega} e^T(x,t)\left[I_N \otimes \left(\frac{\Theta}{2} + \frac{PBB^TP}{2} - \frac{PA+AP}{2}\right) - \sum_{k=1}^{q}\frac{W}{2l_k^2}\right] e(x,t)dx\\
&\quad + \int_{\Omega} \hat{c}\eta^T(x,t)\left(\Lambda \otimes \frac{P\hat{\Gamma} + \hat{\Gamma}P}{2}\right)\eta(x,t)dx.
\end{aligned}
\tag{6.35}
$$

Since $P\hat{\Gamma} + \hat{\Gamma}P$ is a positive definite matrix, we have

$$
\eta^T(x,t)\left(\Lambda \otimes \frac{P\hat{\Gamma} + \hat{\Gamma}P}{2}\right)\eta(x,t) \leqslant \lambda_2\eta^T(x,t)\left(I_N \otimes \frac{P\hat{\Gamma} + \hat{\Gamma}P}{2}\right)\eta(x,t).
$$

Therefore,

$$
\begin{aligned}
\dot{V}_2(t) \leqslant & \int_{\Omega} e^T(x,t)\left[I_N \otimes \left(\frac{\Theta}{2} + \frac{PBB^TP}{2} - \frac{PA+AP}{2}\right) - \sum_{k=1}^{q}\frac{W}{2l_k^2}\right]e(x,t)dx \\
& + \int_{\Omega} \hat{c}\lambda_2 \eta^T(x,t)\left(I_N \otimes \frac{P\hat{\Gamma}+\hat{\Gamma}P}{2}\right)\eta(x,t)dx \\
= & \int_{\Omega} e^T(x,t)\left[I_N \otimes \left(\frac{\Theta}{2} + \frac{PBB^TP}{2} - \frac{PA+AP}{2}\right) - \sum_{k=1}^{q}\frac{W}{2l_k^2}\right]e(x,t)dx \\
& + \int_{\Omega} \hat{c}\lambda_2 e^T(x,t)(\Psi \otimes I_n)\left(I_N \otimes \frac{P\hat{\Gamma}+\hat{\Gamma}P}{2}\right)(\Psi^T \otimes I_n)e(x,t)dx \\
= & \int_{\Omega} e^T(x,t)\left[I_N \otimes \left(\frac{\Theta}{2} + \frac{PBB^TP}{2} - \frac{PA+AP}{2}\right) - \sum_{k=1}^{q}\frac{W}{2l_k^2}\right. \\
& \left. + \hat{c}\lambda_2\left(I_N \otimes \frac{P\hat{\Gamma}+\hat{\Gamma}P}{2}\right)\right]e(x,t)dx. \qquad (6.36)
\end{aligned}
$$

By selecting σ_{ij} sufficiently large such that

$$
\begin{aligned}
& \hat{c}\lambda_2\lambda_m\left(\frac{P\hat{\Gamma}+\hat{\Gamma}P}{2}\right) + \frac{\lambda_M(PBB^TP)}{2} - \lambda_m\left(\sum_{k=1}^{q}\frac{W}{2l_k^2}\right) \\
& + \frac{\max_{i=1,2,\ldots,n}\{\rho_i^2\}}{2} - \frac{\lambda_m(PA+AP)}{2} \leqslant -1, \qquad (6.37)
\end{aligned}
$$

one obtains

$$
\dot{V}_2(t) \leqslant -\|e(\cdot,t)\|_2^2. \qquad (6.38)
$$

Obviously, $V_2(t)$ is non-increasing, and each term of $V_2(t)$ is bounded. Consequently, all the coupling strengths $\hat{G}_{ij}(t)((i,j) \in \hat{\mathcal{E}})$ are bounded, and $\lim_{t\to+\infty} V_2(t)$ exists and is a non-negative real number. Since $\hat{G}_{ij}(t)((i,j) \in \hat{\mathcal{E}})$ is monotonically increasing (see (6.29)), one can conclude that $\hat{G}_{ij}(t)((i,j) \in \hat{\mathcal{E}})$ asymptotically converges to a finite positive value. Therefore, by the definition of $V_2(t)$, we can derive that

$$
\lim_{t\to+\infty}\sum_{i=1}^{N}\int_{\Omega} e_i^T(x,t)Pe_i(x,t)dx
$$

exists and is a non-negative real number. In what follows, we shall prove that

$$
\lim_{t\to+\infty}\sum_{i=1}^{N}\int_{\Omega} e_i^T(x,t)Pe_i(x,t)dx = 0.
$$

If this is not true, we have

$$\lim_{t\to+\infty}\sum_{i=1}^{N}\int_{\Omega}e_i^T(x,t)Pe_i(x,t)dx=\mu>0.$$

Then, there obviously exists a real number $M>0$ such that

$$\sum_{i=1}^{N}\int_{\Omega}e_i^T(x,t)Pe_i(x,t)dx>\frac{\mu}{2}$$

for $t\geqslant M$. From (6.38), we can get

$$\begin{aligned}\dot{V}_2(t)&\leqslant-\frac{\sum_{i=1}^{N}\int_{\Omega}e_i^T(x,t)Pe_i(x,t)dx}{\lambda_M(P)}\\&<\frac{-\mu}{2\lambda_M(P)},\quad t\geqslant M.\end{aligned}\tag{6.39}$$

By integrating (6.39) with respect to t over the time period M to $+\infty$, we can obtain

$$\begin{aligned}-V_2(M)\leqslant V_2(+\infty)-V_2(M)&=\int_M^{+\infty}\dot{V}_2(t)dt\\&<-\int_M^{+\infty}\frac{\mu}{2\lambda_M(P)}dt\\&=-\infty.\end{aligned}$$

This yields a contradiction, and so

$$\lim_{t\to+\infty}\sum_{i=1}^{N}\int_{\Omega}e_i^T(x,t)Pe_i(x,t)dx=0.$$

Then, we can easily obtain

$$\lim_{t\to+\infty}\|e(\cdot,t)\|_2=0.$$

Therefore, the CRDNNs (6.24) achieves synchronization under the adaptive law (6.29). The proof is completed.

When $\hat{\mathcal{E}}=\mathcal{E}$, we can easily obtain the following conclusion.

Corollary 6.8 *If there exists a matrix $P\in\mathbb{R}^{n\times n}>0$ such that*

$$\begin{aligned}I_N\otimes(PD)+I_N\otimes(DP)+cG\otimes(P\Gamma)+cG\otimes(\Gamma P)\geqslant0,\\P\hat{\Gamma}+\hat{\Gamma}P>0,\end{aligned}$$

then the CRDNNs (6.24) is synchronized under the following adaptive law:

$$\begin{aligned}\dot{\hat{G}}_{ij}(t) &= \beta_{ij} \int_{\Omega} (z_i(x,t) - z_j(x,t))^T \frac{P\hat{\Gamma} + \hat{\Gamma}P}{2}(z_i(x,t) \\ &\quad - z_j(x,t))dx, \\ \hat{G}_{ij}(0) &= \hat{G}_{ji}(0) > 0, \ (i,j) \in \mathcal{E},\end{aligned} \tag{6.40}$$

where $\beta_{ij} = \beta_{ji}$ are positive constants.

Remark 6.9 In the past few years, some authors [41, 138–140] have studied the synchronization problem of the traditional complex networks with adaptive coupling, which are described by ordinary differential equations. In Theorem 6.7 and Corollary 6.8, two sufficient conditions are obtained to guarantee the synchronization of the CRDNNs (6.24) by utilizing the Lyapunov functional method and edge-based adaptive strategy. The main difficulty for synchronization analysis of system (6.26) comes from the reaction-diffusion terms, which can not be dealt with by those techniques used in traditional complex networks [41, 138–140].

Remark 6.10 In some realistic systems, complex networks with adaptively adjusted coupling strengths may be more reasonable, such as neural networks, biological systems, etc. [128]. In fact, suitable coupling strengths can significantly enhance synchronizability of the networks. Adaptive technique is a very effective method to tune the coupling strengths of networks. Firstly, coupling strengths can be adjusted very quickly by utilizing the adaptive laws defined by the synchronization errors (see (6.29) and (6.40)). Another important advantage of adaptive technique is to adjust coupling strengths for achieving suitable values in practice [129].

6.3.3 *Adaptive Strategies for $\mathcal{H}_\infty$ Synchronization of CRDNNs*

Let $\bar{z}(x,t) = \frac{1}{N}\sum_{j=1}^{N} z_j(x,t)$ and $e_i(x,t) = z_i(x,t) - \bar{z}(x,t)$. Then, we can derive from (6.25) that

$$\begin{aligned}\frac{\partial e_i(x,t)}{\partial t} &= D\Delta e_i(x,t) - Ae_i(x,t) + Bf(z_i(x,t)) + v_i(x,t) \\ &\quad + c\sum_{j=1}^{N} G_{ij}\Gamma\Delta e_j(x,t) + \hat{c}\sum_{j=1}^{N} \hat{G}_{ij}(t)\hat{\Gamma}e_j(x,t) \\ &\quad - \frac{1}{N}\sum_{j=1}^{N} Bf(z_j(x,t)) - \frac{1}{N}\sum_{j=1}^{N} v_j(x,t), \\ &\quad i = 1, 2, \ldots, N.\end{aligned} \tag{6.41}$$

Similarly, suppose that $\hat{\mathcal{E}}$ is a subset of $\mathcal{E}$, and CRDNNs (6.25) is connected through the undirected edges $\hat{\mathcal{E}}$.

Theorem 6.11 *If there exists a matrix $P \in \mathbb{R}^{n\times n} > 0$ such that*

$$W \geqslant 0,$$
$$P\hat{\Gamma} + \hat{\Gamma}P > 0,$$

where $W = I_N \otimes (PD) + I_N \otimes (DP) + cG \otimes (P\Gamma) + cG \otimes (\Gamma P)$, *then the CRDNNs (6.25) is $\mathcal{H}_\infty$ synchronized under the following adaptive law:*

$$\dot{\hat{G}}_{ij}(t) = \beta_{ij} \int_\Omega (z_i(x,t) - z_j(x,t))^T \frac{P\hat{\Gamma} + \hat{\Gamma}P}{2}(z_i(x,t) - z_j(x,t))dx, \quad \text{if} \ (i,j) \in \hat{\mathcal{E}},$$
$$\dot{\hat{G}}_{ij}(t) = 0, \qquad \text{if} \ (i,j) \in \mathcal{E} - \hat{\mathcal{E}}$$

where $\beta_{ij} = \beta_{ji}$ are positive constants.

Proof Take the same Lyapunov functional $V_2(t)$ as in Theorem 6.7. Define the matrix $\Sigma = (\sigma_{ij})_{N\times N}$, where $\sigma_{ii} = -\sum_{\substack{j=1\\ j\neq i}}^{N} \sigma_{ij}$, $i = 1, 2, \ldots, N$. Then, we can easily derive

$$\begin{aligned}
&\sum_{i=1}^{N} \int_0^{t_f} \int_\Omega (e_i^T(x,t)e_i(x,t) - \gamma^2 v_i^T(x,t)v_i(x,t))dxdt \\
&= \int_0^{t_f} \left[\sum_{i=1}^{N} \int_\Omega (e_i^T(x,t)e_i(x,t) - \gamma^2 v_i^T(x,t)v_i(x,t))dx + \dot{V}_2(t) \right] dt \\
&\quad + V_2(0) - V_2(t_f) \\
&= V_2(0) - V_2(t_f) + \sum_{i=1}^{N} \int_0^{t_f} \int_\Omega e_i^T(x,t)e_i(x,t)dxdt \\
&\quad + \sum_{i=1}^{N} \sum_{(i,j)\in\hat{\mathcal{E}}} \int_0^{t_f} \frac{\hat{c}}{2\beta_{ij}} (\hat{G}_{ij}(t) - \sigma_{ij})\dot{\hat{G}}_{ij}(t)dt \\
&\quad + \sum_{i=1}^{N} \int_0^{t_f} \int_\Omega e_i^T(x,t)P\frac{\partial e_i(x,t)}{\partial t}dxdt \\
&\quad - \gamma^2 \sum_{i=1}^{N} \int_0^{t_f} \int_\Omega v_i^T(x,t)v_i(x,t)dxdt
\end{aligned}$$

$$\leqslant V_2(0) - V_2(t_f) + \int_0^{t_f}\int_\Omega e^T(x,t)e(x,t)dxdt$$
$$- \gamma^2 \int_0^{t_f}\int_\Omega v^T(x,t)v(x,t)dxdt + \int_0^{t_f}\int_\Omega e^T(x,t)\left[I_N \otimes \left(\frac{\Theta}{2} + \frac{PBB^TP}{2}\right.\right.$$
$$\left.\left.-\frac{PA+AP}{2}\right) - \sum_{k=1}^{q}\frac{W}{2l_k^2}\right]e(x,t)dxdt + \int_0^{t_f}\int_\Omega e^T(x,t)(I_N \otimes P)v(x,t)dxdt$$
$$+ \hat{c}\int_0^{t_f}\int_\Omega e^T(x,t)\left(\Sigma \otimes \frac{P\hat{\Gamma} + \hat{\Gamma}P}{2}\right)e(x,t)dxdt$$
$$= V_2(0) - V_2(t_f) - \int_0^{t_f}\int_\Omega \left[\gamma v(x,t) - \frac{1}{2\gamma}(I_N \otimes P)\,e(x,t)\right]^T [\gamma v(x,t)$$
$$- \frac{1}{2\gamma}(I_N \otimes P)e(x,t)]dxdt + \int_0^{t_f}\int_\Omega e^T(x,t)\left[\frac{1}{4\gamma^2}(I_N \otimes P^2)\right.$$
$$+ I_N \otimes \left(\frac{\Theta}{2} + \frac{PBB^TP}{2} - \frac{PA+AP}{2} + I_n\right)$$
$$\left.+\hat{c}\left(\Sigma \otimes \frac{P\hat{\Gamma} + \hat{\Gamma}P}{2}\right) - \sum_{k=1}^{q}\frac{W}{2l_k^2}\right] e(x,t)dxdt$$
$$\leqslant \int_0^{t_f}\int_\Omega e^T(x,t)\left[-\sum_{k=1}^{q}\frac{W}{2l_k^2} + I_N \otimes \left(\frac{\Theta}{2} - \frac{PA+AP}{2} + \frac{PBB^TP}{2} + I_n\right)\right.$$
$$\left.+ \hat{c}\left(\Sigma \otimes \frac{P\hat{\Gamma} + \hat{\Gamma}P}{2}\right) + \frac{1}{4\gamma^2}(I_N \otimes P^2)\right] e(x,t)dxdt + V_2(0) \tag{6.42}$$

where γ is a positive constant, $v(x,t) = (v_1^T(x,t),\ v_2^T(x,t), \ldots, v_N^T(x,t))^T$, $\Theta = \text{diag}(\rho_1^2, \rho_2^2, \ldots, \rho_n^2)$, $e(x,t) = (e_1^T(x,t), e_2^T(x,t), \ldots, e_N^T(x,t))^T$.

Obviously, there exists a unitary matrix $\Psi = (\psi_1, \psi_2, \ldots, \psi_N) \in \mathbb{R}^{N\times N}$ such that $\Psi^T\Sigma\Psi = \Lambda$ with $\Lambda = \text{diag}(\lambda_1, \lambda_2, \ldots, \lambda_N)$. $\lambda_i, i = 1, 2, \ldots, N$, are the eigenvalues of Σ and $0 = \lambda_1 > \lambda_2 \geqslant \lambda_3 \geqslant \ldots \geqslant \lambda_N$. Similar to the proof of Theorem 6.7, we have

$$\int_0^{t_f}\int_\Omega (e^T(x,t)e(x,t) - \gamma^2 v^T(x,t)v(x,t))dxdt$$
$$\leqslant \int_0^{t_f}\int_\Omega e^T(x,t)\left[-\sum_{k=1}^{q}\frac{W}{2l_k^2} + I_N \otimes \left(\frac{\Theta}{2} + \frac{PBB^TP}{2} - \frac{PA+AP}{2} + I_n\right)\right.$$
$$+ \frac{1}{4\gamma^2}(I_N \otimes P^2)$$
$$\left.+\hat{c}\lambda_2\left(I_N \otimes \frac{P\hat{\Gamma} + \hat{\Gamma}P}{2}\right)\right] e(x,t)dxdt + V_2(0).$$

By selecting σ_{ij} sufficiently large such that

$$\hat{c}\lambda_2\lambda_m\left(\frac{P\hat{\Gamma}+\hat{\Gamma}P}{2}\right)+\frac{\lambda_M(PBB^TP)}{2}+1-\lambda_m\left(\sum_{k=1}^{q}\frac{W}{2l_k^2}\right)$$
$$+\frac{\max_{i=1,2,\ldots,n}\{\rho_i^2\}}{2}-\lambda_m\left(\frac{PA+AP}{2}\right)+\frac{\lambda_M^2(P)}{4\gamma^2}\leqslant 0,$$

one obtains

$$\int_0^{t_f}\int_\Omega e^T(x,t)e(x,t)dxdt$$
$$\leqslant V_2(0)+\gamma^2\int_0^{t_f}\int_\Omega v^T(x,t)v(x,t)dxdt.$$

Therefore, the CRDNNs (6.25) is $\mathcal{H}_\infty$ synchronized with the disturbance attenuation level γ. The proof is completed.

When $\hat{\mathcal{E}}=\mathcal{E}$, we can easily obtain the following conclusion.

Corollary 6.12 *If there exists a matrix $P\in\mathbb{R}^{n\times n}>0$ such that*

$$I_N\otimes(PD)+I_N\otimes(DP)+cG\otimes(P\Gamma)+cG\otimes(\Gamma P)\geqslant 0,$$
$$P\hat{\Gamma}+\hat{\Gamma}P>0,$$

then the CRDNNs (6.25) is $\mathcal{H}_\infty$ synchronized under the following adaptive law:

$$\dot{\hat{G}}_{ij}(t)=\beta_{ij}\int_\Omega(z_i(x,t)-z_j(x,t))^T\frac{P\hat{\Gamma}+\hat{\Gamma}P}{2}(z_i(x,t)-z_j(x,t))dx,$$
$$\hat{G}_{ij}(0)=\hat{G}_{ji}(0)>0,\ (i,j)\in\mathcal{E},$$

where $\beta_{ij}=\beta_{ji}$ are positive constants.

Remark 6.13 In [122], Wu and Chen investigated the asymptotical synchronization and $\mathcal{H}_\infty$ synchronization of N-coupled linear partial differential systems via spatial diffusion coupling, and several sufficient conditions to guarantee the asymptotical synchronization and the $\mathcal{H}_\infty$ synchronization were derived. In this chapter, we consider a CRDNNs consisting of N identical RDNNs with hybrid coupling, and some sufficient conditions ensuring the synchronization and $\mathcal{H}_\infty$ synchronization are established. On the other hand, it is well known that CRDNNs cannot synchronize by themselves in many circumstances. Therefore, some adaptive strategies to tune the coupling strengths among network nodes are designed for reaching synchronization and $\mathcal{H}_\infty$ synchronization.

6.4 Numerical Examples

As an application of the above obtained theoretical results, two representative examples are given in this section.

Example 6.14 Consider a CRDNNs consisting of 10 identical nodes with hybrid coupling, in which each node is a 3-dimensional reaction-diffusion neural network described by

$$\frac{\partial w_i(x,t)}{\partial t} = d_i \frac{\partial^2 w_i(x,t)}{\partial x^2} - a_i w_i(x,t) + J_i + \sum_{j=1}^{3} b_{ij} f_j(w_j(x,t))$$

where $i = 1, 2, 3$, $\Omega = \{x \mid -0.5 < x < 0.5\}$, $f_j(\xi) = \frac{|\xi+1|-|\xi-1|}{4}$, $d_1 = 0.3$, $d_2 = 0.4$, $d_3 = 0.3$, $a_1 = 0.3$, $a_2 = 0.5$, $a_3 = 0.4$, $J_1 = 0.2$, $J_2 = 0.4$, $J_3 = 0.2$, and the matrix $B = (b_{ij})_{3\times 3}$ is chosen as

$$B = \begin{pmatrix} 0.3 & 0.2 & 0.1 \\ 0.1 & 0.2 & 0.1 \\ 0.3 & 0.2 & 0.2 \end{pmatrix}.$$

Obviously, $f_j(\cdot)(j = 1, 2, 3)$ satisfies the Lipschitz condition with $\rho_j = \frac{1}{2}$. We take $c = 0.1$, $\hat{c} = 0.2$, $\Gamma = \text{diag}(0.4, 0.5, 0.3)$ and $\hat{\Gamma} = \text{diag}(0.5, 0.6, 0.4)$. The coupling strengths G_{ij} and $\hat{G}_{ij}(i, j = 1, 2, 3, \ldots, 10)$ are chosen as follows: $G_{11} = -0.1$, $G_{16} = G_{61} = 0.1$, $G_{22} = -0.2$, $G_{29} = G_{92} = 0.2$, $G_{33} = -0.1$, $G_{34} = G_{43} = 0.1$, $G_{44} = -0.3$, $G_{410} = G_{104} = 0.2$, $G_{55} = -0.3$, $G_{510} = G_{105} = 0.3$, $G_{66} = -0.5$, $G_{67} = G_{76} = 0.4$, $G_{77} = -0.5$, $G_{78} = G_{87} = 0.1$, $G_{88} = -0.3$, $G_{89} = G_{98} = 0.2$, $G_{99} = -0.7$, $G_{910} = G_{109} = 0.3$, $G_{1010} = -0.8$, $\hat{G}_{11} = -0.2$, $\hat{G}_{16} = \hat{G}_{61} = 0.2$, $\hat{G}_{22} = -0.3$, $\hat{G}_{29} = \hat{G}_{92} = 0.3$, $\hat{G}_{33} = -0.2$, $\hat{G}_{34} = \hat{G}_{43} = 0.2$, $\hat{G}_{44} = -0.5$, $\hat{G}_{410} = \hat{G}_{104} = 0.3$, $\hat{G}_{55} = -0.3$, $\hat{G}_{510} = \hat{G}_{105} = 0.3$, $\hat{G}_{66} = -0.6$, $\hat{G}_{67} = \hat{G}_{76} = 0.4$, $\hat{G}_{77} = -0.6$, $\hat{G}_{78} = \hat{G}_{87} = 0.2$, $\hat{G}_{88} = -0.4$, $\hat{G}_{89} = \hat{G}_{98} = 0.2$, $\hat{G}_{99} = -0.8$, $\hat{G}_{910} = \hat{G}_{109} = 0.3$, $\hat{G}_{1010} = -0.9$. Here, only the non-zero coupling strengths are given.

By using the MATLAB LMI Toolbox, we can find the positive definite matrix

$$P = \begin{pmatrix} 0.7610 & -0.0071 & -0.0190 \\ -0.0071 & 0.5917 & -0.0074 \\ -0.0190 & -0.0074 & 0.7204 \end{pmatrix}$$

satisfying (6.21) and (6.22) with $\gamma = 1.2$. According to Theorem 6.5, the CRDNNs (6.17) with above given parameters is $\mathcal{H}_\infty$ synchronized in the sense of Definition 6.4. Set $v_i(x,t) = (\cos(i\pi x)e^{-0.01t}, \cos(i\pi x)e^{-0.02t}, \cos(i\pi x)\, e^{-0.01t})^T$. The simulation results are shown in Fig. 6.1.

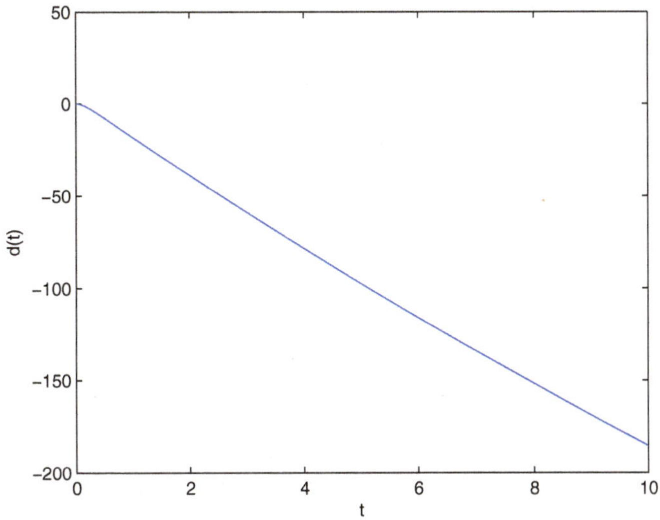

Fig. 6.1 The change process of $d(t)$, in which $d(t) = \int_0^t \int_\Omega e^T(x,s)e(x,s)dxds - 1.44 \int_0^t \int_\Omega v^T(x,s)v(x,s)dxds$

Obviously, if we take

$$P = \begin{pmatrix} 0.7610 & -0.0071 & -0.0190 \\ -0.0071 & 0.5917 & -0.0074 \\ -0.0190 & -0.0074 & 0.7204 \end{pmatrix},$$

then the conditions (6.7) and (6.8) are also satisfied. From Theorem 6.3, the CRDNNs (6.1) with above given parameters is synchronized. The simulation results are shown in Fig. 6.2.

Example 6.15 Consider a CRDNNs consisting of 5 identical nodes with hybrid coupling, in which each node is a 3-dimensional reaction-diffusion neural network described by

$$\frac{\partial w_i(x,t)}{\partial t} = d_i \frac{\partial^2 w_i(x,t)}{\partial x^2} - a_i w_i(x,t) + J_i + \sum_{j=1}^{3} b_{ij} f_j(w_j(x,t))$$

where $i = 1, 2, 3$, $\Omega = \{x \mid -0.5 < x < 0.5\}$, $f_j(\xi) = \frac{|\xi+1|-|\xi-1|}{4}$, $d_1 = 0.4$, $d_2 = 0.3$, $d_3 = 0.4$, $a_1 = 0.2$, $a_2 = 0.1$, $a_3 = 0.1$, $J_1 = 0.3$, $J_2 = 0.1$, $J_3 = 0.1$, and the matrix $B = (b_{ij})_{3\times 3}$ is chosen as

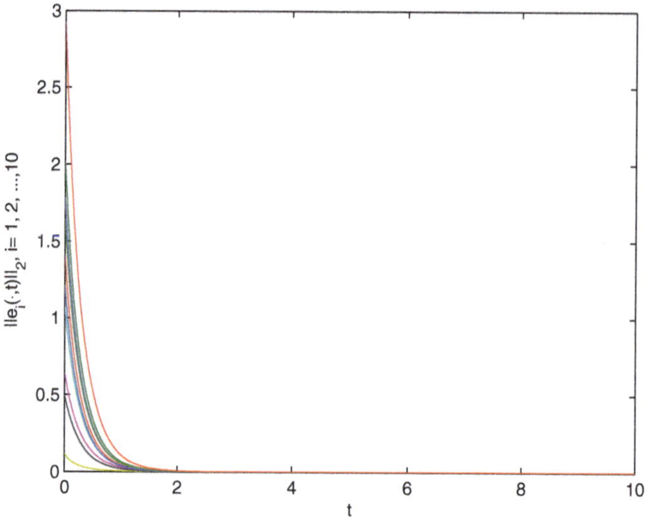

Fig. 6.2 The change processes of $\|e_i(\cdot,t)\|_2$, $i = 1, 2, \ldots, 10$

$$B = \begin{pmatrix} 0.5 & 0.4 & 0.3 \\ 0.6 & 0.5 & 0.4 \\ 0.5 & 0.3 & 0.4 \end{pmatrix}.$$

Obviously, $f_j(\cdot)(j = 1, 2, 3)$ satisfies the Lipschitz condition with $\rho_j = \frac{1}{2}$.

We take $c = 0.1$, $\hat{c} = 0.2$, $\Gamma = \text{diag}(0.2, 0.3, 0.1)$ and $\hat{\Gamma} = \text{diag}(0.3, 0.4, 0.3)$. The matrices G and $\hat{G}(0)$ are chosen as, respectively,

$$G = \begin{pmatrix} -0.4 & 0.1 & 0.2 & 0 & 0.1 \\ 0.1 & -0.3 & 0.2 & 0 & 0 \\ 0.2 & 0.2 & -0.5 & 0.1 & 0 \\ 0 & 0 & 0.1 & -0.3 & 0.2 \\ 0.1 & 0 & 0 & 0.2 & -0.3 \end{pmatrix},$$

$$\hat{G}(0) = \begin{pmatrix} -0.3 & 0.1 & 0.1 & 0 & 0.1 \\ 0.1 & -0.3 & 0.2 & 0 & 0 \\ 0.1 & 0.2 & -0.5 & 0.2 & 0 \\ 0 & 0 & 0.2 & -0.3 & 0.1 \\ 0.1 & 0 & 0 & 0.1 & -0.2 \end{pmatrix}.$$

It is easy to verify that the conditions (6.27) and (6.28) are satisfied if $P = \text{diag}(0.5, 0.5, 0.4)$.

Choose $\beta_{15} = \beta_{51} = 0.4$, $\beta_{23} = \beta_{32} = 0.4$, $\beta_{34} = \beta_{43} = 0.3$, $\beta_{45} = \beta_{54} = 0.4$. According to Theorem 6.7, the CRDNNs (6.24) with above given parameters is synchronized under the adaptive law (6.29). The simulation results are shown in

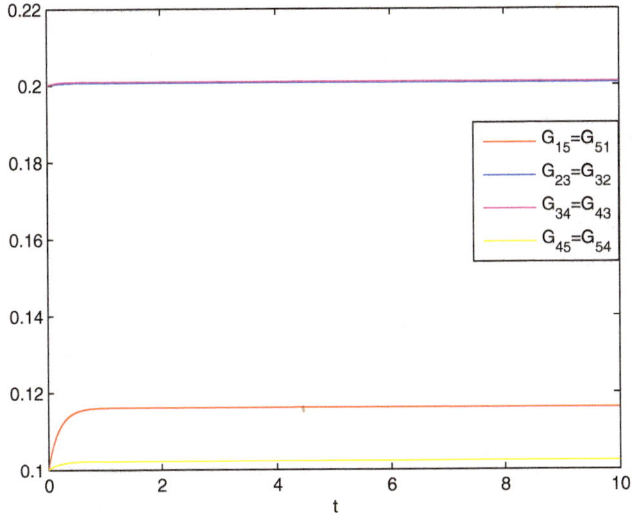

Fig. 6.3 Adaptive coupling strengths

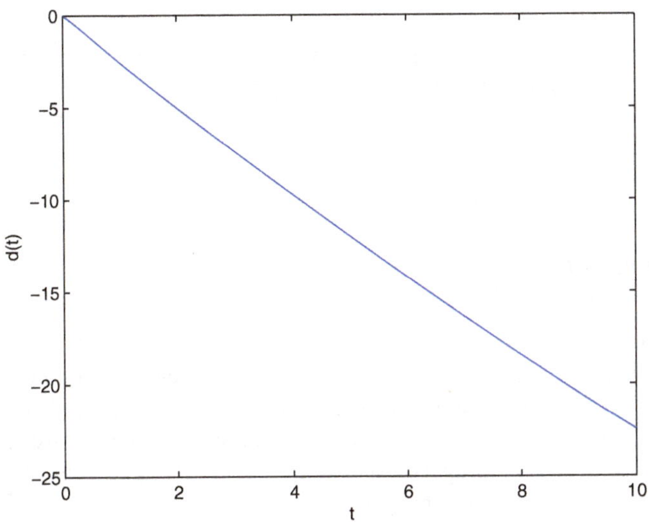

Fig. 6.4 The change process of $d(t)$, in which $d(t) = \int_0^t \int_\Omega e^T(x, s)e(x, s)dxds - 0.36 \int_0^t \int_\Omega v^T(x, s)v(x, s)dxds$

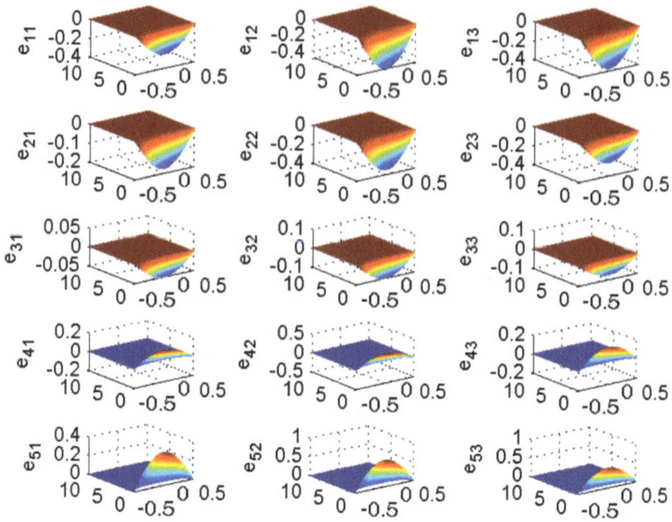

Fig. 6.5 The change processes of $e_{ij}(x, t), i = 1, 2, \ldots, 5, j = 1, 2, 3$

Figs. 6.5 and 6.3. On the other hand, by Theorem 6.11, we know that the CRDNNs (6.25) with above given parameters is $\mathcal{H}_\infty$ synchronized under the adaptive law (6.29). Set $v_i(x, t) = (\cos(i\pi x)e^{-0.01t}, \cos(i\pi x)e^{-0.02t}, \cos(i\pi x)e^{-0.01t})^T$. The simulation results are shown in Fig. 6.4.

6.5 Conclusions

A general array model of CRDNNs with hybrid coupling has first been introduced. Both synchronization and $\mathcal{H}_\infty$ synchronization have been taken into consideration in this chapter, and some sufficient conditions have been derived. Furthermore, several novel edge-based adaptive strategies to tune the coupling strengths among network nodes have been proposed, and some criteria for synchronization and $\mathcal{H}_\infty$ synchronization have been established by using the designed adaptive laws. Illustrative simulations have been provided to verify the correctness and effectiveness of the theoretical results.

Chapter 7
Passivity-Based Synchronization of CRDNNs with Time-Varying Delay

7.1 Introduction

Recently, passivity theory has also received a great deal of attention, and many results on this topic have been reported. The passivity theory was firstly proposed in the circuit analysis [176], and since then has found successful applications in diverse areas such as stability, complexity, signal processing, chaos control and synchronization, fuzzy control, group coordination, power control, flow control, energy management, and so on [177–180]. Although research on passivity has attracted so much attention, little of that had been devoted to the passivity properties of the spatially and temporally complex dynamical networks until Wang, Wu and Guo [29] obtained the conditions for passivity of RDNNs. To the best of our knowledge, the passivity of arrays of CRDNNs has not yet been considered. Therefore, it is important and interesting to study the passivity of CRDNNs. On the other hand, the passivity theory has long been a nice tool for analyzing the synchronization of the complex networks. But in most existing works, it is assumed that the node state is only dependent on the time. Therefore, it is essential to investigate the relationship between passivity and synchronization of the CRDNNs.

The objective of this chapter is to study the synchronization problem of arrays of CRDNNs by using the passivity theory. The main contributions of this chapter are as follows. First, we establish a criterion for the output strict passivity by utilizing some inequality techniques. Second, we reveal the relationship between output strict passivity and synchronization of the proposed network model. Third, by employing the obtained passivity result and the relationship between output strict passivity and synchronization, a sufficient condition for synchronization of the CRDNNs is derived.

J.-L. Wang et al., *Analysis and Control of Coupled Neural Networks with Reaction-Diffusion Terms*, DOI 10.1007/978-981-10-4907-1_7

7.2 Network Model

N mutually coupled RDNNs (2.4) can result in a CRDNNs, which is described by

$$\frac{\partial z_i(x,t)}{\partial t} = D\Delta z_i(x,t) - Az_i(x,t) + Bf(z_i(x,t)) + c_1 \sum_{j=1}^{N} G_{ij}^1 \Gamma_1 z_j(x,t) + J + c_2 \sum_{j=1}^{N} G_{ij}^2 \Gamma_2 z_j(x,t-\tau(t)) + u_i(x,t) \tag{7.1}$$

where $i = 1, 2, \ldots, N$, N is the number of nodes in the network; $\tau(t)$ is the time-varying delay with $0 \leqslant \tau(t) \leqslant \tau$; $z_i(x,t) = (z_{i1}(x,t), z_{i2}(x,t), \ldots, z_{in}(x,t))^T \in \mathbb{R}^n$ is the state vector of node i; $u_i(x,t) \in \mathbb{R}^n$ is the control input; c_1 and c_2 are positive real numbers, which represent the coupling strength for non-delayed configuration and delayed one, respectively; $\Gamma_1 = \text{diag}(\gamma_1^1, \gamma_2^1, \ldots, \gamma_n^1) \in \mathbb{R}^{n \times n}$ and $\Gamma_2 = \text{diag}(\gamma_1^2, \gamma_2^2, \ldots, \gamma_n^2) \in \mathbb{R}^{n \times n}$ are positive definite matrices, which describe the individual coupling between two nodes for non-delayed configuration and delayed one, respectively; $G^1 = (G_{ij}^1)_{N \times N}$ and $G^2 = (G_{ij}^2)_{N \times N}$ represent the topological structure of network and coupling strength between nodes for non-delayed configuration and delayed one, respectively, where G_{ij}^1 (G_{ij}^2 can be defined similarly) is defined as follows: if there exists a connection between node i and node j, then $G_{ij}^1 = G_{ji}^1 > 0$; otherwise, $G_{ij}^1 = G_{ji}^1 = 0 (i \neq j)$, and the diagonal elements of matrix G^1 are defined by

$$G_{ii}^1 = -\sum_{\substack{j=1 \\ j \neq i}}^{N} G_{ij}^1, \quad i = 1, 2, \ldots, N.$$

In this section, the topological structure of CRDNNs (7.1) is undirected and weighted. The initial value and boundary value conditions associated with network (7.1) are given in the form

$$z_i(x,t) = \Phi_i(x,t) \in \mathbb{R}^n, (x,t) \in \Omega \times [-\tau, 0], \tag{7.2}$$

$$z_i(x,t) = 0, (x,t) \in \partial\Omega \times [-\tau, +\infty) \tag{7.3}$$

where $\Phi_i(x,t) (i = 1, 2, \ldots, N)$ is a bounded and continuous function on $\Omega \times [-\tau, 0]$.

Remark 7.1 Many systems in nature and society can be modeled as complex networks, which consist of nodes connected by edges. In a complex network, each node represents a fundamental unit with specific activity, while edges represent the relationship of these fundamental units. In this chapter, we consider a CRDNNs con-

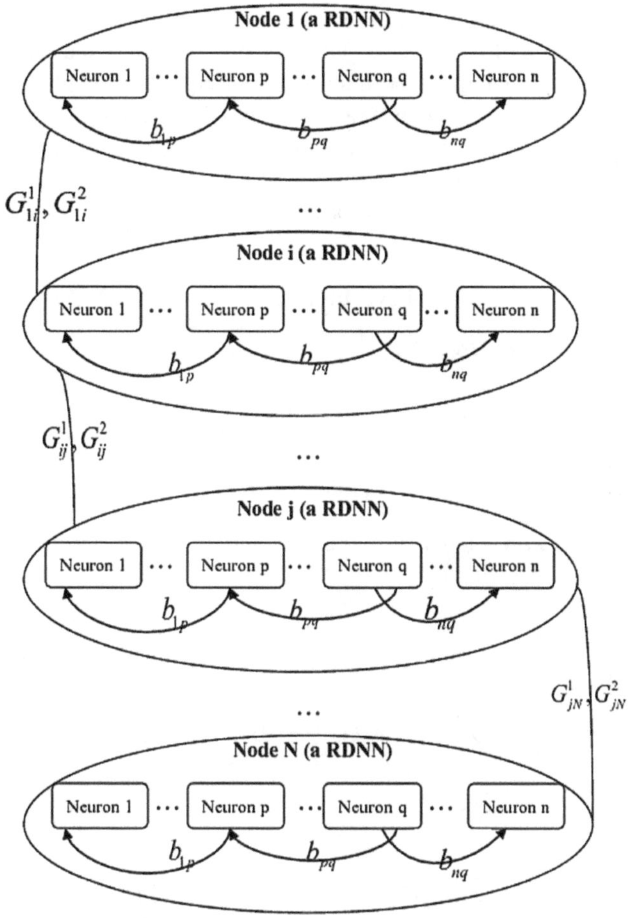

Fig. 7.1 CRDNNs (7.1) ($i, j \in \{1, 2, \ldots, N\}$, $p, q \in \{1, 2, \ldots, n\}$)

sisting of N identical nodes, in which each node is an n-dimensional RDNN (see Fig. 7.1). In CRDNNs (7.1), G^1_{ij} and G^2_{ij} denote the impact strength of the jth node on the ith node for non-delayed case and delayed case, respectively.

7.3 Passivity-Based Synchronization of CRDNNs

Suppose that $w^*(x) = (w^*_1(x), w^*_2(x), \ldots, w^*_n(x))^T$ is an equilibrium solution of the reaction-diffusion neural network (2.4), then it satisfies (2.3) and

$$0 = D\Delta w^*(x) - Aw^*(x) + J + Bf(w^*(x)). \tag{7.4}$$

Remark 7.2 In this section, we assume that $w^*(x)$ is an equilibrium solution of the reaction-diffusion neural network (2.4), but we do not require that the equilibrium solution be unique. Obviously, the uniqueness of equilibrium solution is also an important and interesting problem. Practically, in [31], a class of interval reaction-diffusion Hopfield neural networks with Dirichlet boundary conditions was considered, and reaction-diffusion neural network (2.4) is a special case of this model. The author obtained a sufficient condition for the uniqueness of the equilibrium solution in [31].

Next, we give the definition of synchronization for the CRDNNs (7.1).

Definition 7.3 The CRDNNs (7.1) is said to achieve synchronization if

$$\lim_{t\to+\infty} \|z_i(\cdot,t) - w^*(\cdot)\|_2 = 0 \text{ for all } i = 1, 2, \dots, N$$

under the condition that $u_i(x,t) = 0, i = 1, 2, \dots, N$.

Defining $e_i(x,t) = z_i(x,t) - w^*(x)$, then the dynamics of the error vector $e_i(x,t)$ is governed by the following equation:

$$\begin{aligned}\frac{\partial e_i(x,t)}{\partial t} &= D\Delta e_i(x,t) - Ae_i(x,t) + Bf(z_i(x,t)) - Bf(w^*(x)) \\ &\quad + c_1\sum_{j=1}^{N} G^1_{ij}\Gamma_1 e_j(x,t) + c_2\sum_{j=1}^{N} G^2_{ij}\Gamma_2 e_j(x,t-\tau(t)) \\ &\quad + u_i(x,t)\end{aligned} \tag{7.5}$$

where $i = 1, 2, \dots, N$.

In what follows, we establish the passivity of system (7.5). The output vector $y_i(x,t)$ of system (7.5) is defined as

$$y_i(x,t) = Fe_i(x,t) + Hu_i(x,t) \tag{7.6}$$

for the passivity scheme, where $F, H \in \mathbb{R}^{n\times n}$ are known real matrices, $i=1, 2, \dots, N$.

For the convenience, we denote

$$\begin{aligned}\hat{D} &= \text{diag}(D, D, \dots, D),\\ \hat{A} &= \text{diag}(A, A, \dots, A),\\ \hat{B} &= \text{diag}(B, B, \dots, B),\\ \hat{F} &= \text{diag}(F, F, \dots, F),\\ \hat{H} &= \text{diag}(H, H, \dots, H),\\ \Theta &= \text{diag}(\rho_1^2, \rho_2^2, \dots, \rho_n^2),\\ e(x,t) &= (e_1^T(x,t), e_2^T(x,t), \dots, e_N^T(x,t))^T,\end{aligned}$$

$$u(x,t) = (u_1^T(x,t), u_2^T(x,t), \ldots, u_N^T(x,t))^T,$$
$$y(x,t) = (y_1^T(x,t), y_2^T(x,t), \ \ldots, \ y_N^T(x,t))^T,$$
$$\hat{f}(\hat{w}^*(x)) = (f^T(w^*(x)), f^T(w^*(x)), \ldots, f^T(w^*(x)))^T,$$
$$\hat{f}(z(x,t)) = (f^T(z_1(x,t)), f^T(z_2(x,t)), \ \ldots, \ f^T(z_N(x,t)))^T.$$

By using Kronecker product, we can rewrite systems (7.5) and (7.6) in a compact form as follows:

$$\begin{aligned}\frac{\partial e(x,t)}{\partial t} &= \hat{D}\Delta e(x,t) - \hat{A}e(x,t) + \hat{B}\hat{f}(z(x,t)) - \hat{B}\hat{f}(\hat{w}^*(x))\\ &\quad + c_1(G^1 \otimes \Gamma_1)e(x,t) + c_2(G^2 \otimes \Gamma_2)e(x,t-\tau(t))\\ &\quad + u(x,t)\\ y(x,t) &= \hat{F}e(x,t) + \hat{H}u(x,t). \end{aligned} \tag{7.7}$$

Lemma 7.4 *Let $\dot{\tau}(t) \leqslant \sigma < 1$. If there exist matrices $\hat{P} = \mathrm{diag}(P^1, P^2, \ldots, P^N) \in \mathbb{R}^{nN\times nN} > 0$, $Q \in \mathbb{R}^{nN\times nN} > 0$ and a scalar $\gamma > 0$ such that*

$$\hat{P}\hat{D} + \hat{D}\hat{P} \geqslant 0 \tag{7.8}$$

$$\begin{pmatrix} \chi & \hat{P} - \hat{F}^T + \gamma\hat{F}^T\hat{H} \\ \hat{P} - \hat{F} + \gamma\hat{H}^T\hat{F} & -\hat{H} - \hat{H}^T + \gamma\hat{H}^T\hat{H} \end{pmatrix} \leqslant 0 \tag{7.9}$$

where $\hat{\Theta} = \mathrm{diag}(\Theta, \Theta, \ldots, \Theta)$, $\chi = -\sum_{k=1}^{q} \frac{1}{l_k^2}(\hat{P}\hat{D} + \hat{D}\hat{P}) - \hat{P}\hat{A} - \hat{A}\hat{P} + \hat{\Theta} + \hat{P}\hat{B}\hat{B}^T\hat{P} + c_1\hat{P}(G^1 \otimes \Gamma_1) + c_1(G^1 \otimes \Gamma_1)\hat{P} + \frac{Q}{1-\sigma} + c_2^2\hat{P}(G^2 \otimes \Gamma_2)Q^{-1}(G^2 \otimes \Gamma_2)$ $\hat{P} + \gamma\hat{F}^T\hat{F}$, then system (7.7) is output-strictly passive in the sense of Definition 1.5.

Proof First, construct a Lyapunov functional for system (7.7) as follows:

$$\begin{aligned} V(t) &= \int_\Omega e^T(x,t)\hat{P}e(x,t)dx \\ &\quad + \frac{1}{1-\sigma}\int_{t-\tau(t)}^{t}\int_\Omega e^T(x,s)Qe(x,s)dxds. \end{aligned} \tag{7.10}$$

Calculating the time derivative of $V(t)$ along the trajectory of system (7.7), we can get

$$\begin{aligned} \dot{V}(t) &= 2\int_\Omega e^T(x,t)\hat{P}\frac{\partial e(x,t)}{\partial t}dx + \frac{1}{1-\sigma}\int_\Omega e^T(x,t)Qe(x,t)dx \\ &\quad - \frac{1-\dot{\tau}(t)}{1-\sigma}\int_\Omega e^T(x,t-\tau(t))Qe(x,t-\tau(t))dx \\ &\leqslant \int_\Omega [e^T(x,t)\hat{P}\hat{D}\Delta e(x,t) + (\Delta e(x,t))^T\hat{D}\hat{P}e(x,t) \end{aligned}$$

$$
\begin{aligned}
&+ e^T(x,t)(-\hat{P}\hat{A} - \hat{A}\hat{P} + c_1\hat{P}(G^1 \otimes \Gamma_1) + \frac{Q}{1-\sigma} \\
&+ c_1(G^1 \otimes \Gamma_1)\hat{P})e(x,t) + 2e^T(x,t)\hat{P}\hat{B}(\hat{f}(z(x,t)) \\
&- \hat{f}(\hat{w}^*(x))) + 2e^T(x,t)\hat{P}u(x,t) \\
&+ 2c_2 e^T(x,t)\hat{P}(G^2 \otimes \Gamma_2)e(x,t-\tau(t))]dx \\
&- \int_\Omega e^T(x,t-\tau(t))Qe(x,t-\tau(t))dx.
\end{aligned}
$$

From Green's formula and the boundary condition, we have

$$
\int_\Omega e_{il}(x,t)\Delta e_{ij}(x,t)dx = -\sum_{k=1}^{q}\int_\Omega \frac{\partial e_{il}(x,t)}{\partial x_k}\frac{\partial e_{ij}(x,t)}{\partial x_k}dx
$$

where $l, j \in \{1, 2, \ldots, n\}$, $i = 1, 2, \ldots, N$. Letting $P^i = (p^i_{jl})_{n\times n}$, we can obtain

$$
\begin{aligned}
&\int_\Omega e^T(x,t)\hat{P}\hat{D}\Delta e(x,t)dx \\
&= \sum_{i=1}^{N}\int_\Omega e_i^T(x,t)P^i D\Delta e_i(x,t)dx \\
&= \sum_{i=1}^{N}\sum_{j=1}^{n}\sum_{l=1}^{n} p^i_{jl}d_l \int_\Omega e_{ij}(x,t)\Delta e_{il}(x,t)dx \\
&= -\sum_{k=1}^{q}\sum_{i=1}^{N}\sum_{j=1}^{n}\sum_{l=1}^{n} p^i_{jl}d_l \int_\Omega \frac{\partial e_{ij}(x,t)}{\partial x_k}\frac{\partial e_{il}(x,t)}{\partial x_k}dx \\
&= -\sum_{k=1}^{q}\int_\Omega \left(\frac{\partial e(x,t)}{\partial x_k}\right)^T \hat{P}\hat{D}\frac{\partial e(x,t)}{\partial x_k}dx, \\
&\int_\Omega (\Delta e(x,t))^T\hat{D}\hat{P}e(x,t)dx \\
&= \sum_{i=1}^{N}\int_\Omega (\Delta e_i(x,t))^T DP^i e_i(x,t)dx \\
&= \sum_{i=1}^{N}\sum_{j=1}^{n}\sum_{l=1}^{n} d_j p^i_{jl}\int_\Omega e_{il}(x,t)\Delta e_{ij}(x,t)dx \\
&= -\sum_{k=1}^{q}\int_\Omega \left(\frac{\partial e(x,t)}{\partial x_k}\right)^T \hat{D}\hat{P}\frac{\partial e(x,t)}{\partial x_k}dx.
\end{aligned}
$$

Then, we can get

$$\int_{\Omega}(e^T(x,t)\hat{P}\hat{D}\Delta e(x,t)+(\Delta e(x,t))^T\hat{D}\hat{P}e(x,t))dx$$
$$=-\sum_{k=1}^{q}\int_{\Omega}\left(\frac{\partial e(x,t)}{\partial x_k}\right)^T(\hat{P}\hat{D}+\hat{D}\hat{P})\frac{\partial e(x,t)}{\partial x_k}dx.$$

On the other hand, there obviously exists a real matrix $C\in\mathbb{R}^{nN\times nN}$ such that

$$\hat{P}\hat{D}+\hat{D}\hat{P}=C^TC,$$

then

$$\left(\frac{\partial e(x,t)}{\partial x_k}\right)^T(\hat{P}\hat{D}+\hat{D}\hat{P})\frac{\partial e(x,t)}{\partial x_k}=\left(\frac{\partial e(x,t)}{\partial x_k}\right)^T C^TC\frac{\partial e(x,t)}{\partial x_k}$$
$$=\left(\frac{\partial(Ce(x,t))}{\partial x_k}\right)^T\frac{\partial(Ce(x,t))}{\partial x_k}.$$

Let $\vartheta(x,t)=Ce(x,t)$, for $(x,t)\in\partial\Omega\times[-\tau,+\infty)$ from the boundary condition (7.3), we have $\vartheta(x,t)=Ce(x,t)=0$. In view of Lemma 1.22, one has

$$\sum_{k=1}^{q}\int_{\Omega}\left(\frac{\partial\vartheta(x,t)}{\partial x_k}\right)^T\frac{\partial\vartheta(x,t)}{\partial x_k}dx$$
$$\geqslant\sum_{k=1}^{q}\frac{1}{l_k^2}\int_{\Omega}\vartheta^T(x,t)\vartheta(x,t)dx$$
$$=\sum_{k=1}^{q}\frac{1}{l_k^2}\int_{\Omega}e^T(x,t)(\hat{P}\hat{D}+\hat{D}\hat{P})e(x,t)dx. \tag{7.11}$$

Furthermore, we can easily derive

$$2e^T(x,t)\hat{P}\hat{B}(\hat{f}(z(x,t))-\hat{f}(\hat{w}^*(x)))$$
$$=2\sum_{i=1}^{N}e_i^T(x,t)P^iB(f(z_i(x,t))-f(w^*(x)))$$
$$\leqslant\sum_{i=1}^{N}e_i^T(x,t)P^iBB^TP^ie_i(x,t)+\sum_{i=1}^{N}e_i^T(x,t)\Theta e_i(x,t)$$
$$=e^T(x,t)(\hat{P}\hat{B}\hat{B}^T\hat{P}+\hat{\Theta})e(x,t), \tag{7.12}$$

$$\begin{aligned}
&2e^T(x,t)\hat{P}(G^2\otimes\Gamma_2)e(x,t-\tau(t))\\
&\leqslant c_2e^T(x,t)\hat{P}(G^2\otimes\Gamma_2)Q^{-1}(G^2\otimes\Gamma_2)\hat{P}e(x,t)\\
&+\frac{1}{c_2}e^T(x,t-\tau(t))Qe(x,t-\tau(t)).
\end{aligned}\tag{7.13}$$

It follows from (7.11)–(7.13) that

$$\begin{aligned}
\dot{V}(t)\leqslant&\int_\Omega e^T(x,t)[-\sum_{k=1}^{q}\frac{1}{l_k^2}(\hat{P}\hat{D}+\hat{D}\hat{P})-\hat{P}\hat{A}-\hat{A}\hat{P}\\
&+\hat{\Theta}+\hat{P}\hat{B}\hat{B}^T\hat{P}+c_1\hat{P}(G^1\otimes\Gamma_1)+c_1(G^1\otimes\Gamma_1)\hat{P}\\
&+\frac{Q}{1-\sigma}+c_2^2\hat{P}(G^2\otimes\Gamma_2)Q^{-1}(G^2\otimes\Gamma_2)\hat{P}]e(x,t)dx\\
&+2\int_\Omega e^T(x,t)\hat{P}u(x,t)dx.
\end{aligned}$$

Therefore,

$$\begin{aligned}
&\dot{V}(t)-2\int_\Omega y^T(x,t)u(x,t)dx+\gamma\int_\Omega y^T(x,t)y(x,t)dx\\
\leqslant&\int_\Omega e^T(x,t)\Big[-\sum_{k=1}^{q}\frac{1}{l_k^2}(\hat{P}\hat{D}+\hat{D}\hat{P})-\hat{P}\hat{A}-\hat{A}\hat{P}\\
&+\hat{\Theta}+\hat{P}\hat{B}\hat{B}^T\hat{P}+c_1\hat{P}(G^1\otimes\Gamma_1)+c_1(G^1\otimes\Gamma_1)\hat{P}\\
&+\frac{Q}{1-\sigma}+c_2^2\hat{P}(G^2\otimes\Gamma_2)Q^{-1}(G^2\otimes\Gamma_2)\hat{P}+\gamma\hat{F}^T\hat{F}\Big]\\
&\times e(x,t)dx+2\int_\Omega e^T(x,t)(\hat{P}-\hat{F}^T+\gamma\hat{F}^T\hat{H})u(x,t)dx\\
&+\int_\Omega u^T(x,t)(-\hat{H}-\hat{H}^T+\gamma\hat{H}^T\hat{H})u(x,t)dx\\
=&\int_\Omega \xi^T(x,t)\begin{pmatrix}\chi & \chi_1\\ \chi_1^T & \chi_2\end{pmatrix}\xi(x,t)dx,
\end{aligned}$$

where $\xi(x,t)=(e^T(x,t),u^T(x,t))^T$, $\chi_1=\hat{P}-\hat{F}^T+\gamma\hat{F}^T\hat{H}$, $\chi_2=-\hat{H}-\hat{H}^T+\gamma\hat{H}^T\hat{H}$. From (7.9), we can get

$$\dot{V}(t)+\gamma\int_\Omega y^T(x,t)y(x,t)dx\leqslant 2\int_\Omega y^T(x,t)u(x,t)dx \tag{7.14}$$

By integrating Eq. (7.14) with respect to t over the time period t_0 to t_p, we can obtain

$$2\int_{t_0}^{t_p}\int_{\Omega} y^T(x,t)u(x,t)dxdt \geqslant V(t_p) - V(t_0) + \gamma\int_{t_0}^{t_p}\int_{\Omega} y^T(x,t)y(x,t)dxdt$$

Namely,

$$\int_{t_0}^{t_p}\int_{\Omega} y^T(x,t)u(x,t)dxdt \geqslant \frac{V(t_p) - V(t_0)}{2} + \frac{\gamma}{2}\int_{t_0}^{t_p}\int_{\Omega} y^T(x,t)y(x,t)dxdt$$

for any $t_p, t_0 \in \mathbb{R}^+$ and $t_p \geqslant t_0$. The proof is completed.

Remark 7.5 To our knowledge, very few researchers have discussed the passivity of RDNNs [29], in which the input and output variables are varied with the time and space variables. Especially, the passivity of CRDNNs has not yet been investigated. In this chapter, we consider a CRDNNs consisting of N linearly and diffusively coupled identical RDNNs, which is totally different from the network model considered in [29]. In Lemma 7.4, a sufficient condition ensuring output strict passivity is established, which can be applied to analyze linearly coupled RDNNs (7.1) with output vector defined by (7.6).

Lemma 7.6 *Suppose that $V : \mathbb{R}^+ \times \mathrm{C} \to \mathbb{R}^+$ is continuously differentiable and satisfies the following condition*

$$v_1(\|e(\cdot,t)\|_2) \leqslant V(t, e_t(x,\theta)) \leqslant v_2(\|e(\cdot,t)\|_\tau) \tag{7.15}$$

where $e_t(x,\theta) \triangleq e(x, t+\theta)$, $-\tau \leqslant \theta \leqslant 0$, $v_1, v_2 : \mathbb{R}^+ \to \mathbb{R}^+$ are continuous and strictly monotonically nondecreasing functions, $v_1(s)$ and $v_2(s)$ are positive for $s > 0$ with $v_1(0) = v_2(0) = 0$. The CRDNNs (7.1) is locally synchronized in the sense of Definition 7.3 if system (7.7) is output-strictly passive with respect to storage function V and matrix $\hat{F}$ is nonsingular.

Proof System (7.7) is output-strictly passive with respect to storage function V, then there obviously exists a positive constant γ such that

$$\begin{aligned} V(t+\varepsilon) - V(t) \leqslant &\int_{t}^{t+\varepsilon}\int_{\Omega} y^T(x,s)u(x,s)dxds \\ &-\gamma\int_{t}^{t+\varepsilon}\int_{\Omega} y^T(x,s)y(x,s)dxds \end{aligned} \tag{7.16}$$

for any $t \in \mathbb{R}^+$ and $\varepsilon > 0$. Obviously, we can derive from (7.16) that

$$\frac{V(t+\varepsilon)-V(t)}{\varepsilon} \leqslant \frac{\int_t^{t+\varepsilon}\int_\Omega y^T(x,s)u(x,s)dxds}{\varepsilon} - \gamma\frac{\int_t^{t+\varepsilon}\int_\Omega y^T(x,s)y(x,s)dxds}{\varepsilon} \tag{7.17}$$

By taking limit $\epsilon \to 0$ in (7.17), we have

$$\dot{V}(t) \leqslant \int_\Omega y^T(x,t)u(x,t)dx - \gamma\int_\Omega y^T(x,t)y(x,t)dx \tag{7.18}$$

Letting $u(x,t)=0$, we can get from (7.18) that

$$\dot{V}(t) \leqslant -\gamma\lambda_m(\hat{F}^T\hat{F})\|e(\cdot,t)\|_2^2.$$

From Lemma 1.23, we can conclude that system (7.7) is internally locally asymptotically stable. Therefore, the CRDNNs (7.1) is locally synchronized in the sense of Definition 7.3. The proof is completed.

From Lemmas 7.4 and 7.6, we can obtain the following conclusion.

Theorem 7.7 *Let $\dot{\tau}(t) \leqslant \sigma < 1$. If there exist matrices $F, H \in \mathbb{R}^{n\times n}$, $0 < \hat{P} = \mathrm{diag}(P^1, P^2, \ldots, P^N) \in \mathbb{R}^{nN\times nN}$, $0 < Q \in \mathbb{R}^{nN\times nN}$ and a scalar $\gamma > 0$ such that*

$$\hat{P}\hat{D} + \hat{D}\hat{P} \geqslant 0 \tag{7.19}$$

$$\begin{pmatrix} \chi & \hat{P} - \hat{F}^T + \gamma\hat{F}^T\hat{H} \\ \hat{P} - \hat{F} + \gamma\hat{H}^T\hat{F} & -\hat{H} - \hat{H}^T + \gamma\hat{H}^T\hat{H} \end{pmatrix} \leqslant 0 \tag{7.20}$$

where F is a nonsingular matrix, $\hat{\Theta} = \mathrm{diag}(\Theta, \Theta, \ldots, \Theta)$, $\hat{F} = \mathrm{diag}(F, F, \ldots, F)$, $\hat{H} = \mathrm{diag}(H, H, \ldots, H)$, $\chi = -\sum_{k=1}^{q}\frac{1}{l_k^2}(\hat{P}\hat{D} + \hat{D}\hat{P}) - \hat{P}\hat{A} - \hat{A}\hat{P} + \hat{\Theta} + \hat{P}\hat{B}\hat{B}^T\hat{P} + c_1\hat{P}(G^1\otimes\Gamma_1) + c_1(G^1\otimes\Gamma_1)\hat{P} + \frac{Q}{1-\sigma} + c_2^2\hat{P}(G^2\otimes\Gamma_2)Q^{-1}(G^2\otimes\Gamma_2)\hat{P} + \gamma\hat{F}^T\hat{F}$, then CRDNNs (7.1) is locally synchronized in the sense of Definition 7.3.

Remark 7.8 In [181, 182], the authors studied the output synchronization of multi-agent systems, in which the agents' dynamics are input-output passive and described by affine nonlinear ODE systems. In this chapter, we proposes a complex delayed dynamical network consisting of N identical RDNNs. Firstly, we analyze the output strict passivity of the proposed network model, which can not be dealt with by those techniques used in [181, 182]. Then, we reveal the relationship between output strict passivity and synchronization of the complex delayed dynamical network (7.1). Moreover, by using the obtained passivity result and the relationship between output strict passivity and synchronization, a sufficient condition ensuring synchronization is obtained.

Remark 7.9 To our knowledge, there are very few works on the synchronization of CRDNNs [97, 98], and the asymptotic stability of synchronization error system has not yet been investigated. In Theorem 7.7, a sufficient condition is established to ensure that the synchronization error system is locally asymptotically stable.

7.4 Numerical Example

In this section, an illustrative example is provided to verify the effectiveness of the proposed theoretical results.

Consider a CRDNNs consisting of 5 identical nodes with diffusive coupling, in which each node is a 3-dimensional reaction-diffusion neural network described by

$$\frac{\partial w_i(x,t)}{\partial t} = d_i \Delta w_i(x,t) - a_i w_i(x,t) + J_i + \sum_{j=1}^{3} b_{ij} f_j(w_j(x,t)) \tag{7.21}$$

where $i = 1, 2, 3$, $\Omega = \{x \mid -0.5 < x < 0.5\}$, $f_j(\xi) = \frac{|\xi+1|-|\xi-1|}{4}$, $d_1 = 0.3$, $d_2 = 0.2$, $d_3 = 0.4$, $a_1 = 0.4$, $a_2 = 0.3$, $a_3 = 0.3$, $J_1 = J_2 = J_3 = 0$, and the matrix $B = (b_{ij})_{3\times 3}$ is chosen as

$$B = \begin{pmatrix} 0.2 & 0.2 & 0.3 \\ 0.1 & 0.3 & 0.3 \\ 0.3 & 0.1 & 0.2 \end{pmatrix}.$$

Obviously, $w^*(x) = (0, 0, 0)^T \in \mathbb{R}^3$ is an equilibrium solution of the reaction-diffusion neural network (7.21), and $f_j(\cdot)(j = 1, 2, 3)$ satisfies the Lipschitz condition with $\rho_j = \frac{1}{2}$. In the following, we analyze the asymptotic stability of synchronized state $[(w^*(x))^T, (w^*(x))^T, \ldots, (w^*(x))^T]^T$.

We take $c_1 = 0.2$, $c_2 = 0.1$, $\Gamma_1 = \text{diag}(0.4, 0.2, 0.3)$, $\Gamma_2 = \text{diag}(0.3, 0.1, 0.2)$ and $\tau(t) = 0.1 - 0.1e^{-t}$. The matrices G^1 and G^2 are chosen as, respectively,

$$G^1 = \begin{pmatrix} -0.5 & 0.2 & 0 & 0.3 & 0 \\ 0.2 & -0.8 & 0.2 & 0.2 & 0.2 \\ 0 & 0.2 & -0.5 & 0 & 0.3 \\ 0.3 & 0.2 & 0 & -0.5 & 0 \\ 0 & 0.2 & 0.3 & 0 & -0.5 \end{pmatrix},$$

$$G^2 = \begin{pmatrix} -0.3 & 0.1 & 0 & 0.2 & 0 \\ 0.1 & -0.6 & 0.2 & 0.1 & 0.2 \\ 0 & 0.2 & -0.4 & 0 & 0.2 \\ 0.2 & 0.1 & 0 & -0.3 & 0 \\ 0 & 0.2 & 0.2 & 0 & -0.4 \end{pmatrix}.$$

By using the YALMIP Toolbox of MATLAB, we can find the following matrices

$$F = \begin{pmatrix} 0.8605 & -0.0401 & -0.0185 \\ -0.0427 & 1.0158 & -0.0226 \\ -0.0178 & -0.0204 & 0.7640 \end{pmatrix},$$

$$H = \begin{pmatrix} 0.6442 & 0.0048 & 0.0023 \\ 0.0048 & 0.6406 & 0.0025 \\ 0.0023 & 0.0025 & 0.6471 \end{pmatrix},$$

$$P = I_5 \otimes \begin{pmatrix} 1.1401 & -0.0413 & -0.0153 \\ -0.0413 & 1.3780 & -0.0203 \\ -0.0153 & -0.0203 & 1.0000 \end{pmatrix} > 0,$$

$$Q = I_5 \otimes \begin{pmatrix} 1.2851 & -0.1148 & -0.0773 \\ -0.1148 & 0.8580 & -0.0708 \\ -0.0773 & -0.0708 & 1.4187 \end{pmatrix} > 0$$

satisfying (7.19) and (7.20) with $\gamma = 0.4127$. According to Theorem 7.7, the CRDNNs (7.1) with above given parameters is locally synchronized in the sense of Definition 7.3. The simulation results are shown in Fig. 7.2.

From Fig. 7.2, we clearly see that $\|z_i(\cdot, t)\|_2, i = 1, 2, \ldots, 5$ are very close to 0 when the time t increases gradually to 2.5s, and this state is maintained along with

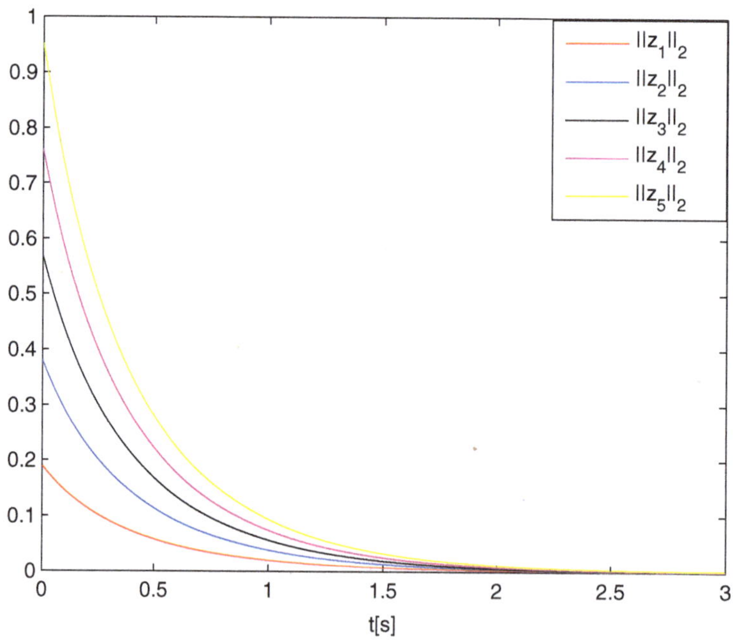

Fig. 7.2 The change processes of $\|z_i(\cdot, t)\|_2, i = 1, 2, \ldots, 5$

the increasing of the time. These results show that $z_i(x,t)(i=1,2,\ldots,5)$ asymptotically converges to $w^*(x)=(0,0,0)^T$, thus the CRDNNs (7.1) is synchronized. Figure 7.2 demonstrates the correctness and effectiveness of the obtained result in Theorem 7.7.

Remark 7.10 It is obvious that the delay has a strong influence on the rate of change of the node state (see (7.1)). The change processes of $\|z_i(\cdot,t)\|_2$, $i=1,2,\ldots,5$, may be different if we take other time-varying delays in Fig. 7.2, although we omit this comparison for brevity. On the other hand, since $\sum_{j=1}^{N} G_{ij}^1=0$ and $\sum_{j=1}^{N} G_{ij}^2=0(i=1,2,\ldots,N)$, $\sum_{j=1}^{N} G_{ij}^1\Gamma_1 z_j(x,t)$ and $\sum_{j=1}^{N} G_{ij}^2\Gamma_2 z_j(x,t-\tau(t))$ will tend to zero asymptotically if the CRDNNs (7.1) is synchronized.

Remark 7.11 YALMIP Toolbox is a very convenient tool to solve linear matrix inequality (LMI), which has been widely used in analyzing the stability of various systems. Using the Schur complement, we can obtain that (7.20) is equivalent to

$$\begin{pmatrix} \mathbb{W}_1+\frac{Q}{1-\sigma} & \hat{P}-\hat{F}^T & \hat{F}^T & \hat{P}\hat{B} & \mathbb{W}_2 \\ \hat{P}-\hat{F} & -\hat{H}-\hat{H}^T & \hat{H}^T & 0 & 0 \\ \hat{F} & \hat{H} & -\frac{1}{\gamma}I_{nN} & 0 & 0 \\ \hat{B}^T\hat{P} & 0 & 0 & -I_{nN} & 0 \\ \mathbb{W}_2^T & 0 & 0 & 0 & -Q \end{pmatrix} \leqslant 0,$$

where $\mathbb{W}_1=-\sum_{k=1}^{q}\frac{1}{l_k^2}(\hat{P}\hat{D}+\hat{D}\hat{P})-\hat{P}\hat{A}-\hat{A}\hat{P}+\hat{\Theta}+c_1\hat{P}(G^1\otimes\Gamma_1)+c_1(G^1\otimes\Gamma_1)\hat{P}$, $\mathbb{W}_2=c_2\hat{P}(G^2\otimes\Gamma_2)$. Then, by utilizing the YALMIP Toolbox of MATLAB, we can find the F, H, $\hat{P}$, Q, γ satisfying (7.19) and the above LMI.

7.5 Conclusion

In this chapter, a general array model of CRDNNs with delay coupling has been introduced. The output strict passivity of the proposed network model has been taken into consideration, and a sufficient condition has been established. Furthermore, the relationship between output strict passivity and synchronization of the proposed network model has been revealed. In addition, a sufficient condition ensuring synchronization has been derived by utilizing the obtained passivity result and the relationship between output strict passivity and synchronization. A numerical example has been provided to verify the correctness and effectiveness of the theoretical results. In future work, we shall study the passivity of CRDNNs with impulsive effects.

Chapter 8
Passivity and Synchronization of CRDNNs with Adaptive Coupling

8.1 Introduction

Passivity is part of a broader and a general theory of dissipativeness [105, 183]. The main point of passivity theory is that the passive properties of systems can keep the systems internally stable. In recent years, as a powerful tool, passivity has played an important role in synchronization, group coordination, flow control, energy management, and so on [179, 180, 184–190]. However, in most existing works on the passivity, it is assumed that the input and output variables are only dependent on the time. But, in reality, the input and output variables are not only dependent on the time, but also intensively dependent on space variable in many circumstances. Thus, it is important and interesting to study the passivity of systems, in which the input and output variables are varied with the time and space variables. In [29], the authors gave the passivity definitions for the case where input and output variables are varied with the time and space variables, and investigated the passivity and stability problems of a reaction-diffusion neural network with Dirichlet boundary conditions. In [167], the robust global exponential stability and robust passivity were studied for a class of parabolic complex networks with spatially and temporally varying input and output variables. To the best of our knowledge, very few researchers have investigated the passivity of CRDNNs.

Motivated by the above discussions, the objective of this chapter is to study the passivity of CRDNNs. However, CRDNNs are not passive in many circumstances, thus some control strategies should be adopted. It is well known that the topological structure and the coupling strength are two key factors impacting the synchronization in complex networks. In recent years, some adaptive strategies to tune the coupling strengths among network nodes have been designed for reaching synchronization [99, 100]. To our knowledge, in many real-world networks, the coupling strength between nodes is adaptively adjusted according to changes in the environment or the network itself (for instance, neural networks, wireless sensor networks, biological networks). Unfortunately, the passivity of CRDNNs with adaptive coupling has not yet been investigated. Therefore, it is important and interesting to study the passivity

J.-L. Wang et al., *Analysis and Control of Coupled Neural Networks with Reaction-Diffusion Terms*, DOI 10.1007/978-981-10-4907-1_8

of CRDNNs with adaptive coupling. On the other hand, the passivity theory has long been a nice tool for analyzing the synchronization of the complex networks. But in most existing works, the coupling strength is fixed and the node state is only dependent on the time. Therefore, it is essential to consider the relationship between passivity and synchronization of the CRDNNs with adaptive coupling.

The main contributions of this chapter are as follows. First, several sufficient conditions ensuring the dissipativity and passivity are derived for the CRDNNs with adaptive coupling by utilizing some inequality techniques. Second, we reveal the relationship between passivity and synchronization of the CRDNNs. Third, a criterion for synchronization is established by employing the obtained passivity results and the relationship between passivity and synchronization.

8.2 Network Model

N mutually coupled RDNNs (2.4) can result in a CRDNNs, which is described by

$$\begin{aligned}\frac{\partial z_i(x,t)}{\partial t} &= D\Delta z_i(x,t) - Az_i(x,t) + Bf(z_i(x,t)) + c_1\sum_{j=1}^{N} G^1_{ij}\Gamma_1 z_j(x,t) + J \\ &\quad + c_2\sum_{j=1}^{N} G^2_{ij}\Gamma_2 z_j(x,t-\tau(t)) + u_i(x,t), \qquad (8.1)\end{aligned}$$

where $i = 1, 2, \ldots, N$, N is the number of nodes in the network; $\tau(t)$ is the time-varying delay with $0 \leqslant \tau(t) \leqslant \tau$; $z_i(x,t) = (z_{i1}(x,t), z_{i2}(x,t), \ldots, z_{in}(x,t))^T \in \mathbb{R}^n$ is the state vector of node i; $u_i(x,t) \in \mathbb{R}^n$ is the control input; c_1 and c_2 are positive real numbers, which represent the coupling strength for non-delayed configuration and delayed one, respectively; $\Gamma_1 = \mathrm{diag}(\gamma^1_1, \gamma^1_2, \ldots, \gamma^1_n) \in \mathbb{R}^{n\times n}$ and $\Gamma_2 = \mathrm{diag}(\gamma^2_1, \gamma^2_2, \ldots, \gamma^2_n) \in \mathbb{R}^{n\times n}$ are positive definite matrices, which describe the individual coupling between two nodes for non-delayed configuration and delayed one, respectively; $G^1 = (G^1_{ij})_{N\times N}$ and $G^2 = (G^2_{ij})_{N\times N}$ represent the topological structure of network and coupling strength between nodes for non-delayed configuration and delayed one, respectively, where G^1_{ij} (G^2_{ij} can be defined similarly) is defined as follows: if there exists a connection between node i and node j, then $G^1_{ij} = G^1_{ji} > 0$; otherwise, $G^1_{ij} = G^1_{ji} = 0 (i \neq j)$, and the diagonal elements of matrix G^1 are defined by

$$G^1_{ii} = -\sum_{\substack{j=1\\ j\neq i}}^{N} G^1_{ij}, \quad i = 1, 2, \ldots, N.$$

In this chapter, we always assume network (8.1) is connected, and the topological structure of network (8.1) is fixed. The initial value and boundary value conditions associated with network (8.1) are given in the form

$$z_i(x,t) = \Phi_i(x,t) \in \mathbb{R}^n, \quad (x,t) \in \Omega \times [-\tau, 0], \tag{8.2}$$

$$z_i(x,t) = 0, \quad (x,t) \in \partial\Omega \times [-\tau, +\infty), \tag{8.3}$$

where $\Phi_i(x,t)(i = 1, 2, \ldots, N)$ is a bounded and continuous function on $\Omega \times [-\tau, 0]$.

Remark 8.1 In [97, 98, 101], the authors investigated the synchronization and synchronization control problems of CRDNNs. Similarly, we propose a general array model of CRDNNs with delay coupling in this chapter.

Let $\mathcal{V} = \{1, 2, \ldots, N\}$ and $\mathcal{E} \subset \mathcal{V} \times \mathcal{V}$ respectively denote the set of nodes and the set of undirected links in the network (8.1). The neighbor set of node i is denoted by $\mathcal{N}_i = \{j \in \mathcal{V} : (i,j) \in \mathcal{E}\}$. Here, consider the case where only the coupling strengths G^1_{ij} are adjusted. In this case, the CRDNNs (8.1) can be described by the following equations:

$$\begin{aligned}\frac{\partial z_i(x,t)}{\partial t} &= D\Delta z_i(x,t) - Az_i(x,t) + Bf(z_i(x,t)) + c_1 \sum_{j=1}^{N} G^1_{ij}(t)\Gamma_1 z_j(x,t) \\ &\quad + J + c_2 \sum_{j=1}^{N} G^2_{ij}\Gamma_2 z_j(x, t - \tau(t)) + u_i(x,t),\end{aligned} \tag{8.4}$$

where $i = 1, 2, \ldots, N$, $G^1(t) = (G^1_{ij}(t))_{N\times N}$ is a time-varying matrix, where $G^1_{ij}(t)$ is defined as follows: if there is an edge between node i and node $j (i \neq j)$ at time t, then the element $G^1_{ij}(t) = G^1_{ji}(t) > 0$; otherwise, $G^1_{ij}(t) = G^1_{ji}(t) = 0 (i \neq j)$, and the diagonal elements of matrix $G^1(t)$ are defined by

$$G^1_{ii}(t) = -\sum_{\substack{j=1 \\ j\neq i}}^{N} G^1_{ij}(t), \quad i = 1, 2, \ldots, N.$$

Let $\bar{z}(x,t) = \frac{1}{N}\sum_{j=1}^{N} z_j(x,t)$. Then, we have

$$\begin{aligned}\frac{\partial \bar{z}(x,t)}{\partial t} &= D\Delta\bar{z}(x,t) - A\bar{z}(x,t) + J + \frac{1}{N}\sum_{j=1}^{N} Bf(z_j(x,t)) \\ &\quad + \frac{1}{N}\sum_{j=1}^{N} u_j(x,t).\end{aligned} \tag{8.5}$$

Define $e_i(x,t) = (e_{i1}(x,t), e_{i2}(x,t), \ldots, e_{in}(x,t))^T = z_i(x,t) - \bar{z}(x,t), i = 1, 2, \ldots, N$. The dynamics of the error vector $e_i(x,t)$ is governed by the following equation:

$$
\begin{aligned}
\frac{\partial e_i(x,t)}{\partial t} &= D\Delta e_i(x,t) - Ae_i(x,t) - \frac{1}{N}\sum_{j=1}^{N} Bf(z_j(x,t)) + Bf(z_i(x,t)) \\
&+ c_1\sum_{j=1}^{N} G^1_{ij}(t)\Gamma_1 e_j(x,t) + c_2\sum_{j=1}^{N} G^2_{ij}\Gamma_2 e_j(x,t-\tau(t)) \\
&+ u_i(x,t) - \frac{1}{N}\sum_{j=1}^{N} u_j(x,t)
\end{aligned}
\tag{8.6}
$$

where $i = 1, 2, \ldots, N$.

8.3 Passivity and Synchronization of CRDNNs

In this section, some adaptive strategies to tune the coupling strengths among network nodes are designed in order to ensure the passivity of the network. Moreover, we reveal the relationship between passivity and synchronization of the CRDNNs with adaptive coupling, and establish a global synchronization criterion by using the obtained passivity results and the relationship between passivity and synchronization.

8.3.1 Passivity Analysis of CRDNNs

In what follows, we establish the passivity of system (8.6). The output vector $y_i(x,t)$ of system (8.6) is defined as

$$
y_i(x,t) = Fe_i(x,t) + Hu_i(x,t) \tag{8.7}
$$

for the passivity scheme, where $F, H \in \mathbb{R}^{n\times n}$ are known real matrices, i=1, $2, \ldots, N$. Let $u(x,t) = (u_1^T(x,t), u_2^T(x,t), \ldots, u_N^T(x,t))^T$, $y(x,t) = (y_1^T(x,t), y_2^T(x,t), \ldots, y_N^T(x,t))^T$.

Theorem 8.2 *Let $\dot{\tau}(t) \leqslant \sigma < 1$. If there exists a scalar $\gamma \geqslant 0$ such that*

$$
H + H^T + \gamma I_n > 0, \tag{8.8}
$$

then system (8.6) satisfies

$$
\int_{t_0}^{t_p}\int_{\Omega} y^T(x,t)u(x,t)dxdt \geqslant V(t_p) - V(t_0) - \frac{\gamma}{2}\int_{t_0}^{t_p}\int_{\Omega} u^T(x,t)u(x,t)dxdt
$$

under the following adaptive law:

$$\dot{G}^1_{ij}(t) = \beta_{ij} \int_\Omega (z_i(x,t) - z_j(x,t))^T \Gamma_1 (z_i(x,t) - z_j(x,t))dx, \ (i,j) \in \mathcal{E} \quad (8.9)$$

where $t_p, t_0 \in \mathbb{R}^+$, $t_p \geqslant t_0$, $\beta_{ij} = \beta_{ji}$ *are positive constants, and* $V : \mathbb{R}^+ \to \mathbb{R}^+$ *is a continuously differentiable function.*

Proof For the system (8.6), let us define the following Lyapunov functional:

$$V_1(t) = \sum_{i=1}^N \int_\Omega e_i^T(x,t) e_i(x,t)dx + + \sum_{i=1}^N \sum_{j \in \mathcal{N}_i} \frac{c_1}{2\beta_{ij}} (G^1_{ij}(t) - \sigma_{ij})^2$$
$$+ \sum_{i=1}^N \sum_{j=1}^N \frac{c_2 |G^2_{ij}|}{1-\sigma} \int_{t-\tau(t)}^t \int_\Omega e_j^T(x,s) \Gamma_2 e_j(x,s) dx ds, \quad (8.10)$$

where $\sigma_{ij} = \sigma_{ji}(i \neq j)$ are nonnegative constants, and $\sigma_{ij} = 0(i \neq j)$ if and only if $G^1_{ij}(t) = 0$. Since $\sum_{i=1}^N e_i^T(x,t) = 0$, we can obtain

$$\dot{V}_1(t) \leqslant \sum_{i=1}^N 2 \int_\Omega e_i^T(x,t) \frac{\partial e_i(x,t)}{\partial t} dx + \sum_{i=1}^N \sum_{j \in \mathcal{N}_i} \frac{c_1}{\beta_{ij}} (G^1_{ij}(t) - \sigma_{ij}) \dot{G}^1_{ij}(t)$$
$$- \sum_{i=1}^N \sum_{j=1}^N c_2 |G^2_{ij}| \int_\Omega e_j^T(x, t-\tau(t)) \Gamma_2 e_j(x, t-\tau(t)) dx$$
$$+ \sum_{i=1}^N \sum_{j=1}^N \frac{c_2 |G^2_{ij}|}{1-\sigma} \int_\Omega e_j^T(x,t) \Gamma_2 e_j(x,t) dx$$
$$= \sum_{i=1}^N 2 \int_\Omega e_i^T(x,t) \Big[D\Delta e_i(x,t) + Bf(\bar{z}(x,t)) - \frac{1}{N} \sum_{j=1}^N Bf(z_j(x,t))$$
$$- Bf(\bar{z}(x,t)) + u_i(x,t) + Bf(z_i(x,t)) + c_1 \sum_{j=1}^N G^1_{ij}(t) \Gamma_1 e_j(x,t)$$
$$- Ae_i(x,t) + c_2 \sum_{j=1}^N G^2_{ij} \Gamma_2 e_j(x, t-\tau(t)) - \frac{1}{N} \sum_{j=1}^N u_j(x,t) \Big] dx$$
$$+ \sum_{i=1}^N \sum_{j \in \mathcal{N}_i} c_1 (G^1_{ij}(t) - \sigma_{ij}) \int_\Omega (z_i(x,t) - z_j(x,t))^T \Gamma_1 (z_i(x,t)$$
$$- z_j(x,t))dx + \sum_{i=1}^N \sum_{j=1}^N \frac{c_2 |G^2_{ij}|}{1-\sigma} \int_\Omega e_j^T(x,t) \Gamma_2 e_j(x,t) dx$$

$$
\begin{aligned}
&-c_2\sum_{i=1}^{N}\sum_{j=1}^{N}|G_{ij}^2|\int_{\Omega}e_j^T(x,t-\tau(t))\Gamma_2 e_j(x,t-\tau(t))dx\\
&=\sum_{i=1}^{N}2\int_{\Omega}e_i^T(x,t)\Big[D\Delta e_i(x,t)-Bf(\bar{z}(x,t))+c_1\sum_{j=1}^{N}G_{ij}^1(t)\Gamma_1 e_j(x,t)\\
&\quad+Bf(z_i(x,t))+u_i(x,t)+c_2\sum_{j=1}^{N}G_{ij}^2\Gamma_2 e_j(x,t-\tau(t))-Ae_i(x,t)\Big]dx\\
&\quad+\sum_{i=1}^{N}\sum_{j\in\mathcal{N}_i}c_1(G_{ij}^1(t)-\sigma_{ij})\int_{\Omega}(e_i(x,t)-e_j(x,t))^T\Gamma_1(e_i(x,t)-e_j(x,t))dx\\
&\quad-c_2\sum_{i=1}^{N}\sum_{j=1}^{N}|G_{ij}^2|\int_{\Omega}e_j^T(x,t-\tau(t))\Gamma_2 e_j(x,t-\tau(t))dx\\
&\quad+\sum_{i=1}^{N}\sum_{j=1}^{N}\frac{c_2|G_{ij}^2|}{1-\sigma}\int_{\Omega}e_j^T(x,t)\Gamma_2 e_j(x,t)dx.
\end{aligned}
$$

From Green's formula and Lemma 1.22, we have

$$
\begin{aligned}
&\int_{\Omega}e_i^T(x,t)D\Delta e_i(x,t)dx\\
&=-\sum_{k=1}^{q}\int_{\Omega}\left(\frac{\partial e_i(x,t)}{\partial x_k}\right)^T D\frac{\partial e_i(x,t)}{\partial x_k}dx\\
&\leqslant-\lambda_m(D)\sum_{k=1}^{q}\int_{\Omega}\left(\frac{\partial e_i(x,t)}{\partial x_k}\right)^T\frac{\partial e_i(x,t)}{\partial x_k}dx\\
&\leqslant\sum_{k=1}^{q}\frac{-\lambda_m(D)}{l_k^2}\int_{\Omega}e_i^T(x,t)e_i(x,t)dx.
\end{aligned}\tag{8.11}
$$

Define the matrix $\Sigma=(\sigma_{ij})_{N\times N}$, where $\sigma_{ii}=-\sum_{\substack{j=1\\ j\neq i}}^{N}\sigma_{ij}$, $i=1,2,\ldots,N$. Then, we can easily derive

$$
\begin{aligned}
&\sum_{i=1}^{N}\sum_{j\in\mathcal{N}_i}(G_{ij}^1(t)-\sigma_{ij})(e_i(x,t)-e_j(x,t))^T\Gamma_1(e_i(x,t)-e_j(x,t))\\
&=-2\sum_{i=1}^{N}\sum_{j=1}^{N}(G_{ij}^1(t)-\sigma_{ij})e_i^T(x,t)\Gamma_1 e_j(x,t).
\end{aligned}\tag{8.12}
$$

Let $\Theta = \mathrm{diag}(\rho_1^2, \rho_2^2, \ldots, \rho_n^2)$. Obviously

$$\begin{aligned} &2e_i^T(x,t)B(f(z_i(x,t)) - f(\bar{z}(x,t))) \\ &\leqslant e_i^T(x,t)BB^T e_i(x,t) + e_i^T(x,t)\Theta e_i(x,t). \end{aligned} \tag{8.13}$$

Furthermore,

$$\begin{aligned} &2c_2\sum_{i=1}^{N}\sum_{j=1}^{N} G_{ij}^2 e_i^T(x,t)\Gamma_2 e_j(x,t-\tau(t)) \\ &\leqslant c_2\sum_{i=1}^{N}\sum_{j=1}^{N} |G_{ij}^2| e_j^T(x,t-\tau(t))\Gamma_2 e_j(x,t-\tau(t)) \\ &\quad + c_2\sum_{i=1}^{N}\sum_{j=1}^{N} |G_{ij}^2| e_i^T(x,t)\Gamma_2 e_i(x,t). \end{aligned} \tag{8.14}$$

It follows from (8.11)–(8.14) that

$$\begin{aligned} \dot{V}_1(t) \leqslant &\sum_{i=1}^{N}\int_{\Omega} e_i^T(x,t)\Big[2\sum_{k=1}^{q}\frac{-\lambda_m(D)}{l_k^2}I_n - 2A + BB^T + \frac{2c_2|G_{ii}^2|(2-\sigma)}{1-\sigma}\Gamma_2 \\ &+ \Theta\Big]e_i(x,t)dx + 2\sum_{i=1}^{N}\int_{\Omega} e_i^T(x,t)u_i(x,t)dx \\ &+ 2\sum_{i=1}^{N}\sum_{j=1}^{N} c_1\sigma_{ij}\int_{\Omega} e_i^T(x,t)\Gamma_1 e_j(x,t)dx. \end{aligned}$$

Therefore,

$$\begin{aligned} &\dot{V}_1(t) - 2\int_{\Omega} y^T(x,t)u(x,t)dx - \gamma\int_{\Omega} u^T(x,t)u(x,t)dx \\ &\leqslant \sum_{i=1}^{N}\int_{\Omega} e_i^T(x,t)\Big[2\sum_{k=1}^{q}\frac{-\lambda_m(D)}{l_k^2}I_n - 2A + BB^T + \frac{2c_2|G_{ii}^2|(2-\sigma)}{1-\sigma}\Gamma_2 \\ &+ \Theta\Big]e_i(x,t)dx + 2\sum_{i=1}^{N}\int_{\Omega} e_i^T(x,t)(I_n - F^T)u_i(x,t)dx \\ &- \sum_{i=1}^{N}\int_{\Omega} u_i^T(x,t)(H + H^T + \gamma I_n)u_i(x,t)dx \\ &+ 2\sum_{i=1}^{N}\sum_{j=1}^{N} c_1\sigma_{ij}\int_{\Omega} e_i^T(x,t)\Gamma_1 e_j(x,t)dx \end{aligned}$$

$$
\begin{aligned}
&\leqslant \sum_{i=1}^{N}\int_{\Omega} e_i^T(x,t)\Big[2\sum_{k=1}^{q}\frac{-\lambda_m(D)}{l_k^2}I_n - 2A + BB^T + \Theta + \frac{2c_2|G_{ii}^2|(2-\sigma)}{1-\sigma}\Gamma_2 \\
&+ (I_n - F^T)(H + H^T + \gamma I_n)^{-1}(I_n - F)\Big]e_i(x,t)dx \\
&+ 2\sum_{i=1}^{N}\sum_{j=1}^{N} c_1\sigma_{ij}\int_{\Omega} e_i^T(x,t)\Gamma_1 e_j(x,t)dx \\
&= \int_{\Omega} e^T(x,t)[I_N \otimes \Upsilon + 2c_1(\Sigma \otimes \Gamma_1)]e(x,t)dx
\end{aligned}
\tag{8.15}
$$

where $e(x,t) = (e_1^T(x,t), e_2^T(x,t), \ldots, e_N^T(x,t))^T$, $\Upsilon = 2\sum_{k=1}^{q}\frac{-\lambda_m(D)}{l_k^2}I_n - 2A + BB^T + \Theta + \frac{2c_2|G_{ii}^2|(2-\sigma)}{1-\sigma}\Gamma_2 + (I_n - F^T)(H + H^T + \gamma I_n)^{-1}(I_n - F)$.

According to Lemma 1.17, there obviously exists a unitary matrix $\Psi = (\psi_1, \psi_2, \ldots, \psi_N) \in \mathbb{R}^{N\times N}$ such that

$$
\Psi^T \Sigma \Psi = \Lambda
$$

with $\Lambda = \mathrm{diag}(\lambda_1, \lambda_2, \ldots, \lambda_N)$. $\lambda_i, i = 1, 2, \ldots, N$, are the eigenvalues of Σ and $0 = \lambda_1 > \lambda_2 \geqslant \lambda_3 \geqslant \ldots \geqslant \lambda_N$. Let $\xi(x,t) = (\xi_1^T(x,t), \xi_2^T(x,t), \ldots, \xi_N^T(x,t))^T = (\Psi^T \otimes I_n)e(x,t)$. Since $\psi_1 = \frac{1}{\sqrt{N}}(1, 1, \ldots, 1)^T$, one has

$$
\xi_1(x,t) = (\psi_1^T \otimes I_n)e(x,t) = 0.
$$

Then, we can derive from (8.15) that

$$
\begin{aligned}
&\dot{V}_1(t) - 2\int_{\Omega} y^T(x,t)u(x,t)dx - \gamma\int_{\Omega} u^T(x,t)u(x,t)dx \\
&\leqslant \int_{\Omega} e^T(x,t)[2c_1(\Psi \otimes I_n)(\Lambda \otimes \Gamma_1)(\Psi^T \otimes I_n) + I_N \otimes \Upsilon]e(x,t)dx \\
&= \int_{\Omega} e^T(x,t)(I_N \otimes \Upsilon)e(x,t)dx + 2c_1\int_{\Omega} \xi^T(x,t)(\Lambda \otimes \Gamma_1)\xi(x,t)dx \\
&\leqslant \int_{\Omega} e^T(x,t)(I_N \otimes \Upsilon)e(x,t)dx + 2c_1\lambda_2\int_{\Omega} \xi^T(x,t)(I_N \otimes \Gamma_1)\xi(x,t)dx \\
&= \int_{\Omega} e^T(x,t)[I_N \otimes \Upsilon + 2c_1\lambda_2(I_N \otimes \Gamma_1)]e(x,t)dx.
\end{aligned}
$$

By selecting σ_{ij} sufficiently large such that

$$
2c_1\lambda_2\hat{\gamma} + \lambda_M(\Upsilon) \leqslant 0,
$$

one obtains

$$I_N \otimes \Upsilon + 2c_1\lambda_2(I_N \otimes \Gamma_1) \leqslant 0$$

where $\hat{\gamma} = \min_{i=1,2,\ldots,n}\{\gamma_i^1\}$. Therefore,

$$\frac{\dot{V}_1(t)}{2} - \frac{\gamma}{2}\int_\Omega u^T(x,t)u(x,t)dx \leqslant \int_\Omega y^T(x,t)u(x,t)dx \tag{8.16}$$

Let $V(t) = \frac{V_1(t)}{2}$. By integrating Equation (8.16) with respect to t over the time period t_0 to t_p, we can obtain

$$\int_{t_0}^{t_p}\int_\Omega y^T(x,t)u(x,t)dxdt \geqslant V(t_p) - V(t_0) - \frac{\gamma}{2}\int_{t_0}^{t_p}\int_\Omega u^T(x,t)u(x,t)dxdt.$$

The proof is completed.

Remark 8.3 On the one hand, system (8.6) is dissipative under the adaptive law (8.9) with respect to

$$\vartheta(u,y) = \int_\Omega y^T(x,t)u(x,t)dx + \frac{\gamma}{2}\int_\Omega u^T(x,t)u(x,t)dx$$

if condition (8.8) holds. On the other hand, from Theorem 8.2, system (8.6) is passive under the adaptive law (8.9) in the sense of Definition 1.4 if $\gamma = 0$.

In Theorem 8.2, all the coupling strengths G_{ij}^1 are updated. In the following, consider the case where only a small fraction of the coupling strengths is adjusted. Suppose that $\hat{\mathcal{E}}$ is a subset of $\mathcal{E}$, and CRDNNs (8.1) is connected through the undirected edges $\hat{\mathcal{E}}$. Use the following adaptive law:

$$\dot{G}_{ij}^1(t) = \hat{\beta}_{ij}\int_\Omega (z_i(x,t) - z_j(x,t))^T\Gamma_1(z_i(x,t) - z_j(x,t))dx,\ (i,j) \in \hat{\mathcal{E}} \tag{8.17}$$

where $\hat{\beta}_{ij} = \hat{\beta}_{ji}$ are positive constants.

Construct a Lyapunov functional for system (8.6) as follows:

$$V_2(t) = \sum_{i=1}^N \int_\Omega e_i^T(x,t)e_i(x,t)dx + \sum_{i=1}^N \sum_{(i,j)\in\hat{\mathcal{E}}} \frac{c_1}{2\hat{\beta}_{ij}}(G_{ij}^1(t) - \hat{\sigma}_{ij})^2$$
$$+ \sum_{i=1}^N\sum_{j=1}^N \frac{c_2|G_{ij}^2|}{1-\sigma}\int_{t-\tau(t)}^t \int_\Omega e_j^T(x,s)\Gamma_2 e_j(x,s)dxds \tag{8.18}$$

where $\hat{\sigma}_{ij}$ is defined as follows: if $(i,j) \in \hat{\mathcal{E}}$, then $\hat{\sigma}_{ij} = \hat{\sigma}_{ji} > 0$; otherwise, $\hat{\sigma}_{ij} = 0 (i \neq j)$.

By a minor modification of the proof of Theorem 8.2, we can easily get the following conclusion.

Theorem 8.4 *Let $\dot{\tau}(t) \leqslant \sigma < 1$. If there exists a scalar $\gamma \geqslant 0$ such that*

$$H + H^T + \gamma I_n > 0,$$

then system (8.6) satisfies

$$\int_{t_0}^{t_p} \int_{\Omega} y^T(x,t)u(x,t)dxdt \geqslant \frac{V_2(t_p)}{2} - \frac{V_2(t_0)}{2} - \frac{\gamma}{2} \int_{t_0}^{t_p} \int_{\Omega} u^T(x,t)u(x,t)dxdt$$

under the adaptive law (8.17), where $t_p, t_0 \in \mathbb{R}^+, t_p \geqslant t_0$.

By the similar proofs of Theorems 8.2 and 8.4, we can obtain the following conclusions. Here we omit their proofs to avoid the repetition.

Theorem 8.5 *Let $\dot{\tau}(t) \leqslant \sigma < 1$. If there exists a scalar $\gamma > 0$ such that*

$$H + H^T - \gamma I_n > 0,$$

then system (8.6) is input-strictly passive under the adaptive law (8.9) (or (8.17)).

Theorem 8.6 *Let $\dot{\tau}(t) \leqslant \sigma < 1$. If there exists a scalar $\gamma > 0$ such that*

$$H + H^T - \gamma H^T H > 0,$$

then system (8.6) is output-strictly passive under the adaptive law (8.9) (or (8.17)).

Remark 8.7 To our knowledge, very few researchers have discussed the dissipativity and passivity of RDNNs [29], in which the input and output variables are varied with the time and space variables. Especially, the dissipativity and passivity of CRDNNs with adaptive coupling has not yet been investigated. In this chapter, we consider a CRDNNs consisting of N linearly and diffusively coupled identical RDNNs, which is totally different from the network model considered in [29]. Therefore, we generalize the existing passivity results to some extent, and this can serve as a stepping stone to study the passivity problem of CRDNNs with adaptive coupling.

8.3.2 The Relationship Between Passivity and Synchronization of CRDNNs

Next, we give the definition of synchronization for the CRDNNs (8.4).

Definition 8.8 The CRDNNs (8.4) is said to achieve synchronization if

$$\lim_{t\to+\infty} \|z_i(\cdot,t) - \frac{1}{N}\sum_{j=1}^{N} z_j(\cdot,t)\|_2 = 0, \text{ for all } i = 1, 2, \ldots, N$$

under the condition that $u_i(x,t) = 0, i = 1, 2, \ldots, N$.

Theorem 8.9 *Suppose that F is a nonsingular matrix.*

(i) *Under the adaptive law (8.9), CRDNNs (8.4) is synchronized in the sense of Definition 8.8 if system (8.6) is output-strictly passive with respect to storage function $\hat{V}_1(t) = \frac{V_1(t)}{2}$ (see (8.10));*

(ii) *Under the adaptive law (8.17), CRDNNs (8.4) is synchronized in the sense of Definition 8.8 if system (8.6) is output-strictly passive with respect to storage function $\hat{V}_2(t) = \frac{V_2(t)}{2}$ (see (8.18)).*

Proof If system (8.6) is output-strictly passive with respect to storage function $\hat{V}_1(t)$, then there exists a positive constant γ such that

$$\begin{aligned}\frac{\hat{V}_1(t+\varepsilon) - \hat{V}_1(t)}{\varepsilon} \leqslant & \frac{\int_t^{t+\varepsilon}\int_\Omega y^T(x,s)u(x,s)dxds}{\varepsilon} \\ & -\gamma\frac{\int_t^{t+\varepsilon}\int_\Omega y^T(x,s)y(x,s)dxds}{\varepsilon}\end{aligned} \tag{8.19}$$

for any $t \in \mathbb{R}^+$ and $\varepsilon > 0$. By taking limit $\epsilon \to 0$ in (8.19), we have

$$\dot{\hat{V}}_1(t) \leqslant \int_\Omega y^T(x,t)u(x,t)dx - \gamma\int_\Omega y^T(x,t)y(x,t)dx. \tag{8.20}$$

Letting $u(x,t) = 0$, we can get from (8.20) that

$$\dot{\hat{V}}_1(t) \leqslant -\gamma\lambda_m(F^TF)\|e(\cdot,t)\|_2^2. \tag{8.21}$$

Obviously, $\hat{V}_1(t)$ is non-increasing, and each term of $\hat{V}_1(t)$ is bounded. Consequently, all the coupling strengths $G_{ij}^1(t)((i,j) \in \mathcal{E})$ are bounded, and $\lim_{t\to+\infty}\hat{V}_1(t)$ exists and is a non-negative real number. Since $G_{ij}^1(t)$ is monotonically increasing (see (8.9)), one can conclude that $G_{ij}^1(t)((i,j) \in \mathcal{E})$ asymptotically converges to a finite positive value. Moreover, by (8.21), we have

$$\|e(\cdot,t)\|_2^2 \leqslant \frac{-\dot{\hat{V}}_1(t)}{\gamma\lambda_m(F^TF)}. \tag{8.22}$$

It implies $\lim_{t\to+\infty}\int_0^t \|e(\cdot,s)\|_2^2 ds$ exists and is finite. Since $0 \leqslant \tau(t) \leqslant \tau$, we can get

$$\lim_{t\to+\infty}\int_{t-\tau(t)}^{t}\int_{\Omega} e_j^T(x,s)\Gamma_2 e_j(x,s)dxds = 0,$$

where $j = 1, 2, \ldots, N$. Therefore, by the definition of $\hat{V}_1(t)$, we can derive that $\lim_{t\to+\infty}\|e(\cdot,t)\|_2^2$ exists and is a non-negative real number. In what follows, we shall prove that

$$\lim_{t\to+\infty}\|e(\cdot,t)\|_2^2 = 0.$$

If this is not true, we have

$$\lim_{t\to+\infty}\|e(\cdot,t)\|_2^2 = \mu > 0.$$

Then, there obviously exists a real number $M > 0$ such that

$$\|e(\cdot,t)\|_2^2 > \frac{\mu}{2}$$

for $t \geqslant M$. From (8.21), we can get

$$\dot{\hat{V}}_1(t) < \frac{-\gamma\lambda_m(F^T F)\mu}{2}, \quad t \geqslant M. \tag{8.23}$$

By integrating (8.23) with respect to t over the time period M to $+\infty$, we can obtain

$$\begin{aligned}
-\hat{V}_1(M) &\leqslant \hat{V}_1(+\infty) - \hat{V}_1(M) \\
&= \int_M^{+\infty} \dot{\hat{V}}_1(t)dt \\
&< -\int_M^{+\infty} \frac{\gamma\lambda_m(F^T F)\mu}{2}dt \\
&= -\infty.
\end{aligned}$$

This yields a contradiction, and so

$$\lim_{t\to+\infty}\|e(\cdot,t)\|_2 = 0.$$

Therefore, CRDNNs (8.4) is synchronized under the adaptive law (8.9). Conclusion (ii) can be easily proved in a way similar to conclusion (i). The proof is completed.

Remark 8.10 In this chapter, the CRDNNs model (8.4) and output equation (8.7) have the same control input $u_i(x,t)$. Therefore, if $u_i(x,t)$ is set to be zero in network (8.4), the control input $u_i(x,t) = 0$ in the output equation (8.7).

8.3.3 Synchronization of CRDNNs

From Theorems 8.6 and 8.9, we can obtain the following conclusion.

Theorem 8.11 *Let $\dot{\tau}(t) \leqslant \sigma < 1$. CRDNNs (8.4) is synchronized in the sense of Definition 8.8 under the adaptive law*

$$\dot{G}^1_{ij}(t) = \beta_{ij} \int_\Omega (z_i(x,t) - z_j(x,t))^T \Gamma_1 (z_i(x,t) - z_j(x,t))dx, \ (i,j) \in \mathcal{E},$$

or

$$\dot{G}^1_{ij}(t) = \hat{\beta}_{ij} \int_\Omega (z_i(x,t) - z_j(x,t))^T \Gamma_1 (z_i(x,t) - z_j(x,t))dx, \ (i,j) \in \hat{\mathcal{E}},$$

where $\beta_{ij} = \beta_{ji}$ and $\hat{\beta}_{ij} = \hat{\beta}_{ji}$ are positive constants, $\hat{\mathcal{E}}$ is a subset of $\mathcal{E}$, and CRDNNs (8.1) is connected through the undirected edges $\hat{\mathcal{E}}$.

Remark 8.12 In recent years, as a powerful tool, passivity has played an important role in synchronization, group coordination, flow control, energy management, and so on [179, 180, 184–190]. Unfortunately, in most existing works on the passivity, it is assumed that the state, input and output variables are only dependent on the time. In Theorem 8.11, a sufficient condition for synchronization of the CRDNNs with adaptive coupling is established by employing the obtained passivity results and the relationship between passivity and synchronization.

Remark 8.13 On the one hand, the dissipativity, passivity, input strict passivity and output strict passivity are not dependent on the node state (see Definitions 1.3–1.5). On the other hand, the objective of this chapter is to establish a global synchronization criterion by using the passivity theory. Therefore, we assume that the function $f_j(\cdot)(j = 1, 2, \ldots, n)$ satisfies the global Lipschitz condition. Practically, this assumption has been widely used in analyzing the stability of RDNNs [100, 117].

8.4 Numerical Example

In this section, an illustrative example is provided to verify the effectiveness of the proposed theoretical results.

Consider a CRDNNs consisting of 5 identical nodes with diffusive coupling, in which each node is a 3-dimensional reaction-diffusion neural network described by

$$\frac{\partial w_i(x,t)}{\partial t} = d_i \Delta w_i(x,t) - a_i w_i(x,t) + J_i + \sum_{j=1}^{3} b_{ij} f_j(w_j(x,t))$$

where $i = 1, 2, 3$, $\Omega = \{x \mid -0.5 < x < 0.5\}$, $f_j(\xi) = |\xi + 1| - |\xi - 1|$, $d_1 = 0.1$, $d_2 = 0.2$, $d_3 = 0.1$, $a_1 = 0.2$, $a_2 = 0.3$, $a_3 = 0.1$, $J_1 = 0.5$, $J_2 = 0.3$, $J_3 = 0.6$, and the matrix $B = (b_{ij})_{3\times 3}$ is chosen as

$$B = \begin{pmatrix} 0.4 & 0.2 & 0.1 \\ 0.1 & 0.3 & 0.3 \\ 0.3 & 0.5 & 0.2 \end{pmatrix}.$$

Obviously, $f_j(\cdot)(j = 1, 2, 3)$ satisfies the Lipschitz condition with $\rho_j = 2$.

We take $c_1 = 0.2, c_2 = 0.3, \Gamma_1 = \text{diag}(0.3, 0.2, 0.4), \Gamma_2 = \text{diag}(0.4, 0.3, 0.5)$ and $\tau(t) = 0.3 - 0.3e^{-t}$. The matrices $G^1(0)$ and G^2 are chosen as, respectively,

$$G^1(0) = \begin{pmatrix} -0.4 & 0.1 & 0 & 0.3 & 0 \\ 0.1 & -0.8 & 0.2 & 0.2 & 0.3 \\ 0 & 0.2 & -0.6 & 0 & 0.4 \\ 0.3 & 0.2 & 0 & -0.5 & 0 \\ 0 & 0.3 & 0.4 & 0 & -0.7 \end{pmatrix},$$

$$G^2 = \begin{pmatrix} -0.5 & 0.2 & 0 & 0.3 & 0 \\ 0.2 & -1 & 0.3 & 0.2 & 0.3 \\ 0 & 0.3 & -0.8 & 0 & 0.5 \\ 0.3 & 0.2 & 0 & -0.5 & 0 \\ 0 & 0.3 & 0.5 & 0 & -0.8 \end{pmatrix}.$$

Case 1: Choose $\beta_{12} = \beta_{21} = 0.3, \beta_{14} = \beta_{41} = 0.2, \beta_{23} = \beta_{32} = 0.2, \beta_{24} = \beta_{42} = 0.4, \beta_{25} = \beta_{52} = 0.3, \beta_{35} = \beta_{53} = 0.1$. According to Theorem 8.11, the CRDNNs (8.4) with above given parameters is synchronized in the sense of Definition 8.8. The simulation results are shown in Figs. 8.1 and 8.2.

Case 2: Choose $\hat{\beta}_{12} = \hat{\beta}_{21} = 0.3, \hat{\beta}_{23} = \hat{\beta}_{32} = 0.2, \hat{\beta}_{24} = \hat{\beta}_{42} = 0.4, \hat{\beta}_{25} = \hat{\beta}_{52} = 0.3$. It follows from Theorem 8.11 that the CRDNNs (8.4) with above given parameters is synchronized in the sense of Definition 8.8. The simulation results are shown in Figs. 8.3 and 8.4.

From Figs. 8.1 and 8.3, we clearly see that synchronization errors $e_{ij}(x,t)$, $i = 1, 2, \ldots, 5$, $j = 1, 2, 3$, are very close to 0 when the time t increases gradually to 5, and this state is maintained along with the increasing of the time. These results show that $z_i(x,t) - \bar{z}(x,t) = e_i(x,t) = (e_{i1}(x,t), e_{i2}(x,t), e_{i3}(x,t))^T (i = 1, 2, \ldots, 5)$ asymptotically converges to 0, thus the complex network is synchronized. Figures 8.2 and 8.4 visualize the change processes of the coupling strengths in time interval [0, 10]. The numerical results clearly show that each coupling strength asymptotically

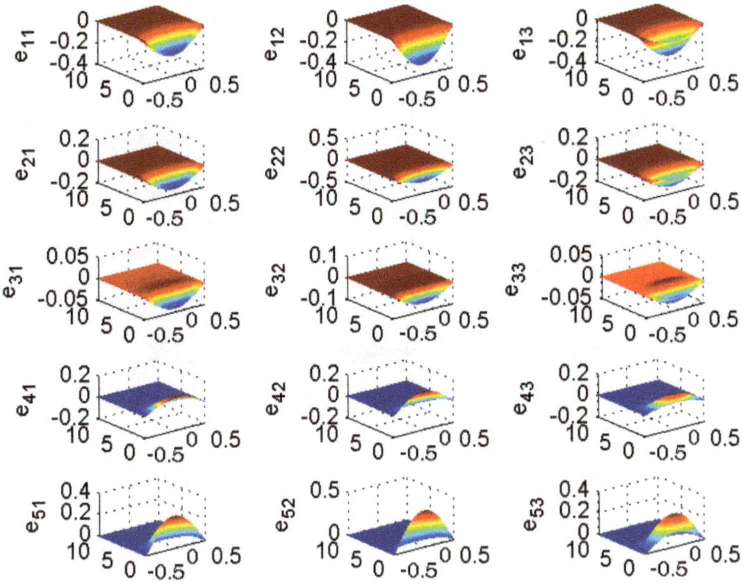

Fig. 8.1 The change processes of $e_{ij}(x, t), i = 1, 2, \ldots, 5, j = 1, 2, 3$

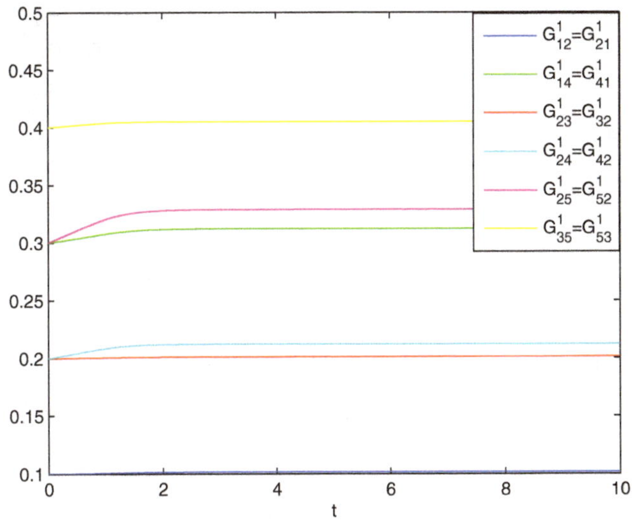

Fig. 8.2 Adaptive coupling strengths

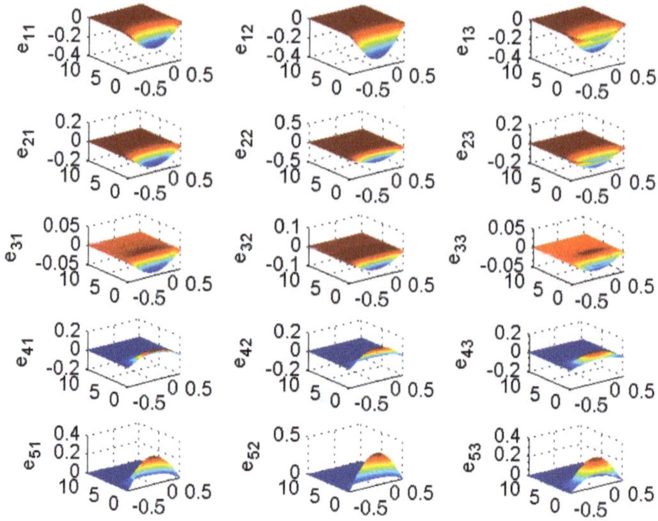

Fig. 8.3 The change processes of $e_{ij}(x,t)$, $i = 1, 2, \ldots, 5$, $j = 1, 2, 3$

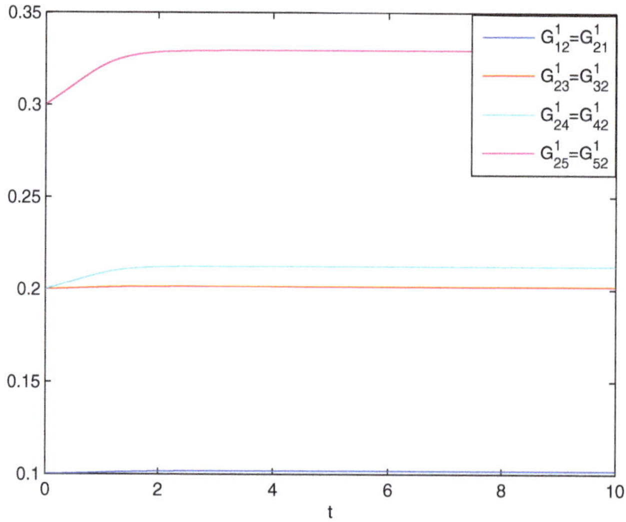

Fig. 8.4 Adaptive coupling strengths

converges to some finite positive value, which has been proved in Theorem 8.9. Figures 8.1, 8.2, 8.3 and 8.4 demonstrate the correctness and effectiveness of the obtained results in Theorems 8.9 and 8.11.

Remark 8.14 From (8.9) and (8.17), the time derivative of $G^1_{ij}(t)$ at time t depends on $z_i(\cdot, t)$ and $z_j(\cdot, t)$. Obviously, the time derivative of $z_i(x, t)$ at time t also depends on the value of $G^1_{ij}(t)$ (see (8)). In our simulations, $G^1_{ij}(t)(0 \leqslant t \leqslant 10)$ is discretized by utilizing the finite-difference method. Namely,

$$G^1_{ij}(t_{k+1}) = G^1_{ij}(t_k) + \epsilon\beta_{ij}\int_{\Omega}(z_i(x, t_k) - z_j(x, t_k))^T \Gamma_1(z_i(x, t_k) - z_j(x, t_k))dx,$$

where $\beta_{ij} = \beta_{ji}$ are positive constants, $t_k - t_{k-1} = \frac{10}{s} = \epsilon, t_0 = 0, t_s = 10$, $k \in \{1, 2, \ldots, s\}$. Similarly, $z_i(x, t)$ can be discretized, and $z_i(x, t_k)$ depends on the value of $G^1_{ij}(t_{k-1})$.

8.5 Conclusion

A general array model of CRDNNs with adaptive coupling has been investigated. Both passivity and synchronization have been taken into consideration in this chapter. Several novel adaptive strategies to tune the coupling strengths among network nodes have been proposed, and some criteria for passivity have been established by using the designed adaptive laws. Furthermore, the relationship between passivity and synchronization of the CRDNNs has been revealed. In addition, a global synchronization criterion has been derived by utilizing the obtained passivity results and the relationship between passivity and synchronization. Illustrative simulations have been provided to verify the correctness and effectiveness of the theoretical results. In future work, we shall study the passivity and robust passivity of the CRDNNs with switching topology.

Chapter 9
Passivity Analysis of CRDNNs with Dirichlet Boundary Conditions

9.1 Introduction

In last few decades, much investigative attention has been focused on passivity problem of complex dynamical networks [188]. The main reason for this is that the passive system is internally stable. For instance, Yao et al. [43] considered passivity of a switched complex network, including the presence of communication time-delays. Based on Lyapunov–Krasovskii functional method, several criteria for passivity were obtained. In [50], the authors respectively discussed passivity of complex networks with presence or lack of communication time delays. In [74], Wang et al. proposed a complex dynamical network model with nonsymmetric, nonlinear and delayed coupling, and gave some conditions for input strict passivity and output strict passivity of the proposed network model. It should be pointed out that time-dependent node state is considered in these existing works [43, 50, 74]. In fact, the state variable of node is time and space-dependent in some complex networks (e.g., RDNNs and food webs). In [29], several passivity and robust passivity criteria for RDNNs with spatially and temporally varying input and output vectors were obtained. In [95, 96], some passivity criteria were derived for CRDNNs with fixed and adaptive coupling weights. Moreover, they analyzed the synchronization of passive CRDNNs. Unfortunately, the input vector has the same dimension as the output vector in the network models considered in [29, 43, 50, 74, 95, 96]. Practically, input and output vectors in some networks have different dimensions. But, for the systems with different dimensions of input and output vectors, we can not deal with by traditional passivity definitions. However, very few authors have discussed passivity of CRDNNs with different dimensions of input and output.

Based on the above discussions, this chapter presents two CRDNNs with different dimensions of input and output. The only difference between them is whether time-varying delay is incorporated in the network model. In this chapter, we analyze dissipativity and passivity of these CRDNNs with different dimensions of input and output. The main contributions of this chapter are as follows. First, several dissipativity and passivity criteria for CRDNNs without time-varying delay are established

J.-L. Wang et al., *Analysis and Control of Coupled Neural Networks with Reaction-Diffusion Terms*, DOI 10.1007/978-981-10-4907-1_9

by exploiting some inequality techniques. Second, we analyze stability of passive CRDNNs without time-varying delay. Third, these results are extended to the case when coupling delay is considered in CRDNNs.

9.2 Passivity of CRDNNs

9.2.1 *Network Model*

Consider the following CRDNNs:

$$\frac{\partial z_i(x,t)}{\partial t} = D\Delta z_i(x,t) - Az_i(x,t) + J + Bf(z_i(x,t)) + c\sum_{j=1}^{N} G_{ij}\Gamma z_j(x,t) + Eu_i(x,t), \tag{9.1}$$

$$z_i(x,0) = \Phi_i(x) \in \mathbb{R}^n, \qquad x \in \Omega, \tag{9.2}$$

$$z_i(x,t) = 0, \qquad (x,t) \in \partial\Omega \times [0,+\infty), \tag{9.3}$$

where $x = (x_1, x_2, \ldots, x_q)^T \in \Omega \subset \mathbb{R}^q, i = 1, 2, \ldots, N$; $\Phi_i(x)$ is continuous on $\Omega \cup \partial\Omega$. $\Delta = \sum_{k=1}^{q} \frac{\partial^2}{\partial x_k^2}$ is Laplace diffusion operator; $z_i(x,t) = (z_{i1}(x,t), z_{i2}(x,t), \ldots, z_{in}(x,t))^T \in \mathbb{R}^n$ is the state vector of node i; $u_i(x,t) \in \mathbb{R}^p (1 \leqslant p \leqslant n)$ is the control input; $D = \mathrm{diag}(d_1, d_2, \ldots, d_n) > 0$, and d_i is the transmission diffusion coefficient along the ith neuron; $A = \mathrm{diag}(a_1, a_2, \ldots, a_n) > 0$, and a_i is the rate with which the ith neuron will reset its potential to the resting state when disconnected from the neural network and external input; $J = (J_1, J_2, \ldots, J_n)^T$ is a constant external input vector; $B = (b_{ij})_{n\times n} \in \mathbb{R}^{n\times n}$, and b_{ij} is the strength of the jth neuron on the ith neuron; E is a known constant matrix with appropriate dimensions; $f(z_i(x,t)) = (f_1(z_{i1}(x,t)), f_2(z_{i2}(x,t)), \ldots, f_n(z_{in}(x,t)))^T$, and $f_j(\cdot)$ is the activation function of the jth neuron; c is a positive real number, which represents the overall coupling strength; $\Gamma \in \mathbb{R}^{n\times n} > 0$ is inner coupling matrix; $G = (G_{ij})_{N\times N}$ is the outer coupling matrix, where G_{ij} is defined as follows: if there exists a connection from node i to node $j (i \neq j)$, then $G_{ij} > 0$; otherwise, $G_{ij} = 0 (i \neq j)$, and the diagonal elements of matrix G are defined by

$$G_{ii} = -\sum_{\substack{j=1\\ j\neq i}}^{N} G_{ij}, i = 1, 2, \ldots, N.$$

9.2.2 Passivity Analysis of CRDNNs

Suppose that $w^*(x) = (w_1^*(x), w_2^*(x), \dots, w_n^*(x))^T \in \mathbb{R}^n$ is an equilibrium solution of an isolated node of the CRDNNs (9.1), then it satisfies (9.3) and

$$0 = D\Delta w^*(x) - Aw^*(x) + J + Bf(w^*(x)). \tag{9.4}$$

Define $e_i(x,t) = z_i(x,t) - w^*(x)$. From (9.1) and (9.4), we can get

$$\begin{aligned}\frac{\partial e_i(x,t)}{\partial t} &= D\Delta e_i(x,t) - Ae_i(x,t) + Bf(z_i(x,t)) - Bf(w^*(x))\\ &\quad + c\sum_{j=1}^{N} G_{ij}\Gamma e_j(x,t) + Eu_i(x,t), \quad i = 1, 2, \dots, N.\end{aligned} \tag{9.5}$$

The output vector $y_i(x,t) \in \mathbb{R}^m (1 \leqslant m \leqslant n, i = 1, 2, \dots, N)$ of CRDNNs (9.5) is chosen as

$$y_i(x,t) = Fe_i(x,t) + Hu_i(x,t), \tag{9.6}$$

where $F \in \mathbb{R}^{m\times n}$ and $H \in \mathbb{R}^{m\times p}$ are known constant matrices.

For convenient analysis, we let

$$\begin{aligned}
z(x,t) &= (z_1^T(x,t), z_2^T(x,t), \dots, z_N^T(x,t))^T,\\
e(x,t) &= (e_1^T(x,t), e_2^T(x,t), \dots, e_N^T(x,t))^T,\\
y(x,t) &= (y_1^T(x,t), y_2^T(x,t), \dots, y_N^T(x,t))^T,\\
u(x,t) &= (u_1^T(x,t), u_2^T(x,t), \dots, u_N^T(x,t))^T,\\
\hat{w}^*(x) &= ((w^*(x))^T, (w^*(x))^T, \dots, (w^*(x))^T)^T,\\
\hat{f}(\hat{w}^*(x)) &= (f^T(w^*(x)), f^T(w^*(x)), \dots, f^T(w^*(x)))^T,\\
\hat{f}(z(x,t)) &= (f^T(z_1(x,t)), f^T(z_2(x,t)), \dots, f^T(z_N(x,t)))^T.
\end{aligned}$$

From (9.5) and (9.6), one has

$$\begin{aligned}\frac{\partial e(x,t)}{\partial t} &= (I_N \otimes D)\Delta e(x,t) - (I_N \otimes A)e(x,t) + c(G \otimes \Gamma)e(x,t)\\ &\quad + (I_N \otimes B)\big(\hat{f}(z(x,t)) - \hat{f}(\hat{w}^*(x))\big) + (I_N \otimes E)u(x,t),\\ y(x,t) &= (I_N \otimes F)e(x,t) + (I_N \otimes H)u(x,t).\end{aligned} \tag{9.7}$$

Theorem 9.1 *If there are* $0 < P \in \mathbb{R}^{nN\times nN}$, $0 \leqslant W \in \mathbb{R}^{pN\times pN}$ *and* $Q \in \mathbb{R}^{mN\times pN}$ *such that*

$$P(I_N \otimes D) + (I_N \otimes D)P \geqslant 0, \tag{9.8}$$

$$\begin{pmatrix} \chi & \chi_1 \\ \chi_1^T & -Q^T(I_N \otimes H) - W - (I_N \otimes H^T)Q \end{pmatrix} \leqslant 0, \tag{9.9}$$

where $\Theta = \mathrm{diag}(\rho_1^2, \rho_2^2, \ldots, \rho_n^2)$, $\chi = -\sum_{k=1}^{q} \frac{1}{l_k^2}\big[P(I_N \otimes D) + (I_N \otimes D)P\big] - P(I_N \otimes A) - (I_N \otimes A)P + cP(G \otimes \Gamma) + c(G^T \otimes \Gamma)P + P\left(I_N \otimes (BB^T)\right)P + I_N \otimes \Theta$, $\chi_1 = P(I_N \otimes E) - (I_N \otimes F^T)Q$, *then system* (9.7) *is dissipative with respect to*

$$\vartheta(u, y) = \int_\Omega y^T(x,t)Qu(x,t)dx + \int_\Omega u^T(x,t)\frac{W}{2}u(x,t)dx.$$

Proof Choose the following Lyapunov functional:

$$V_1(t) = \int_\Omega e^T(x,t)Pe(x,t)dx. \tag{9.10}$$

The time derivative of $V_1(t)$ is given by

$$\begin{aligned}
\dot{V}_1(t) &= 2\int_\Omega e^T(x,t)P\frac{\partial e(x,t)}{\partial t}dx \\
&= \int_\Omega \Big[e^T(x,t)P(I_N \otimes D)\Delta e(x,t) + (\Delta e(x,t))^T(I_N \otimes D)Pe(x,t)\Big]dx \\
&\quad + \int_\Omega e^T(x,t)\Big[-P(I_N \otimes A) - (I_N \otimes A)P + c(G^T \otimes \Gamma)P \\
&\quad + cP(G \otimes \Gamma)\Big]e(x,t)dx + 2\int_\Omega e^T(x,t)P(I_N \otimes E)u(x,t)dx \\
&\quad + 2\int_\Omega e^T(x,t)P(I_N \otimes B)\Big(\hat{f}(z(x,t)) - \hat{f}(\hat{w}^*(x))\Big)dx.
\end{aligned} \tag{9.11}$$

From the Dirichlet boundary conditions (9.3), one obtains

$$\int_\Omega e_{il}(x,t)\Delta e_{jh}(x,t)dx = -\sum_{k=1}^{q}\int_\Omega \frac{\partial e_{il}(x,t)}{\partial x_k}\frac{\partial e_{jh}(x,t)}{\partial x_k}dx,$$

where $i, j \in \{1, 2, \ldots, N\}$, $l, h \in \{1, 2, \ldots, n\}$.

Let

$$P^{ij} = (p_{lh}^{ij})_{n\times n} \text{ and } P = \begin{pmatrix} P^{11} & P^{12} & \ldots & P^{1N} \\ P^{21} & P^{22} & \ldots & P^{2N} \\ \vdots & \vdots & \vdots & \vdots \\ P^{N1} & P^{N2} & \ldots & P^{NN} \end{pmatrix}.$$

Then, we can get

$$\begin{aligned}
&\int_{\Omega} e^T(x,t)P(I_N \otimes D)\Delta e(x,t)dx \\
&= \sum_{i=1}^{N}\sum_{j=1}^{N}\int_{\Omega} e_i^T(x,t)P^{ij}D\Delta e_j(x,t)dx \\
&= \sum_{i=1}^{N}\sum_{j=1}^{N}\sum_{l=1}^{n}\sum_{h=1}^{n} p_{lh}^{ij}d_h \int_{\Omega} e_{il}(x,t)\Delta e_{jh}(x,t)dx \\
&= -\sum_{k=1}^{q}\sum_{i=1}^{N}\sum_{j=1}^{N}\sum_{l=1}^{n}\sum_{h=1}^{n} p_{lh}^{ij}d_h \int_{\Omega} \frac{\partial e_{il}(x,t)}{\partial x_k}\frac{\partial e_{jh}(x,t)}{\partial x_k}dx \\
&= -\sum_{k=1}^{q}\int_{\Omega}\left(\frac{\partial e(x,t)}{\partial x_k}\right)^T P(I_N \otimes D)\frac{\partial e(x,t)}{\partial x_k}dx, \\
&\int_{\Omega} (\Delta e(x,t))^T(I_N \otimes D)Pe(x,t)dx \\
&= \sum_{i=1}^{N}\sum_{j=1}^{N}\int_{\Omega} (\Delta e_i(x,t))^T DP^{ij}e_j(x,t)dx \\
&= \sum_{i=1}^{N}\sum_{j=1}^{N}\sum_{l=1}^{n}\sum_{h=1}^{n} p_{lh}^{ij}d_l \int_{\Omega} e_{jh}(x,t)\Delta e_{il}(x,t)dx \\
&= -\sum_{k=1}^{q}\sum_{i=1}^{N}\sum_{j=1}^{N}\sum_{l=1}^{n}\sum_{h=1}^{n} p_{lh}^{ij}d_l \int_{\Omega} \frac{\partial e_{il}(x,t)}{\partial x_k}\frac{\partial e_{jh}(x,t)}{\partial x_k}dx \\
&= -\sum_{k=1}^{q}\int_{\Omega}\left(\frac{\partial e(x,t)}{\partial x_k}\right)^T (I_N \otimes D)P\frac{\partial e(x,t)}{\partial x_k}dx. \qquad (9.12)
\end{aligned}$$

Therefore,

$$\begin{aligned}
&\int_{\Omega}\left[e^T(x,t)P(I_N \otimes D)\Delta e(x,t) + (\Delta e(x,t))^T(I_N \otimes D)Pe(x,t)\right]dx \\
&= -\sum_{k=1}^{q}\int_{\Omega}\left(\frac{\partial e(x,t)}{\partial x_k}\right)^T \left[P(I_N \otimes D) + (I_N \otimes D)P\right]\frac{\partial e(x,t)}{\partial x_k}dx.
\end{aligned}$$

By (9.8), we can find the matrix $M \in \mathbb{R}^{nN \times nN}$ satisfying

$$P(I_N \otimes D) + (I_N \otimes D)P = M^T M.$$

Thus,

$$\begin{aligned}&\left(\frac{\partial e(x,t)}{\partial x_k}\right)^T\left[P(I_N\otimes D)+(I_N\otimes D)P\right]\frac{\partial e(x,t)}{\partial x_k}\\&=\left(\frac{\partial e(x,t)}{\partial x_k}\right)^T M^T M\frac{\partial e(x,t)}{\partial x_k}\\&=\left(\frac{\partial (Me(x,t))}{\partial x_k}\right)^T\frac{\partial (Me(x,t))}{\partial x_k}.\end{aligned}$$

According to Lemma 1.22, we can easily obtain

$$\begin{aligned}&\sum_{k=1}^{q}\int_{\Omega}\left(\frac{\partial (Me(x,t))}{\partial x_k}\right)^T\frac{\partial (Me(x,t))}{\partial x_k}dx\\&\geqslant\sum_{k=1}^{q}\frac{1}{l_k^2}\int_{\Omega}(Me(x,t))^T(Me(x,t))dx\\&=\sum_{k=1}^{q}\frac{1}{l_k^2}\int_{\Omega}e^T(x,t)\mathbb{W}e(x,t)dx,\end{aligned}\tag{9.13}$$

where $\mathbb{W}=P(I_N\otimes D)+(I_N\otimes D)P$.

On the other hand,

$$\begin{aligned}&2e^T(x,t)P(I_N\otimes B)\big(\hat{f}(z(x,t))-\hat{f}(\hat{w}^*(x))\big)\\&\leqslant e^T(x,t)\big[P\left(I_N\otimes(BB^T)\right)P+I_N\otimes\Theta\big]e(x,t).\end{aligned}\tag{9.14}$$

By (9.11)–(9.14), one obtains

$$\begin{aligned}\dot{V}_1(t)\leqslant&\int_{\Omega}e^T(x,t)\Big[-\sum_{k=1}^{q}\frac{1}{l_k^2}\mathbb{W}-P(I_N\otimes A)-(I_N\otimes A)P+cP(G\otimes\Gamma)\\&+c(G^T\otimes\Gamma)P+P\left(I_N\otimes(BB^T)\right)P+I_N\otimes\Theta\Big]e(x,t)dx\\&+2\int_{\Omega}e^T(x,t)P(I_N\otimes E)u(x,t)dx.\end{aligned}$$

Thus,

$$\begin{aligned}&\dot{V}_1(t)-2\int_{\Omega}y^T(x,t)Qu(x,t)dx-\int_{\Omega}u^T(x,t)Wu(x,t)dx\\&\leqslant\int_{\Omega}e^T(x,t)\Big[-\sum_{k=1}^{q}\frac{1}{l_k^2}\mathbb{W}-P(I_N\otimes A)-(I_N\otimes A)P+cP(G\otimes\Gamma)\end{aligned}$$

$$+ c(G^T \otimes \Gamma)P + P\left(I_N \otimes (BB^T)\right)P + I_N \otimes \Theta\Big]e(x,t)dx$$
$$+ \int_\Omega u^T(x,t)\big[-Q^T(I_N \otimes H) - W - (I_N \otimes H^T)Q\big]u(x,t)dx$$
$$-2\int_\Omega e^T(x,t)(I_N \otimes F^T)Qu(x,t)dx$$
$$+2\int_\Omega e^T(x,t)P(I_N \otimes E)u(x,t)dx$$
$$= \int_\Omega \vartheta^T(x,t)\begin{pmatrix} \chi & \chi_1 \\ \chi_1^T & \chi_2 \end{pmatrix}\vartheta(x,t)dx,$$

where $\chi_2 = -Q^T(I_N \otimes H) - (I_N \otimes H^T)Q - W, \vartheta(x,t) = (e^T(x,t), u^T(x,t))^T$.

From (9.9), one has

$$\int_\Omega y^T(x,t)Qu(x,t)dx$$
$$\geqslant \dot{S}(t) - \int_\Omega u^T(x,t)\frac{W}{2}u(x,t)dx, \tag{9.15}$$

where $S(t) = \frac{V_1(t)}{2}$.

By (9.15), we can obtain

$$\int_{t_0}^{t_p}\int_\Omega y^T(x,t)Qu(x,t)dxdt$$
$$\geqslant S(t_p) - S(t_0) - \int_{t_0}^{t_p}\int_\Omega u^T(x,t)\frac{W}{2}u(x,t)dxdt$$

for any $t_p, t_0 \in [0, +\infty)$ and $t_p \geqslant t_0$.

By utilizing the same method as in proving Theorem 9.1, we can get the following conclusions.

Theorem 9.2 *If there are* $0 < P \in \mathbb{R}^{nN\times nN}$ *and* $Q \in \mathbb{R}^{mN\times pN}$ *such that*

$$P(I_N \otimes D) + (I_N \otimes D)P \geqslant 0, \tag{9.16}$$

$$\begin{pmatrix} \chi & \chi_1 \\ \chi_1^T & -Q^T(I_N \otimes H) - (I_N \otimes H^T)Q \end{pmatrix} \leqslant 0, \tag{9.17}$$

where $\Theta = \mathrm{diag}(\rho_1^2, \rho_2^2, \ldots, \rho_n^2), \chi = -\sum_{k=1}^q \frac{1}{l_k^2}\big[P(I_N \otimes D) + (I_N \otimes D)P\big] - P(I_N \otimes A) - (I_N \otimes A)P + cP(G \otimes \Gamma) + c(G^T \otimes \Gamma)P + P\left(I_N \otimes (BB^T)\right)P + I_N \otimes \Theta, \chi_1 = P(I_N \otimes E) - (I_N \otimes F^T)Q$*, then system (9.7) is passive.*

Theorem 9.3 *If there are* $0 < P \in \mathbb{R}^{nN\times nN}$, $0 < W \in \mathbb{R}^{pN\times pN}$ *and* $Q \in \mathbb{R}^{mN\times pN}$ *such that*

$$P(I_N \otimes D) + (I_N \otimes D)P \geqslant 0, \tag{9.18}$$

$$\begin{pmatrix} \chi & \chi_1 \\ \chi_1^T & -Q^T(I_N \otimes H) + W - (I_N \otimes H^T)Q \end{pmatrix} \leqslant 0, \tag{9.19}$$

where $\Theta = \text{diag}(\rho_1^2, \rho_2^2, \ldots, \rho_n^2)$, $\chi = -\sum_{k=1}^{q} \frac{1}{l_k^2}\left[P(I_N \otimes D) + (I_N \otimes D)P\right] - P(I_N \otimes A) - (I_N \otimes A)P + cP(G \otimes \Gamma) + c(G^T \otimes \Gamma)P + P\left(I_N \otimes (BB^T)\right)P + I_N \otimes \Theta$, $\chi_1 = P(I_N \otimes E) - (I_N \otimes F^T)Q$, *then system* (9.7) *is input-strictly passive.*

Theorem 9.4 *If there are* $0 < P \in \mathbb{R}^{nN \times nN}$, $0 < W \in \mathbb{R}^{mN \times mN}$ *and* $Q \in \mathbb{R}^{mN \times pN}$ *such that*

$$P(I_N \otimes D) + (I_N \otimes D)P \geqslant 0, \tag{9.20}$$

$$\begin{pmatrix} \hat{\chi} & \chi_3 \\ \chi_3^T & \chi_4 \end{pmatrix} \leqslant 0, \tag{9.21}$$

where $\Theta = \text{diag}(\rho_1^2, \rho_2^2, \ldots, \rho_n^2)$, $\hat{\chi} = -\sum_{k=1}^{q} \frac{1}{l_k^2}\left[P(I_N \otimes D) + (I_N \otimes D)P\right] - P(I_N \otimes A) - (I_N \otimes A)P + cP(G \otimes \Gamma) + c(G^T \otimes \Gamma)P + P\left(I_N \otimes (BB^T)\right)P + I_N \otimes \Theta + (I_N \otimes F^T)W(I_N \otimes F)$, $\chi_3 = P(I_N \otimes E) - (I_N \otimes F^T)Q + (I_N \otimes F^T)W(I_N \otimes H)$, $\chi_4 = -Q^T(I_N \otimes H) - (I_N \otimes H^T)Q + (I_N \otimes H^T)W(I_N \otimes H)$, *then system* (9.7) *is output-strictly passive.*

Remark 9.5 It is well known that input and output vectors in many systems have different dimensions. But, for the systems with different dimensions of input and output vectors, we can not deal with by traditional passivity definitions in [29, 43, 50, 74, 95, 96]. In this subsection, we analyze dissipativity and passivity of the CRDNNs with different dimensions of input and output. Several dissipativity and passivity criteria for the CRDNNs are established by exploiting some inequality techniques [see Theorems 9.1–9.4].

9.2.3 The Stability of Passive CRDNNs

More recently, some authors have discussed stability of passive complex dynamical networks under the assumption that input and output vectors have the same dimension [95]. However, in many real networks, the dimensions of input and output vectors are different. Therefore, the stability of passive CRDNNs with different dimensions of input and output is considered in this subsection.

Theorem 9.6 $S : [0, +\infty) \to [0, +\infty)$ *is continuously differentiable and satisfies*

$$v_1(\|e(\cdot, t)\|_2) \leqslant S(t) \leqslant v_2(\|e(\cdot, t)\|_2), \tag{9.22}$$

where $v_1, v_2 : [0, +\infty) \to [0, +\infty)$ are continuous and strictly monotonically increasing functions, $v_1(\beta)$ and $v_2(\beta)$ are positive for $\beta > 0$ with $v_1(0) = v_2(0) = 0$. Then

(i) If system (9.7) is passive with respect to $S(t)$, CRDNNs (9.1) is internally stable;

(ii) If system (9.7) is input-strictly passive with respect to $S(t)$, CRDNNs (9.1) is internally stable;

(iii) If system (9.7) is output-strictly passive with respect to $S(t)$ and $F^T F > 0$, CRDNNs (9.1) is internally asymptotically stable.

Proof In what follows, we complete the proof in two steps. In Step 1, we prove the stability of the CRDNNs (9.1); in Step 2, we prove the attractivity of the CRDNNs (9.1).

Step 1: We shall prove its stability. Firstly, according to the passivity definition [see Definition 1.8], we can get

$$S(t+\varepsilon) - S(t) \leqslant \int_t^{t+\varepsilon} \int_\Omega y^T(x,s) Q u(x,s) dx ds \tag{9.23}$$

for any $t \in [0, +\infty)$ and $\varepsilon > 0$, where $Q \in \mathbb{R}^{mN \times pN}$ is a constant matrix.

From (9.23), one has

$$\frac{S(t+\varepsilon) - S(t)}{\varepsilon} \leqslant \frac{\int_t^{t+\varepsilon} \int_\Omega y^T(x,s) Q u(x,s) dx ds}{\varepsilon}. \tag{9.24}$$

Obviously, by (9.24), we can derive

$$\dot{S}(t) \leqslant \int_\Omega y^T(x,t) Q u(x,t) dx,$$

where $t \in [0, +\infty)$.

Let $u(x,t) = 0$, we then have

$$\dot{S}(t) \leqslant 0, \quad t \in [0, +\infty).$$

Furthermore, for any $\epsilon > 0$, we can find a $\delta = \delta(\epsilon)(0 < \delta < \epsilon)$ satisfying $v_2(\delta) < v_1(\epsilon)$. When $\|e(\cdot, 0)\|_2 < \delta$,

$$v_1(\|e(\cdot,t)\|_2) \leqslant S(t) \leqslant S(0) \leqslant v_2(\|e(\cdot,0)\|_2) < v_2(\delta) < v_1(\epsilon), \ t \geqslant 0. \tag{9.25}$$

By (9.25), one obtains

$$\|e(\cdot,t)\|_2 < \epsilon, \quad t \geqslant 0.$$

Therefore, CRDNNs (9.1) is internally stable.

Similarly, we can prove that CRDNNs (9.1) is internally stable when system (9.7) is input-strictly passive or output-strictly passive. The internal stability of the CRDNNs (9.1) is proven.

Step 2: We shall prove the attractivity of the CRDNNs (9.1) when system (9.7) is output-strictly passive. By utilizing the same method as in Step 1, we can get

$$\dot{S}(t) \leqslant -\alpha\|e(\cdot,t)\|_2^2, \quad t \in [0,+\infty), \tag{9.26}$$

where $W \in \mathbb{R}^{mN\times mN} > 0$, $\alpha = \lambda_m(F^TF)\lambda_m(W) > 0$. Obviously, $\lim_{t\to+\infty} S(t)$ is a finite nonnegative value.

For any $\epsilon > 0$ and $\|e(\cdot,0)\|_2 < b_0 = \delta(1)$, we shall prove that there exists $T\big(e(x,0),\epsilon\big) > 0$ such that $\|e(\cdot,t)\|_2 < \epsilon$ for all $t \geqslant T\big(e(x,0),\epsilon\big)$. If this is not true, we can find ϵ^*, $\Psi(x) \in \mathbb{R}^{nN}(\|\Psi(\cdot)\|_2 < b_0)$ and sequence $\{t_k\}$ such that

$$\|e(\cdot,t_k,\Psi)\|_2 \geqslant \delta(\epsilon^*), \ \xi_{k-1}+\varpi < t_k < \xi_k, \ k \in \mathbb{N},$$

where $\xi_0 = 0$, $\lim_{k\to+\infty}\xi_k = +\infty$, ϖ is a positive real number, $\delta(\epsilon^*)$ and $\delta(1)$ are the above constants for stability.

By (9.3), one obtains

$$\begin{aligned}
&\frac{d\|e(\cdot,t,\Psi)\|_2^2}{dt}\\
&= 2\int_\Omega e^T(x,t,\Psi)\frac{\partial e(x,t,\Psi)}{\partial t}dx\\
&\leqslant \int_\Omega e^T(x,t,\Psi)\big[I_N\otimes(BB^T+\Theta-2A)\big]e(x,t,\Psi)dx\\
&\quad -2\sum_{k=1}^q\int_\Omega\left(\frac{\partial e(x,t,\Psi)}{\partial x_k}\right)^T(I_N\otimes D)\frac{\partial e(x,t,\Psi)}{\partial x_k}dx\\
&\quad +2c\int_\Omega e^T(x,t,\Psi)(G\otimes\Gamma)e(x,t,\Psi)dx\\
&\leqslant \int_\Omega e^T(x,t,\Psi)\Big[I_N\otimes(BB^T+\Theta-2A)+c(G+G^T)\otimes\Gamma\Big]e(x,t,\Psi)dx\\
&\leqslant \mu\|e(\cdot,t,\Psi)\|_2^2\\
&< \mu, \ t\geqslant 0,
\end{aligned} \tag{9.27}$$

where $\Theta = \mathrm{diag}(\rho_1^2,\rho_2^2,\ldots,\rho_n^2)$, $\mu = \max\Big\{1, \lambda_M\big(I_N\otimes(BB^T+\Theta-2A)+c(G+G^T)\otimes\Gamma\big)\Big\} > 0$.

In the following, we shall prove that

$$\|e(\cdot,t,\Psi)\|_2 > \frac{\delta(\epsilon^*)}{2}, \quad t_k - \frac{\delta^2(\epsilon^*)}{2\mu} \leqslant t \leqslant t_k. \tag{9.28}$$

If (9.28) does not hold, then we can find a $t^* \in \left[t_k - \frac{\delta^2(\epsilon^*)}{2\mu}, t_k\right]$ satisfying

$$\|e(\cdot,t_k,\Psi)\|_2^2 - \|e(\cdot,t^*,\Psi)\|_2^2 \geqslant \frac{3\delta^2(\epsilon^*)}{4}. \tag{9.29}$$

Furthermore, from (9.27), one has

$$\|e(\cdot,t_k,\Psi)\|_2^2 - \|e(\cdot,t^*,\Psi)\|_2^2 < \mu \times \frac{\delta^2(\epsilon^*)}{2\mu} = \frac{\delta^2(\epsilon^*)}{2}.$$

This contradicts (9.29), and thus (9.28) holds.

Then, we can derive from (9.26) and (9.28) that

$$\dot{S}(t) \leqslant -\frac{\alpha\delta^2(\epsilon^*)}{4}, \quad t_k - \frac{\delta^2(\epsilon^*)}{2\mu} \leqslant t \leqslant t_k.$$

Define $\varpi = \frac{\delta^2(\epsilon^*)}{\mu}$. Obviously, these intervals do not overlap. Therefore,

$$-S(0) \leqslant S(t_k) - S(0) = \int_0^{t_k} \dot{S}(t)dt \leqslant -\frac{\alpha\delta^2(\epsilon^*)}{4}\frac{\delta^2(\epsilon^*) \times k}{2\mu}. \tag{9.30}$$

Letting $k \to +\infty$ in (9.30), we have

$$-S(0) \leqslant S(+\infty) - S(0) \leqslant -\lim_{k\to+\infty} \frac{\alpha\delta^2(\epsilon^*)}{4}\frac{\delta^2(\epsilon^*) \times k}{2\mu} = -\infty, \tag{9.31}$$

which is a contradiction. This proves the attractivity of the CRDNNs (9.1).

Remark 9.7 In Theorem (9.6), we not only prove that passive (or input-strictly passive) system (9.7) is internally stable, but also prove that output-strictly passive system (9.7) is internally asymptotically stable under the condition that $F^T F > 0$. This can serve as a stepping stone to investigate stability of passive CRDNNs with different dimensions of input and output.

9.3 Passivity of CRDNNs with Time-Varying Delay

9.3.1 Network Model

In what follows, a more general CRDNNs model is considered

$$\frac{\partial z_i(x,t)}{\partial t} = D\Delta z_i(x,t) - Az_i(x,t) + J + Bf(z_i(x,t)) + Eu_i(x,t)$$
$$+ c\sum_{j=1}^{N} G_{ij}\Gamma z_j(x, t-\tau(t)),\ i = 1, 2, \ldots, N, \tag{9.32}$$
$$z_i(x,t) = \Phi_i(x,t) \in \mathbb{R}^n, \quad (x,t) \in \Omega \times [-\tau, 0], \tag{9.33}$$
$$z_i(x,t) = 0, \quad (x,t) \in \partial\Omega \times [-\tau, +\infty), \tag{9.34}$$

where $\tau(t)$ is time-varying delay with $0 \leqslant \tau(t) \leqslant \tau$, $\Phi_i(x,t)$ is continuous on $\Omega \cup \partial\Omega \times [-\tau, 0]$.

Remark 9.8 Practically, time delay always exists in biological and physical networks due to the finite speeds of transmission and spreading as well as traffic congestion. Moreover, time delay could not be ignored in many circumstances, which might lead to undesirable dynamic network behaviors (e.g., instability and oscillation). On the other hand, absolute constant time delay may be scarce and time delay is frequently varied with time. Therefore, a delayed CRDNNs is also studied.

9.3.2 Passivity Analysis of CRDNNs with Time-Varying Delay

Suppose that $w^*(x) \in \mathbb{R}^n$ is an equilibrium solution of an isolated node of the CRDNNs (9.32), then it satisfies (9.34) and

$$0 = D\Delta w^*(x) - Aw^*(x) + J + Bf(w^*(x)). \tag{9.35}$$

Define $e_i(x,t) = z_i(x,t) - w^*(x)$. By (9.32) and (9.35), one obtains

$$\frac{\partial e_i(x,t)}{\partial t} = D\Delta e_i(x,t) + Bf(z_i(x,t)) - Bf(w^*(x)) - Ae_i(x,t) + Eu_i(x,t)$$
$$+ c\sum_{j=1}^{N} G_{ij}\Gamma e_j(x, t-\tau(t)), \quad i = 1, 2, \ldots, N. \tag{9.36}$$

The output vector $y_i(x,t) \in \mathbb{R}^m (1 \leqslant m \leqslant n, i = 1, 2, \ldots, N)$ of CRDNNs (9.36) is chosen as

$$y_i(x,t) = Fe_i(x,t) + Hu_i(x,t),$$

where $F \in \mathbb{R}^{m\times n}$ and $H \in \mathbb{R}^{m\times p}$ are known constant matrices.

Thus,

$$\begin{aligned}\frac{\partial e(x,t)}{\partial t} &= (I_N \otimes D)\Delta e(x,t) - (I_N \otimes A)e(x,t) + (I_N \otimes E)u(x,t)\\ &\quad + (I_N \otimes B)\big(\hat{f}(z(x,t)) - \hat{f}(\hat{w}^*(x))\big)\\ &\quad + c(G \otimes \Gamma)e(x,t-\tau(t)),\\ y(x,t) &= (I_N \otimes F)e(x,t) + (I_N \otimes H)u(x,t).\end{aligned} \tag{9.37}$$

Theorem 9.9 *Let $\dot{\tau}(t) \leqslant \sigma < 1$. If there exist matrices $0 < P \in \mathbb{R}^{nN\times nN}$, $0 < Z \in \mathbb{R}^{nN\times nN}$, $0 \leqslant W \in \mathbb{R}^{pN\times pN}$ and $Q \in \mathbb{R}^{mN\times pN}$ such that*

$$P(I_N \otimes D) + (I_N \otimes D)P \geqslant 0, \tag{9.38}$$

$$\begin{pmatrix} \chi_5 & \chi_1 \\ \chi_1^T & -Q^T(I_N \otimes H) - W - (I_N \otimes H^T)Q \end{pmatrix} \leqslant 0, \tag{9.39}$$

where $\Theta = \text{diag}(\rho_1^2, \rho_2^2, \ldots, \rho_n^2)$, $\chi_5 = -\sum_{k=1}^{q} \frac{1}{l_k^2}\big[P(I_N \otimes D) + (I_N \otimes D)P\big] - P(I_N \otimes A) - (I_N \otimes A)P + \frac{cZ}{1-\sigma} + cP(G \otimes \Gamma)Z^{-1}(G^T \otimes \Gamma)P + P(I_N \otimes (BB^T))P + I_N \otimes \Theta$, $\chi_1 = P(I_N \otimes E) - (I_N \otimes F^T)Q$, *then system* (9.37) *is dissipative with respect to*

$$\vartheta(u,y) = \int_\Omega y^T(x,t)Qu(x,t)dx + \int_\Omega u^T(x,t)\frac{W}{2}u(x,t)dx.$$

Proof Choose the following Lyapunov functional:

$$V_2(t) = \int_\Omega e^T(x,t)Pe(x,t)dx + \frac{c}{1-\sigma}\int_{t-\tau(t)}^{t}\int_\Omega e^T(x,s)Ze(x,s)dxds.$$

The time derivative of $V_2(t)$ is given by

$$\begin{aligned}\dot{V}_2(t) &\leqslant 2\int_\Omega e^T(x,t)P\frac{\partial e(x,t)}{\partial t}dx + \frac{c}{1-\sigma}\int_\Omega e^T(x,t)Ze(x,t)dx\\ &\quad - c\int_\Omega e^T(x,t-\tau(t))Ze(x,t-\tau(t))dx\\ &= \int_\Omega \Big[e^T(x,t)P(I_N \otimes D)\Delta e(x,t) + (\Delta e(x,t))^T(I_N \otimes D)Pe(x,t)\Big]dx\\ &\quad + \int_\Omega e^T(x,t)[-P(I_N \otimes A) - (I_N \otimes A)P]e(x,t)dx\end{aligned}$$

$$
\begin{aligned}
&+\frac{c}{1-\sigma}\int_{\Omega} e^T(x,t)Ze(x,t)dx - c\int_{\Omega} e^T(x,t-\tau(t))Ze(x,t-\tau(t))dx \\
&+2\int_{\Omega} e^T(x,t)P(I_N\otimes B)\Big(\hat{f}(z(x,t)) - \hat{f}(\hat{w}^*(x))\Big)dx \\
&+2c\int_{\Omega} e^T(x,t)P(G\otimes\Gamma)e(x,t-\tau(t))dx \\
&+2\int_{\Omega} e^T(x,t)P(I_N\otimes E)u(x,t)dx \\
\leqslant &\int_{\Omega} e^T(x,t)\Bigg[-\sum_{k=1}^{q}\frac{1}{l_k^2}\mathbb{W} - P(I_N\otimes A) + \frac{cZ}{1-\sigma} - (I_N\otimes A)P \\
&+cP(G\otimes\Gamma)Z^{-1}(G^T\otimes\Gamma)P + P\left(I_N\otimes(BB^T)\right)P \\
&+I_N\otimes\Theta\Bigg]e(x,t)dx + 2\int_{\Omega} e^T(x,t)P(I_N\otimes E)u(x,t)dx,
\end{aligned}
\tag{9.40}
$$

where $\mathbb{W} = P(I_N\otimes D) + (I_N\otimes D)P$.

From (9.40), we have

$$
\begin{aligned}
&\dot{V}_2(t) - 2\int_{\Omega} y^T(x,t)Qu(x,t)dx - \int_{\Omega} u^T(x,t)Wu(x,t)dx \\
\leqslant &\int_{\Omega} e^T(x,t)\Bigg[-\sum_{k=1}^{q}\frac{1}{l_k^2}\mathbb{W} - P(I_N\otimes A) + \frac{cZ}{1-\sigma} - (I_N\otimes A)P + I_N\otimes\Theta \\
&+cP(G\otimes\Gamma)Z^{-1}(G^T\otimes\Gamma)P + P\left(I_N\otimes(BB^T)\right)P\Bigg]e(x,t)dx \\
&+\int_{\Omega} u^T(x,t)\big[-Q^T(I_N\otimes H) - W - (I_N\otimes H^T)Q\big]u(x,t)dx \\
&+2\int_{\Omega} e^T(x,t)P(I_N\otimes E)u(x,t)dx \\
&-2\int_{\Omega} e^T(x,t)(I_N\otimes F^T)Qu(x,t)dx \\
= &\int_{\Omega} \vartheta^T(x,t)\begin{pmatrix}\chi_5 & \chi_1\\ \chi_1^T & \chi_2\end{pmatrix}\vartheta(x,t)dx,
\end{aligned}
$$

where $\vartheta(x,t) = (e^T(x,t), u^T(x,t))^T$, $\chi_2 = -Q^T(I_N\otimes H) - (I_N\otimes H^T)Q - W$.

By (9.39), one obtains

$$
\int_{\Omega} y^T(x,t)Qu(x,t)dx \geqslant \dot{S}(t) - \int_{\Omega} u^T(x,t)\frac{W}{2}u(x,t)dx, \tag{9.41}
$$

where $S(t) = \frac{V_2(t)}{2}$.

From (9.41), one has

$$\int_{t_0}^{t_p}\int_{\Omega} y^T(x,t)Qu(x,t)dxdt \geqslant S(t_p) - S(t_0) - \int_{t_0}^{t_p}\int_{\Omega} u^T(x,t)\frac{W}{2}u(x,t)dxdt$$

for any $t_p, t_0 \in [0, +\infty)$ and $t_p \geqslant t_0$.

By utilizing the same method as in proving Theorem 9.9, we can get the following conclusions.

Theorem 9.10 *Let* $\dot{\tau}(t) \leqslant \sigma < 1$. *If there exist matrices* $0 < P \in \mathbb{R}^{nN\times nN}$, $0 < Z \in \mathbb{R}^{nN\times nN}$ *and* $Q \in \mathbb{R}^{mN\times pN}$ *such that*

$$P(I_N \otimes D) + (I_N \otimes D)P \geqslant 0, \tag{9.42}$$

$$\begin{pmatrix} \chi_5 & \chi_1 \\ \chi_1^T & -Q^T(I_N \otimes H) - (I_N \otimes H^T)Q \end{pmatrix} \leqslant 0, \tag{9.43}$$

where $\Theta = \mathrm{diag}(\rho_1^2, \rho_2^2, \ldots, \rho_n^2)$, $\chi_5 = -\sum_{k=1}^{q} \frac{1}{l_k^2}\big[P(I_N \otimes D) + (I_N \otimes D)P\big] - P(I_N \otimes A) + \frac{cZ}{1-\sigma} - (I_N \otimes A)P + cP(G \otimes \Gamma)Z^{-1}(G^T \otimes \Gamma)P + P(I_N \otimes (BB^T))P + I_N \otimes \Theta$, $\chi_1 = P(I_N \otimes E) - (I_N \otimes F^T)Q$, *then system* (9.37) *is passive.*

Theorem 9.11 *Let* $\dot{\tau}(t) \leqslant \sigma < 1$. *If there exist matrices* $0 < P \in \mathbb{R}^{nN\times nN}$, $0 < Z \in \mathbb{R}^{nN\times nN}$, $0 < W \in \mathbb{R}^{pN\times pN}$ *and* $Q \in \mathbb{R}^{mN\times pN}$ *such that*

$$P(I_N \otimes D) + (I_N \otimes D)P \geqslant 0, \tag{9.44}$$

$$\begin{pmatrix} \chi_5 & \chi_1 \\ \chi_1^T & -Q^T(I_N \otimes H) + W - (I_N \otimes H^T)Q \end{pmatrix} \leqslant 0, \tag{9.45}$$

where $\Theta = \mathrm{diag}(\rho_1^2, \rho_2^2, \ldots, \rho_n^2)$, $\chi_5 = -\sum_{k=1}^{q} \frac{1}{l_k^2}\big[P(I_N \otimes D) + (I_N \otimes D)P\big] - P(I_N \otimes A) + \frac{cZ}{1-\sigma} - (I_N \otimes A)P + cP(G \otimes \Gamma)Z^{-1}(G^T \otimes \Gamma)P + P(I_N \otimes (BB^T))P + I_N \otimes \Theta$, $\chi_1 = P(I_N \otimes E) - (I_N \otimes F^T)Q$, *then system* (9.37) *is input-strictly passive.*

Theorem 9.12 *Let* $\dot{\tau}(t) \leqslant \sigma < 1$. *If there exist matrices* $0 < P \in \mathbb{R}^{nN\times nN}$, $0 < Z \in \mathbb{R}^{nN\times nN}$, $0 < W \in \mathbb{R}^{mN\times mN}$ *and* $Q \in \mathbb{R}^{mN\times pN}$ *such that*

$$P(I_N \otimes D) + (I_N \otimes D)P \geqslant 0, \tag{9.46}$$

$$\begin{pmatrix} \chi_6 & \chi_3 \\ \chi_3^T & \chi_4 \end{pmatrix} \leqslant 0, \tag{9.47}$$

where $\Theta = \mathrm{diag}(\rho_1^2, \rho_2^2, \ldots, \rho_n^2)$, $\chi_6 = -\sum_{k=1}^{q} \frac{1}{l_k^2}\big[P(I_N \otimes D) + (I_N \otimes D)P\big] - P(I_N \otimes A) + \frac{cZ}{1-\sigma} - (I_N \otimes A)P + cP(G \otimes \Gamma)Z^{-1}(G^T \otimes \Gamma)P + P(I_N \otimes (BB^T))P + I_N \otimes \Theta + (I_N \otimes F^T)W(I_N \otimes F)$, $\chi_3 = P(I_N \otimes E) - (I_N \otimes F^T)Q + (I_N \otimes$

$F^T)W(I_N \otimes H)$, $\chi_4 = -Q^T(I_N \otimes H) - (I_N \otimes H^T)Q + (I_N \otimes H^T)W(I_N \otimes H)$, *then system* (9.37) *is output-strictly passive.*

Remark 9.13 In Theorems 9.9–9.12, several sufficient conditions ensuring dissipativity and passivity of the delayed CRDNNs with different dimensions of input and output are obtained. This is the first to analyze dissipativity and passivity of delayed CRDNNs with different dimensions of input and output, which is a very important and interesting problem.

9.3.3 The Stability of Passive Delayed CRDNNs

Next, we analyze stability of passive delayed CRDNNs with different dimensions of input and output.

Theorem 9.14 $S : [0, +\infty) \rightarrow [0, +\infty)$ *is continuously differentiable and satisfies*

$$v_1(\|e(\cdot, t)\|_2) \leqslant S(t) \leqslant v_2(\|e(\cdot, t)\|_\tau), \tag{9.48}$$

where $v_1, v_2 : [0, +\infty) \rightarrow [0, +\infty)$ *are continuous and strictly monotonically increasing functions,* $v_1(\beta)$ *and* $v_2(\beta)$ *are positive for* $\beta > 0$ *with* $v_1(0) = v_2(0) = 0$. *Then*

(i) If system (9.37) *is passive with respect to* $S(t)$, *CRDNNs* (9.32) *is internally stable;*

(ii) If system (9.37) *is input-strictly passive with respect to* $S(t)$, *CRDNNs* (9.32) *is internally stable;*

(iii) If system (9.37) *is output-strictly passive with respect to* $S(t)$ *and* $F^T F > 0$, *CRDNNs* (9.32) *is internally asymptotically stable.*

Proof By utilizing the same method as in proving Theorem 9.6, we can get these conclusions.

Remark 9.15 To our knowledge, the stability of passive delayed CRDNNs with different dimensions of input and output has not yet been studied. In this subsection, we prove that passive (or input-strictly passive) system (9.37) is internally stable, and output-strictly passive system (9.37) is internally asymptotically stable under the condition that $F^T F > 0$ [see Theorem 9.14].

9.4 Numerical Examples

Two examples and their simulations are given in this section to show the correctness of the proposed output strict passivity and asymptotic stability criteria for CRDNNs with and without coupling delay.

Example 9.16 Consider the following CRDNNs:

$$\frac{\partial z_i(x,t)}{\partial t} = D\frac{\partial^2 z_i(x,t)}{\partial x^2} - Az_i(x,t) + Bf(z_i(x,t)) + 0.3\sum_{j=1}^{5} G_{ij}\Gamma z_j(x,t) + Eu_i(x,t), \tag{9.49}$$

$$z_i(x,0) = i \times \begin{pmatrix} 0.5\cos(\pi x) \\ 0.3\cos(\pi x) \\ 0.6\cos(\pi x) \end{pmatrix}, x \in \Omega,$$

$$z_i(x,t) = 0, \qquad (x,t) \in \partial\Omega \times [0,+\infty),$$

where $i = 1, 2, \ldots, 5$, $f_k(\eta) = \frac{|\eta+1|-|\eta-1|}{4}$, $k = 1, 2, 3$, $\Omega = \{x \mid -0.5 < x < 0.5\}$, and D, A, Γ, E, B, G are chosen as

$$D = \begin{pmatrix} 0.3 & 0 & 0 \\ 0 & 0.2 & 0 \\ 0 & 0 & 0.4 \end{pmatrix}, A = \begin{pmatrix} 0.5 & 0 & 0 \\ 0 & 0.6 & 0 \\ 0 & 0 & 0.4 \end{pmatrix}, \Gamma = \begin{pmatrix} 0.3 & 0 & 0 \\ 0 & 0.1 & 0 \\ 0 & 0 & 0.2 \end{pmatrix},$$

$$E = \begin{pmatrix} 0.5 & 0.2 \\ 0.6 & 0.3 \\ 0.1 & 0.4 \end{pmatrix}, B = \begin{pmatrix} 0.2 & 0.4 & 0.5 \\ 0.3 & 0.6 & 0.4 \\ 0.1 & 0.3 & 0.3 \end{pmatrix},$$

$$G = \begin{pmatrix} -0.1 & 0 & 0.1 & 0 & 0 \\ 0.2 & -0.5 & 0.3 & 0 & 0 \\ 0 & 0 & -0.4 & 0 & 0.4 \\ 0.2 & 0 & 0 & -0.5 & 0.3 \\ 0 & 0 & 0 & 0.5 & -0.5 \end{pmatrix}.$$

It is obvious that $(0,0,0)^T \in \mathbb{R}^3$ is an equilibrium solution of an isolated node of the network (9.49). Moreover, $f_k(\cdot)$, $k = 1, 2, 3$, are Lipschitz continuous with the Lipschitz constant $\rho_k = 0.5$.

The output vector $y_i(x,t) \in \mathbb{R}^3$ of network (9.49) is chosen as

$$y_i(x,t) = Fz_i(x,t) + Hu_i(x,t), \ i = 1, 2, \ldots, 5,$$

where

$$F = \begin{pmatrix} 0.2 & 0 & 0 \\ 0 & 0.8 & 0 \\ 0 & 0 & 0.4 \end{pmatrix}, H = \begin{pmatrix} 0.2 & 0.3 \\ 0.5 & 0.1 \\ 0.6 & 0.2 \end{pmatrix}.$$

Employing MATLAB YALMIP Toolbox to solve (9.20) and (9.21), one obtains

$$P = I_5 \otimes \begin{pmatrix} 0.4898 & -0.0764 & -0.0151 \\ -0.0764 & 0.7477 & -0.0509 \\ -0.0151 & -0.0509 & 0.4844 \end{pmatrix} > 0,$$
$$W = I_5 \otimes \begin{pmatrix} 1.3888 & -0.2869 & -0.1554 \\ -0.2869 & 0.9812 & -0.1731 \\ -0.1554 & -0.1731 & 1.4362 \end{pmatrix} > 0,$$
$$Q = I_5 \otimes \begin{pmatrix} -0.5295 & 1.9331 \\ 0.9066 & 0.0295 \\ 0.9543 & -0.0347 \end{pmatrix}. \tag{9.50}$$

By Theorem 9.4, CRDNNs (9.49) is output-strictly passive. The evolutions of states and outputs of five nodes are depicted in Figs. 9.1 and 9.2, respectively.

By the definition of $V_1(t)$ [see (9.10)], one has

$$\lambda_m(P)\|e(\cdot,t)\|_2^2 \leqslant V_1(t) \leqslant \lambda_M(P)\|e(\cdot,t)\|_2^2.$$

From Theorem 9.4, CRDNNs (9.49) is output-strictly passive with respect to $S(t) = \frac{V_1(t)}{2}$. On the other hand, it is obvious that $F^T F$ is positive definite. Therefore, by

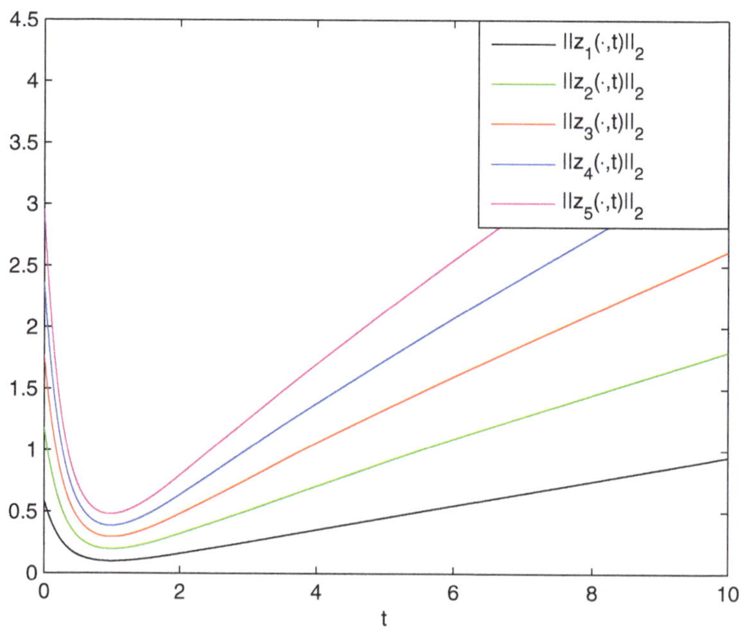

Fig. 9.1 Time evolutions of $\|z_i(\cdot,t)\|_2$, where $u_{i1}(x,t) = 0.2 \times i \times t\cos(\pi x)$, $u_{i2}(x,t) = 0.4 \times i \times t\cos(\pi x)$, $i = 1, 2, \ldots, 5$

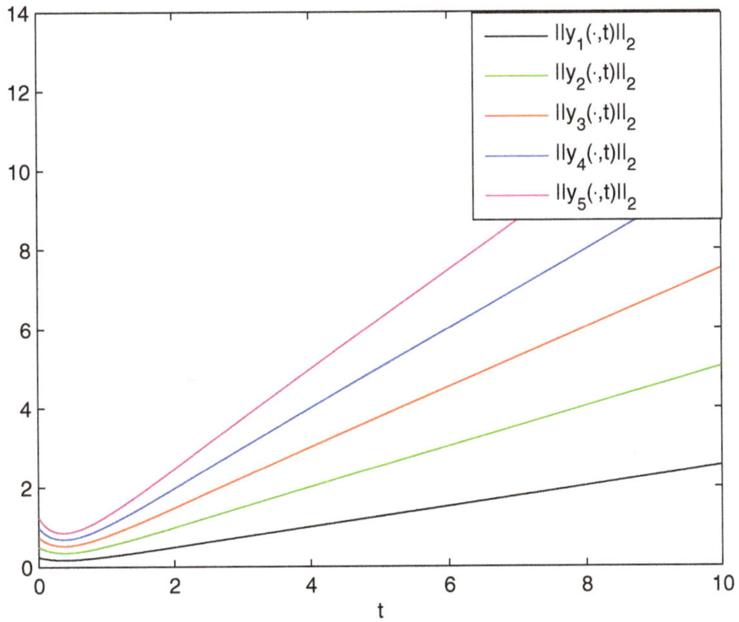

Fig. 9.2 Time evolutions of $\|y_i(\cdot,t)\|_2$, where $u_{i1}(x,t) = 0.2 \times i \times t\cos(\pi x)$, $u_{i2}(x,t) = 0.4 \times i \times t\cos(\pi x)$, $i = 1, 2, \ldots, 5$

Theorem 9.6, CRDNNs (9.49) is internally asymptotically stable. The evolutions of states and outputs of five nodes are depicted in Figs. 9.3 and 9.4, respectively.

Example 9.17 Consider the following delayed CRDNNs:

$$\begin{aligned}
\frac{\partial z_i(x,t)}{\partial t} &= D\frac{\partial^2 z_i(x,t)}{\partial x^2} - Az_i(x,t) + Bf(z_i(x,t)) + Eu_i(x,t) \\
&\quad + 0.2\sum_{j=1}^{5} G_{ij}\Gamma z_j(x,t-\tau(t)),\ i = 1, 2, \ldots, 5, \qquad (9.51)\\
z_i(x,t) &= i \times \begin{pmatrix} 0.4\cos(\pi x) \\ 0.6\cos(\pi x) \\ 0.2\cos(\pi x) \end{pmatrix},\ (x,t) \in \Omega \times [-0.3, 0], \\
z_i(x,t) &= 0, \qquad (x,t) \in \partial\Omega \times [-0.3, +\infty),
\end{aligned}$$

where $f_k(\eta) = \frac{|\eta+1|-|\eta-1|}{4}$, $k = 1, 2, 3$, $\Omega = \{x \mid -0.5 < x < 0.5\}$, $\tau(t) = 0.3 - 0.3e^{-0.4t}$, and the matrices D, A, Γ, E, B, G are chosen as

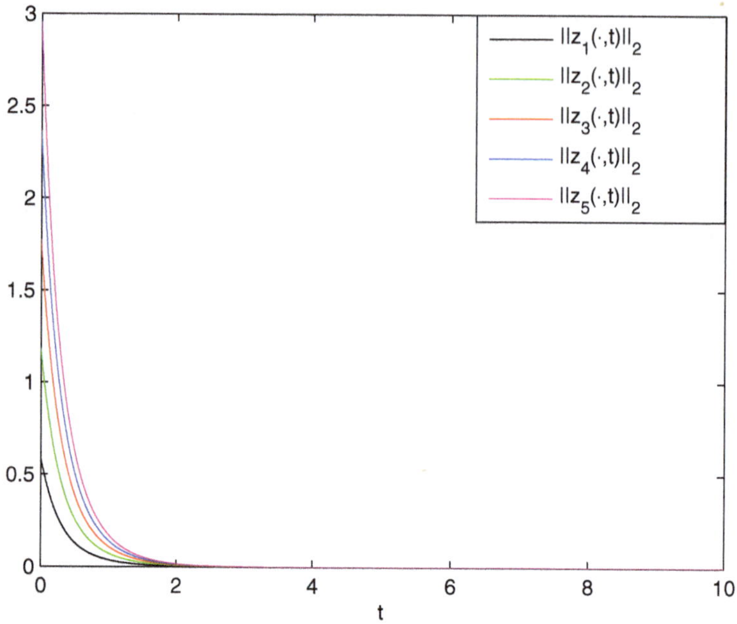

Fig. 9.3 Time evolutions of $\|z_i(\cdot, t)\|_2$, where $u_{i1}(x, t) = u_{i2}(x, t) = 0, i = 1, 2, \ldots, 5$

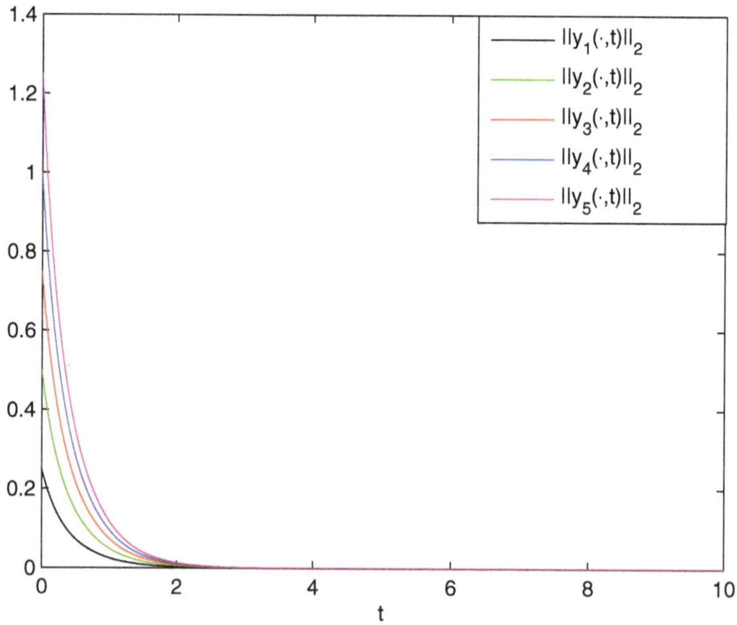

Fig. 9.4 Time evolutions of $\|y_i(\cdot, t)\|_2$, where $u_{i1}(x, t) = u_{i2}(x, t) = 0, i = 1, 2, \ldots, 5$

$$D=\begin{pmatrix}0.4&0&0\\0&0.2&0\\0&0&0.3\end{pmatrix},\ A=\begin{pmatrix}0.6&0&0\\0&0.4&0\\0&0&0.5\end{pmatrix},\ \Gamma=\begin{pmatrix}0.4&0&0\\0&0.1&0\\0&0&0.6\end{pmatrix},$$

$$E=\begin{pmatrix}0.3&0.6\\0.1&0.3\\0.5&0.2\end{pmatrix},\ B=\begin{pmatrix}0.1&0.3&0.6\\0.4&0.2&0.5\\0.6&0.4&0.1\end{pmatrix},$$

$$G=\begin{pmatrix}-0.2&0&0.2&0&0\\0.4&-0.7&0.3&0&0\\0&0&-0.3&0&0.3\\0.4&0&0&-0.6&0.2\\0&0&0&0.5&-0.5\end{pmatrix}.$$

It is obvious that $(0,0,0)^T\in\mathbb{R}^3$ is an equilibrium solution of an isolated node of the network (9.51), $0\leqslant\tau(t)\leqslant\tau=0.3$ and $\dot{\tau}(t)\leqslant\sigma=0.12<1$. Moreover, $f_k(\cdot), k=1,2,3,$ are Lipschitz continuous with the Lipschitz constant $\rho_k=0.5$.

The output vector $y_i(x,t)\in\mathbb{R}^3$ of network (9.51) is chosen as

$$y_i(x,t)=Fz_i(x,t)+Hu_i(x,t),\ i=1,2,\ldots,5,$$

where

$$F=\begin{pmatrix}0.3&0&0\\0&0.5&0\\0&0&0.6\end{pmatrix},\ H=\begin{pmatrix}0.1&0.4\\0.4&0.3\\0.5&0.2\end{pmatrix}.$$

Employing MATLAB YALMIP Toolbox to solve (9.46) and (9.47), one obtains

$$P=I_5\otimes\begin{pmatrix}0.5614&-0.0645&-0.0430\\-0.0645&0.9983&-0.1036\\-0.0430&-0.1036&0.8182\end{pmatrix}>0,$$

$$Z=I_5\otimes\begin{pmatrix}1.6288&-0.0387&-0.0245\\-0.0387&1.5726&-0.0444\\-0.0245&-0.0444&1.6013\end{pmatrix}>0,$$

$$W=I_5\otimes\begin{pmatrix}1.6660&-0.2695&-0.3024\\-0.2695&1.3421&-0.3165\\-0.3024&-0.3165&1.6347\end{pmatrix}>0,$$

$$Q=I_5\otimes\begin{pmatrix}-0.6798&1.9111\\0.4933&0.4068\\1.5267&-0.0521\end{pmatrix}. \tag{9.52}$$

By Theorem 9.12, CRDNNs (9.51) is output-strictly passive. The evolutions of states and outputs of five nodes are depicted in Figs. 9.5 and 9.6, respectively.

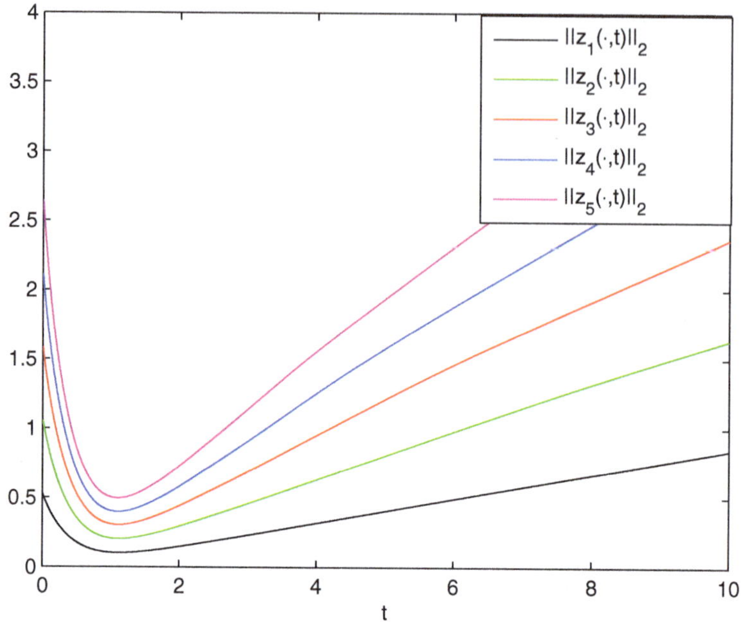

Fig. 9.5 Time evolutions of $\|z_i(\cdot, t)\|_2$, where $u_{i1}(x, t) = 0.2 \times i \times t\cos(\pi x)$, $u_{i2}(x, t) = 0.4 \times i \times t\cos(\pi x)$, $i = 1, 2, \ldots, 5$

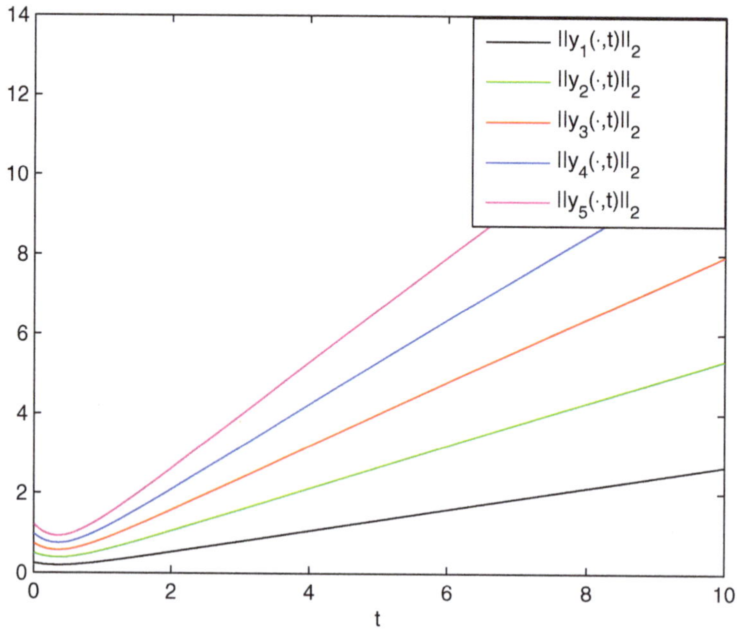

Fig. 9.6 Time evolutions of $\|y_i(\cdot, t)\|_2$, where $u_{i1}(x, t) = 0.2 \times i \times t\cos(\pi x)$, $u_{i2}(x, t) = 0.4 \times i \times t\cos(\pi x)$, $i = 1, 2, \ldots, 5$

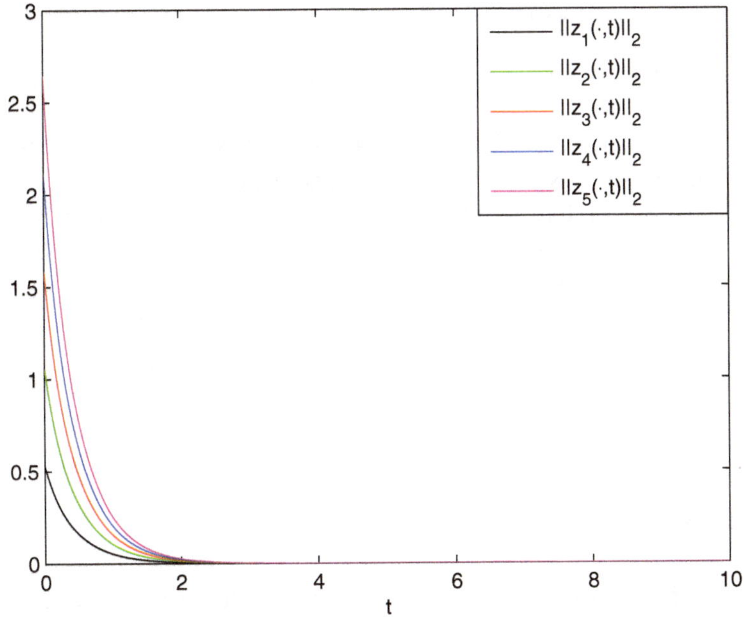

Fig. 9.7 Time evolutions of $\|z_i(\cdot, t)\|_2$, where $u_{i1}(x, t) = u_{i2}(x, t) = 0, i = 1, 2, \ldots, 5$

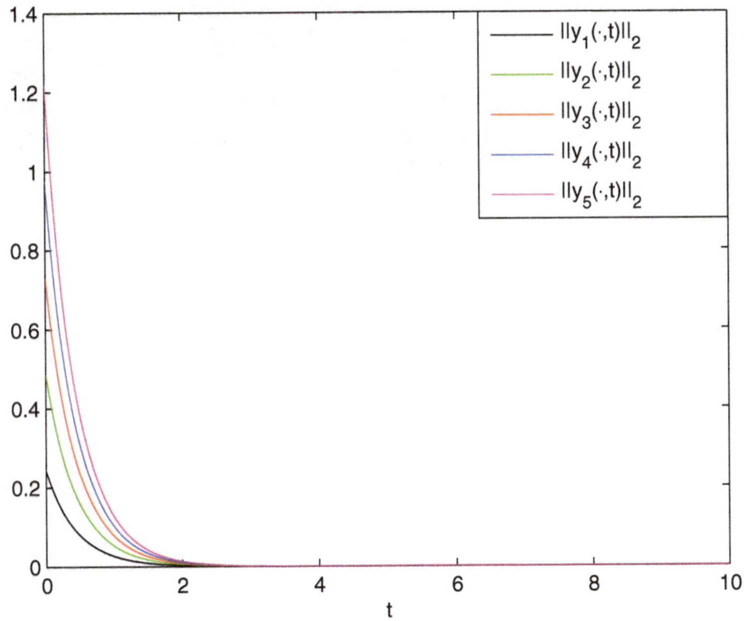

Fig. 9.8 Time evolutions of $\|y_i(\cdot, t)\|_2$, where $u_{i1}(x, t) = u_{i2}(x, t) = 0, i = 1, 2, \ldots, 5$

By the definition of $V_2(t)$, one has

$$\lambda_m(P)\|e(\cdot,t)\|_2^2 \leqslant V_2(t) \leqslant \kappa\|e(\cdot,t)\|_\tau^2,$$

where $\kappa = \lambda_M(P) + \frac{c\tau\lambda_M(Z)}{1-\sigma}$. From Theorem 9.12, CRDNNs (9.51) is output-strictly passive with respect to $S(t) = \frac{V_2(t)}{2}$. On the other hand, it is obvious that $F^T F$ is positive definite. Therefore, by Theorem 9.14, CRDNNs (9.51) is internally asymptotically stable. The evolutions of states and outputs of five nodes are depicted in Figs. 9.7 and 9.8, respectively.

From Figs. 9.3, 9.4, 9.7 and 9.8, we clearly see that $\|z_i(\cdot,t)\|_2$ and $\|y_i(\cdot,t)\|_2 (i = 1, 2, \ldots, 5)$ asymptotically converge to zero when $u_{i1}(x,t) = u_{i2}(x,t) = 0, i = 1, 2, \ldots, 5$. These results show that output-strictly passive CRDNNs (9.49) and (9.51) are internally asymptotically stable.

9.5 Conclusion

Two kinds of CRDNNs have been proposed, in which input and output vectors have different dimensions. On the one hand, we have analyzed the dissipativity and passivity of the proposed network models. By exploiting some inequality techniques, several sufficient conditions ensuring dissipativity and passivity of the CRDNNs have been established. On the other hand, the stability of passive CRDNNs has been studied. We have proved that passive (or input-strictly passive) CRDNNs is internally stable, and output-strictly passive CRDNNs is internally asymptotically stable under the condition that $F^T F > 0$. Finally, two numerical examples have been presented to verify the correctness of the proposed criteria.

Chapter 10
Passivity of Directed and Undirected CRDNNs with Adaptive Coupling Weights

10.1 Introduction

In 1972, Willems [105, 183] initially proposed the concept of dissipativity for ordinary differential equation systems, which is described by supply rate and storage function. In [106, 107, 177, 191, 192], Moylan, Hill, Byrnes, Zhao et al. further discussed the dissipativity of systems, and some important results on dissipativity were established. Especially, passivity [176] has received much attention in recent years, which is defined by dissipativity through an appropriate choice of the supply rate. More recently, the passivity theory has found successful applications in diverse areas such as stability [193], signal processing [178], complexity [194], group coordination [180], power control [186], chaos control and synchronization [179, 195], energy management [187], fuzzy control [196], flow control [185], and so on. Therefore, study of passivity is very important and interesting both in theory and application. So far, a great many interesting results on passivity have been derived for various ordinary differential equation systems.

Many systems in nature and society (for instance, cellular networks, social networks, power grids, communication networks, World Wide Web, food webs, disease transmission networks, metabolic systems, and so on [34, 35]) can be described by complex networks. Therefore, it is important and interesting to study the dynamical behavior of complex networks. Especially, the passivity properties for complex dynamical networks have received much of the focus. In [43], the passivity of a complex spatio-temporal switching network was considered, and several criteria for passivity were obtained. Yao et al. [50] respectively discussed the passivity problem of complex dynamical networks with and without coupling delays. In [74], Wang et al. analyzed the passivity for a complex network with general topology and multiple time-varying delays. Su and Shen [197] considered the passivity of complex dynamical networks with time-varying delay and sampled-data control. Moreover, by employing the passivity property of each node, several criteria for synchronization of complex dynamical networks were derived by designing appropriate control inputs [181, 182, 198–200]. In [181, 182, 198, 199], the authors studied output

J.-L. Wang et al., *Analysis and Control of Coupled Neural Networks with Reaction-Diffusion Terms*, DOI 10.1007/978-981-10-4907-1_10

synchronization of multi-agent systems under the assumption that all the agents in the network are passive systems. In [200], some sufficient conditions for synchronization in networks of neuronal oscillators were established by using the theory of semi-passive and passive systems.

However, the complex network models considered in these existing works [43, 50, 74, 181, 182, 197–202] are described by ordinary differential equations. Practically, some realistic networks should be modeled by partial differential equation systems, such as the food webs, neural networks, coupled neural networks, etc. To our knowledge, very few researchers have discussed the passivity of such complex dynamical networks [29, 95, 96, 167]. In [167], the authors analyzed the passivity of a parabolic complex network model with time-varying delays and parametric uncertainties. Wang et al. [29] respectively investigated the passivity of RDNNs with and without parametric uncertainties. In [95], the authors respectively discussed the synchronization and output strict passivity of the CRDNNs. However, in many circumstances, CRDNNs with known coupling weights are not passive. In [96], the authors investigated the passivity of the CRDNNs with adaptive coupling weights under the assumption that outer coupling matrix is symmetric.

It is well known that input and output vectors in many systems have different dimensions [203, 204]. But in most existing works on the passivity, they always assume that the input vector has the same dimension as the output vector. To the best of our knowledge, there are few works on the passivity of complex dynamical networks with different dimensions of input and output. Especially, the passivity of spatially and temporally complex dynamical networks with different dimensions of input and output has not yet been investigated. On the other hand, in many real-world networks, the coupling weights are adaptively adjusted according to changes in the environment or the network itself (e.g., neural networks, biological networks, wireless sensor networks) [128, 139]. In [205], a direct adaptive neural network control strategy was introduced by utilizing the backstepping method. In order to guarantee a desired system performance, a adaptive neural network tracking controller was designed in [206]. Hence, it is important to investigate the passivity of spatially and temporally complex networks with adaptive coupling weights, in which the input and output vectors have different dimensions. Moreover, the passivity theory [207] has long been a nice tool for analyzing synchronization [208, 209] of complex networks. Unfortunately, for spatially and temporally complex dynamical networks with different dimensions of input and output, the relationship between passivity and synchronization has not yet been considered.

In this chapter, we investigate the adaptive passivity of coupled neural networks with reaction-diffusion terms, in which the input and output have different dimensions. The main contributions of this chapter are as follows. First, we give the passivity definitions for the case where input and output vectors have different dimensions, which generalize some existing concepts of passivity. Second, several sufficient conditions ensuring passivity of the CRDNNs with directed and undirected topologies are obtained by utilizing the designed adaptive laws. Third, we reveal the relationship between output strict passivity and synchronization of the CRDNNs with adaptive coupling weights.

10.2 Network Model

In this section, we consider the following CRDNNs model:

$$\frac{\partial z_i(x,t)}{\partial t} = D\Delta z_i(x,t) - Az_i(x,t) + J + Bf(z_i(x,t)) + c\sum_{j=1}^{N} G_{ij}(t)\Gamma z_j(x,t) + Eu_i(x,t), \tag{10.1}$$

where $i = 1, 2, \ldots, N$, N is the number of RDNNs in the network; $x = (x_1, x_2, \ldots, x_q)^T \in \Omega \subset \mathbb{R}^q$; $\Delta = \sum_{k=1}^{q} \frac{\partial^2}{\partial x_k^2}$ is the Laplace diffusion operator on Ω; $z_i(x,t) = (z_{i1}(x,t), z_{i2}(x,t), \ldots, z_{in}(x,t))^T \in \mathbb{R}^n$ is the state vector of the ith neural network at time t and in space x; $u_i(x,t) \in \mathbb{R}^p$ is the control input; $D = \text{diag}(d_1, d_2, \ldots, d_n) > 0$, and d_i is the transmission diffusion coefficient along the ith neuron; $A = \text{diag}(a_1, a_2, \ldots, a_n) > 0$, and a_i is the rate with which the ith neuron will reset its potential to the resting state when disconnected from the neural network and external input; $J = (J_1, J_2, \ldots, J_n)^T$ is a constant external input vector; $B = (b_{ij})_{n\times n} \in \mathbb{R}^{n\times n}$, and b_{ij} is the strength of the jth neuron on the ith neuron; E is a known constant matrix with appropriate dimensions; $f(z_i(x,t)) = (f_1(z_{i1}(x,t)), f_2(z_{i2}(x,t)), \ldots, f_n(z_{in}(x,t)))^T$, and $f_j(\cdot)$ is the activation function of the jth neuron; c is a positive real number, which represents the overall coupling strength; $\Gamma \in \mathbb{R}^{n\times n} > 0$ is inner coupling matrix; $G(t) = (G_{ij}(t))_{N\times N}$ is a time-varying matrix, where $G_{ij}(t)$ is defined as follows: if there exists a connection from neural network i to neural network $j (i \neq j)$ at time t, then $G_{ij}(t) > 0$; otherwise, $G_{ij}(t) = 0 (i \neq j)$, and the diagonal elements of matrix $G(t)$ are defined by

$$G_{ii}(t) = -\sum_{\substack{j=1 \\ j\neq i}}^{N} G_{ij}(t), i = 1, 2, \ldots, N.$$

The initial value and boundary value conditions associated with network (10.1) are given in the form

$$z_i(x,0) = \Phi_i(x) \in \mathbb{R}^n, \quad x \in \Omega, \tag{10.2}$$

$$z_i(x,t) = 0, \quad (x,t) \in \partial\Omega \times [0, +\infty), \tag{10.3}$$

where $\Phi_i(x)$ is bounded and continuous on Ω.

Remark 10.1 In recent years, the synchronization problem of coupled neural networks has stirred much research interest due to its fruitful applications in various fields [76–79]. Unfortunately, in most existing works, the diffusion effects have not been considered [210]. Strictly speaking, diffusion effects cannot be avoided in neural networks when electrons are moving in asymmetric electromagnetic fields.

However, very few researchers have investigated the coupled neural networks with reaction-diffusion terms.

Remark 10.2 In this section, we always assume CRDNNs (10.1) is strongly connected and the topological structure of CRDNNs (10.1) is fixed. That is, if there is no connection from neural network i to neural network $j (i \neq j)$, then $G_{ij}(t) = 0$ for all t.

Next, we give the definition of synchronization for the CRDNNs (10.1).

Definition 10.3 The CRDNNs (10.1) is said to achieve synchronization if

$$\lim_{t \to +\infty} \left\| z_i(\cdot, t) - z_j(\cdot, t) \right\|_2 = 0 \ \text{ for all } i, j = 1, 2, \ldots, N$$

under the condition that $u_i(x, t) = 0, i = 1, 2, \ldots, N$.

10.3 Passivity of CRDNNs with Directed Topology and Adaptive Coupling Weights

According to Lemmas 1.11 and 1.19, there obviously exists a positive vector $\theta = (\theta_1, \theta_2, \ldots, \theta_N)^T \in \mathbb{R}^N$ such that $\theta^T G(0) = 0$ and

$$\sum_{j=1}^{N} \hat{G}_{ij} = \sum_{j=1}^{N} \hat{G}_{ji} = 0 \text{ for all } i = 1, 2, \ldots, N,$$

where $\Xi = \operatorname{diag}(\theta_1, \theta_2, \ldots, \theta_N)$, $\hat{G} = (\hat{G}_{ij})_{N \times N} = \Xi G(0) + G^T(0)\Xi$.

Let $\xi_i = \frac{\theta_i}{\theta_1 + \theta_2 + \cdots + \theta_N}, i = 1, 2, \ldots, N$. Then, we have $(\xi_1, \xi_2, \ldots, \xi_N)G(0) = 0$ and

$$G_{ii}(0) = -\frac{\sum_{\substack{j=1 \\ j \neq i}}^{N} \left(\xi_i G_{ij}(0) + \xi_j G_{ji}(0) \right)}{2\xi_i}, i = 1, 2, \ldots, N. \tag{10.4}$$

Use the following adaptive law:

$$\dot{G}_{ij}(t) = \begin{cases} \beta_{ij}\xi_j \int_{\Omega} \left(z_i(x, t) - z_j(x, t) \right)^T \Gamma \left(z_i(x, t) \right. \\ \left. -z_j(x, t) \right) dx, \text{ if } (i, j) \in \mathcal{E} \text{ and } (j, i) \in \mathcal{E}, \\ -\sum_{\substack{l=1 \\ l \neq i}}^{N} \dot{G}_{il}(t), \qquad \text{if } i = j, \\ 0, \qquad \text{otherwise}, \end{cases} \tag{10.5}$$

where $\beta_{ij} = \beta_{ji}$ are positive constants.

By (10.5), we can get

$$\sum_{j=1}^{N} \xi_j \dot{G}_{ji}(t) = \sum_{j=1}^{N} \xi_i \dot{G}_{ij}(t) = 0, \tag{10.6}$$

$$\dot{G}_{ii}(t) = -\frac{\sum_{\substack{j=1\\ j\neq i}}^{N} \left(\xi_i \dot{G}_{ij}(t) + \xi_j \dot{G}_{ji}(t)\right)}{2\xi_i}. \tag{10.7}$$

Since $(\xi_1, \xi_2, \ldots, \xi_N)G(0) = 0$, we can obtain

$$\sum_{j=1}^{N} \xi_j G_{ji}(t) = 0$$

for all t and $i \in \{1, 2, ..., N\}$.

Furthermore, we can derive from (10.4) and (10.7) that

$$G_{ii}(t) = -\frac{\sum_{\substack{j=1\\ j\neq i}}^{N} \left(\xi_i G_{ij}(t) + \xi_j G_{ji}(t)\right)}{2\xi_i} \tag{10.8}$$

for all t and $i \in \{1, 2, \ldots, N\}$.

Let $\bar{z}(x,t) = \sum_{i=1}^{N} \xi_i z_i(x,t)$. Then, we have

$$\begin{aligned}
\frac{\partial \bar{z}(x,t)}{\partial t} &= \sum_{i=1}^{N} \xi_i \frac{\partial z_i(x,t)}{\partial t} \\
&= \sum_{i=1}^{N} \xi_i D\Delta z_i(x,t) - \sum_{i=1}^{N} \xi_i A z_i(x,t) + \sum_{i=1}^{N} \xi_i J + \sum_{i=1}^{N} \xi_i E u_i(x,t) \\
&\quad + \sum_{i=1}^{N} \xi_i B f(z_i(x,t)) + c \sum_{i=1}^{N} \xi_i \sum_{j=1}^{N} G_{ij}(t) \Gamma z_j(x,t) \\
&= D\Delta \bar{z}(x,t) - A\bar{z}(x,t) + B \sum_{i=1}^{N} \xi_i f(z_i(x,t)) + E \sum_{i=1}^{N} \xi_i u_i(x,t) \\
&\quad + J + c \sum_{j=1}^{N} \left(\sum_{i=1}^{N} \xi_i G_{ij}(t) \right) \Gamma z_j(x,t) \\
&= D\Delta \bar{z}(x,t) - A\bar{z}(x,t) + B \sum_{i=1}^{N} \xi_i f(z_i(x,t)) \\
&\quad + J + E \sum_{i=1}^{N} \xi_i u_i(x,t).
\end{aligned}$$

Defining $e_i(x,t) = (e_{i1}(x,t), e_{i2}(x,t), \ldots, e_{in}(x,t))^T = z_i(x,t) - \bar{z}(x,t), i = 1, 2, \ldots, N$, then we have

$$\frac{\partial e_i(x,t)}{\partial t} = D\Delta e_i(x,t) - Ae_i(x,t) + Bf(z_i(x,t)) - B\sum_{i=1}^{N}\xi_i f(z_i(x,t)) + c\sum_{j=1}^{N} G_{ij}(t)\Gamma e_j(x,t) + Eu_i(x,t) - E\sum_{i=1}^{N}\xi_i u_i(x,t), \tag{10.9}$$

where $i = 1, 2, \ldots, N$.

In the following, we consider the passivity of system (10.9). The output vector $y_i(x, t)$ of system (10.9) is defined as

$$y_i(x,t) = Fe_i(x,t) + Hu_i(x,t),\ i = 1, 2, \ldots, N, \tag{10.10}$$

where $F \in \mathbb{R}^{m\times n}$ and $H \in \mathbb{R}^{m\times p}$ are known matrices.

For the convenience, we denote

$$A_{ij}(t) = \begin{cases} \xi_i G_{ij}(t) + \xi_j G_{ji}(t), \text{ if } (i,j) \text{ and } (j,i) \in \mathcal{E}, \\ -\sum_{\substack{l=1\\ l\neq i}}^{N} A_{il}(t), & \text{if } i = j, \\ 0, & \text{otherwise,} \end{cases}$$

$$M_{ij} = \begin{cases} \xi_i G_{ij}(0), \text{ if } (i,j) \in \mathcal{E} \text{ and } (j,i) \notin \mathcal{E}, \\ \xi_j G_{ji}(0), \text{ if } (j,i) \in \mathcal{E} \text{ and } (i,j) \notin \mathcal{E}, \\ -\sum_{\substack{l=1\\ l\neq i}}^{N} M_{il}, & \text{if } i = j, \\ 0, & \text{otherwise.} \end{cases}$$

$$\xi = \mathrm{diag}(\xi_1, \xi_2, \ldots, \xi_N),\ \Theta = \mathrm{diag}(\rho_1^2, \rho_2^2, \ldots, \rho_n^2).$$

10.3.1 Passivity Analysis of CRDNNs

Theorem 10.4 *The system (10.9) is passive under the adaptive law (10.5) if there exist matrices $Q \in \mathbb{R}^{mN\times pN}$ and $\kappa = (\kappa_{ij})_{N\times N}$ such that*

$$\begin{pmatrix} \chi_1 & \chi_2 \\ \chi_2^T & -Q^T(I_N \otimes H) - (I_N \otimes H^T)Q \end{pmatrix} \leqslant 0, \tag{10.11}$$

where $M = (M_{ij})_{N\times N}$, $\chi_1 = \xi\otimes\big(-\sum_{k=1}^{q}\frac{2}{l_k^2}D - 2A + BB^T + \Theta\big) + c(M+\kappa)\otimes\Gamma$, $\chi_2 = \xi\otimes E - (I_N\otimes F^T)Q$, $\kappa_{ii} = -\sum_{\substack{j=1\\ j\neq i}}^{N}\kappa_{ij}$, $\kappa_{ij} = \kappa_{ji}(i\neq j)$ are nonnegative constants, and $\kappa_{ij} = 0(i\neq j)$ if and only if $A_{ij}(t) = 0$.

Proof Define the following Lyapunov functional for system (10.9):

$$V_1(\hat{e}(x,t)) = \sum_{i=1}^{N} \xi_i \int_{\Omega} e_i^T(x,t) e_i(x,t) dx + c \sum_{i=1}^{N} \sum_{\substack{(i,j)\in\mathcal{E},\\(j,i)\in\mathcal{E}}} \frac{(\xi_i G_{ij}(t) + \xi_j G_{ji}(t) - \kappa_{ij})^2}{8\beta_{ij}\xi_i\xi_j},$$

where $\hat{e}(x,t) = (e^T(x,t), \underbrace{\ldots, G_{ij}(t), \ldots}_{(i,j)\text{ and }(j,i)\in\mathcal{E}})^T$, $e(x,t) = (e_1^T(x,t), e_2^T(x,t), \ldots, e_N^T(x,t))^T$.

Calculating the time derivative of $V_1(\hat{e}(x,t))$ along the trajectory of system (10.9), we can get

$$\begin{aligned}
\dot{V}_1(\hat{e}(x,t)) &= c \sum_{i=1}^{N} \sum_{\substack{(i,j)\in\mathcal{E},\\(j,i)\in\mathcal{E}}} \frac{\xi_i \dot{G}_{ij}(t) + \xi_j \dot{G}_{ji}(t)}{4\beta_{ij}\xi_i\xi_j} \big(\xi_i G_{ij}(t) + \xi_j G_{ji}(t) - \kappa_{ij}\big) \\
&\quad + 2 \sum_{i=1}^{N} \xi_i \int_{\Omega} e_i^T(x,t) \frac{\partial e_i(x,t)}{\partial t} dx \\
&= 2 \sum_{i=1}^{N} \xi_i \int_{\Omega} e_i^T(x,t) \Big[D\Delta e_i(x,t) - A e_i(x,t) + Bf(z_i(x,t)) \\
&\quad - Bf(\bar{z}(x,t)) + Bf(\bar{z}(x,t)) - B \sum_{i=1}^{N} \xi_i f(z_i(x,t)) + E u_i(x,t) \\
&\quad + c \sum_{j=1}^{N} G_{ij}(t) \Gamma e_j(x,t) - E \sum_{i=1}^{N} \xi_i u_i(x,t) \Big] dx \\
&\quad + \frac{c}{2} \sum_{i=1}^{N} \sum_{\substack{(i,j)\in\mathcal{E},\\(j,i)\in\mathcal{E}}} \big(\xi_i G_{ij}(t) + \xi_j G_{ji}(t) - \kappa_{ij}\big) \int_{\Omega} \Big(e_i(x,t) \\
&\quad - e_j(x,t)\Big)^T \Gamma \Big(e_i(x,t) - e_j(x,t)\Big) dx.
\end{aligned} \tag{10.12}$$

From Green's formula and the boundary condition, we have

$$\int_{\Omega} e_{il}(x,t) \Delta e_{il}(x,t) dx = - \sum_{k=1}^{q} \int_{\Omega} \left(\frac{\partial e_{il}(x,t)}{\partial x_k} \right)^2 dx,$$

where $i = 1, 2, \ldots, N, l = 1, 2, \ldots, n$.

According to Lemma 1.22, we can obtain

$$\begin{aligned}\int_{\Omega} e_i^T(x,t)D\Delta e_i(x,t)dx &= \sum_{l=1}^{n} d_l \int_{\Omega} e_{il}(x,t)\Delta e_{il}(x,t)dx \\ &= -\sum_{l=1}^{n} d_l \sum_{k=1}^{q} \int_{\Omega} \left(\frac{\partial e_{il}(x,t)}{\partial x_k}\right)^2 dx \\ &\leqslant -\sum_{k=1}^{q} \frac{1}{l_k^2} \sum_{l=1}^{n} d_l \int_{\Omega} e_{il}^2(x,t)dx \\ &= -\sum_{k=1}^{q} \frac{1}{l_k^2} \int_{\Omega} e_i^T(x,t)De_i(x,t)dx. \end{aligned} \tag{10.13}$$

Obviously

$$\begin{aligned}&2\sum_{i=1}^{N} \xi_i e_i^T(x,t)B\left[f(z_i(x,t)) - f(\bar{z}(x,t))\right] \\ &\leqslant \sum_{i=1}^{N} \xi_i e_i^T(x,t)(BB^T + \Theta)e_i(x,t). \end{aligned} \tag{10.14}$$

Furthermore

$$\begin{aligned}\sum_{i=1}^{N} \xi_i e_i(x,t) &= \sum_{i=1}^{N} \xi_i \Big(z_i(x,t) - \sum_{j=1}^{N} \xi_j z_j(x,t)\Big) \\ &= \sum_{i=1}^{N} \xi_i z_i(x,t) - \sum_{j=1}^{N} \left(\sum_{i=1}^{N} \xi_i\right) \xi_j z_j(x,t) \\ &= \sum_{i=1}^{N} \xi_i z_i(x,t) - \sum_{j=1}^{N} \xi_j z_j(x,t) \\ &= 0. \end{aligned} \tag{10.15}$$

From (10.15), we can get

$$\sum_{i=1}^{N} \xi_i e_i^T(x,t)\Big(Bf(\bar{z}(x,t)) - B\sum_{i=1}^{N} \xi_i f(z_i(x,t)) - E\sum_{i=1}^{N} \xi_i u_i(x,t)\Big) = 0. \tag{10.16}$$

On the other hand, we can easily derive

$$\sum_{i=1}^{N}\sum_{\substack{(i,j)\in\mathcal{E},\\(j,i)\in\mathcal{E}}}(\xi_i G_{ij}(t)+\xi_j G_{ji}(t)-\kappa_{ij})\Big(e_i(x,t)-e_j(x,t)\Big)^T\Gamma\Big(e_i(x,t)-e_j(x,t)\Big)$$
$$=-2\sum_{i=1}^{N}\sum_{j=1}^{N}(A_{ij}(t)-\kappa_{ij})e_i^T(x,t)\Gamma e_j(x,t). \tag{10.17}$$

It follows from (10.12)–(10.14), (10.16) and (10.17) that:

$$\begin{aligned}
&\dot{V}_1(\hat{e}(x,t))-2\int_{\Omega}y^T(x,t)Qu(x,t)dx\\
&\leqslant\sum_{i=1}^{N}\xi_i\int_{\Omega}e_i^T(x,t)\Big[-\sum_{k=1}^{q}\frac{2}{l_k^2}D-2A+BB^T+\Theta\Big]e_i(x,t)dx\\
&\quad+\int_{\Omega}u^T(x,t)\big[-Q^T(I_N\otimes H)-(I_N\otimes H^T)Q\big]u(x,t)dx\\
&\quad+c\sum_{i=1}^{N}\sum_{j=1}^{N}(\xi_i G_{ij}(t)+\xi_j G_{ji}(t))\int_{\Omega}e_i^T(x,t)\Gamma e_j(x,t)dx\\
&\quad-c\sum_{i=1}^{N}\sum_{j=1}^{N}(A_{ij}(t)-\kappa_{ij})\int_{\Omega}e_i^T(x,t)\Gamma e_j(x,t)dx\\
&\quad-2\int_{\Omega}e^T(x,t)(I_N\otimes F^T)Qu(x,t)dx\\
&\quad+2\sum_{i=1}^{N}\xi_i\int_{\Omega}e_i^T(x,t)Eu_i(x,t)dx\\
&=\sum_{i=1}^{N}\xi_i\int_{\Omega}e_i^T(x,t)\Big[-\sum_{k=1}^{q}\frac{2}{l_k^2}D-2A+BB^T+\Theta\Big]e_i(x,t)dx\\
&\quad+\int_{\Omega}u^T(x,t)\big[-Q^T(I_N\otimes H)-(I_N\otimes H^T)Q\big]u(x,t)dx\\
&\quad+c\sum_{i=1}^{N}\sum_{j=1}^{N}(M_{ij}+\kappa_{ij})\int_{\Omega}e_i^T(x,t)\Gamma e_j(x,t)dx\\
&\quad-2\int_{\Omega}e^T(x,t)(I_N\otimes F^T)Qu(x,t)dx\\
&\quad+2\sum_{i=1}^{N}\xi_i\int_{\Omega}e_i^T(x,t)Eu_i(x,t)dx\\
&=\int_{\Omega}\zeta^T(x,t)\begin{pmatrix}\chi_1 & \chi_2\\ \chi_2^T & -Q^T(I_N\otimes H)-(I_N\otimes H^T)Q\end{pmatrix}\zeta(x,t)dx,
\end{aligned}$$

where $\zeta(x,t) = (e^T(x,t), u^T(x,t))^T$, $u(x,t) = (u_1^T(x,t), u_2^T(x,t), \ldots, u_N^T(x,t))^T$, $y(x,t) = (y_1^T(x,t), y_2^T(x,t), \ldots, y_N^T(x,t))^T$.

Therefore, we can derive from (10.11) that

$$2\int_\Omega y^T(x,t)Qu(x,t)dx \geqslant \dot{V}_1(\hat{e}(x,t)). \tag{10.18}$$

By (10.18), we can obtain

$$\int_{t_0}^{t_p}\int_\Omega y^T(x,t)Qu(x,t)dxdt \geqslant V(t_p) - V(t_0)$$

for any $t_p, t_0 \in \mathbb{R}^+$ and $t_p \geqslant t_0$, where $V(t) = \frac{V_1(\hat{e}(x,t))}{2}$. The proof is completed.

By the similar proof of Theorem 10.4, we can obtain the following conclusions.

Theorem 10.5 *The system (10.9) is input-strictly passive under the adaptive law (10.5) if there exist matrices* $0 < W \in \mathbb{R}^{pN\times pN}$, $Q \in \mathbb{R}^{mN\times pN}$ *and* $\kappa = (\kappa_{ij})_{N\times N}$ *such that*

$$\begin{pmatrix} \chi_1 & \chi_2 \\ \chi_2^T & -Q^T(I_N\otimes H) - (I_N\otimes H^T)Q + W \end{pmatrix} \leqslant 0, \tag{10.19}$$

where $M = (M_{ij})_{N\times N}$, $\chi_1 = \xi\otimes\big(-\sum_{k=1}^q \frac{2}{l_k^2}D - 2A + BB^T + \Theta\big) + c(M+\kappa)\otimes\Gamma$, $\chi_2 = \xi\otimes E - (I_N\otimes F^T)Q$, $\kappa_{ii} = -\sum_{\substack{j=1\\ j\neq i}}^N \kappa_{ij}$, $\kappa_{ij} = \kappa_{ji} (i\neq j)$ *are nonnegative constants, and* $\kappa_{ij} = 0 (i \neq j)$ *if and only if* $A_{ij}(t) = 0$.

Theorem 10.6 *The system (10.9) is output-strictly passive under the adaptive law (10.5) if there exist matrices* $0 < W \in \mathbb{R}^{mN\times mN}$, $Q \in \mathbb{R}^{mN\times pN}$ *and* $\kappa = (\kappa_{ij})_{N\times N}$ *such that*

$$\begin{pmatrix} \chi_3 & \chi_5 \\ \chi_5^T & \chi_4 \end{pmatrix} \leqslant 0, \tag{10.20}$$

where $M = (M_{ij})_{N\times N}$, $\chi_3 = \xi\otimes\big(-\sum_{k=1}^q \frac{2}{l_k^2}D - 2A + BB^T + \Theta\big) + c(M+\kappa)\otimes\Gamma + (I_N\otimes F^T)W(I_N\otimes F)$, $\chi_4 = -Q^T(I_N\otimes H) - (I_N\otimes H^T)Q + (I_N\otimes H^T)W(I_N\otimes H)$, $\chi_5 = \xi\otimes E - (I_N\otimes F^T)Q + (I_N\otimes F^T)W(I_N\otimes H)$, $\kappa_{ii} = -\sum_{\substack{j=1\\ j\neq i}}^N \kappa_{ij}$, $\kappa_{ij} = \kappa_{ji}(i\neq j)$ *are nonnegative constants, and* $\kappa_{ij} = 0(i\neq j)$ *if and only if* $A_{ij}(t) = 0$.

Remark 10.7 To our knowledge, very few researchers have studied the passivity of CRDNNs with adaptive coupling weights [96]. Especially, the passivity of CRDNNs with directed topology and adaptive coupling weights has not yet been investigated. In this section, several sufficient conditions ensuring the passivity, input strict passivity and output strict passivity are derived for the CRDNNs (10.1) by utilizing the designed adaptive law (10.5) and constructing appropriate Lyapunov functional (see Theorems 10.4–10.6).

10.3.2 The Relationship Between Output Strict Passivity and Synchronization of CRDNNs

Theorem 10.8 *The CRDNNs (10.1) is synchronized under the adaptive law (10.5) if F is a nonsingular matrix and system (10.9) under the adaptive law (10.5) is output-strictly passive with respect to storage function*

$$V(t)=\frac{1}{2}\sum_{i=1}^{N}\xi_i\int_{\Omega}e_i^T(x,t)e_i(x,t)dx + c\sum_{i=1}^{N}\sum_{\substack{(i,j)\in\mathcal{E},\\(j,i)\in\mathcal{E}}}\frac{(\xi_iG_{ij}(t)+\xi_jG_{ji}(t)-\kappa_{ij})^2}{16\beta_{ij}\xi_i\xi_j},$$

where $\kappa_{ij}=\kappa_{ji}$ are positive constants.

Proof If system (10.9) is output-strictly passive with respect to storage function $V(t)$, then there exist matrices $0<Q\in\mathbb{R}^{nN\times nN}$ and $P\in\mathbb{R}^{nN\times pN}$ such that

$$\begin{aligned}\frac{V(t+\varepsilon)-V(t)}{\varepsilon}\leqslant&\frac{\int_t^{t+\varepsilon}\int_{\Omega}y^T(x,s)Pu(x,s)dxds}{\varepsilon}\\&-\frac{\int_t^{t+\varepsilon}\int_{\Omega}y^T(x,s)Qy(x,s)dxds}{\varepsilon}\end{aligned}\tag{10.21}$$

for any $t\in\mathbb{R}^+$ and $\varepsilon>0$, where $y(x,t)=(y_1^T(x,t),y_2^T(x,t),\dots,y_N^T(x,t))^T$, $u(x,t)=(u_1^T(x,t),u_2^T(x,t),\dots,u_N^T(x,t))^T$.

By taking limit $\varepsilon\to0$ in (10.21), we obtain

$$\begin{aligned}\dot V(t)&\leqslant\int_{\Omega}y^T(x,t)Pu(x,t)dx-\int_{\Omega}y^T(x,t)Qy(x,t)dx\\&\leqslant\int_{\Omega}y^T(x,t)Pu(x,t)dx-\lambda_m(Q)\|y(\cdot,t)\|_2^2.\end{aligned}\tag{10.22}$$

Letting $u(x,t)=0$, we have

$$\begin{aligned}\dot V(t)&\leqslant-\lambda_m(Q)\sum_{i=1}^{N}\int_{\Omega}e_i^T(x,t)F^TFe_i(x,t)dx\\&\leqslant-\gamma\sum_{i=1}^{N}\int_{\Omega}e_i^T(x,t)e_i(x,t)dx,\end{aligned}\tag{10.23}$$

where $\gamma=\lambda_m(Q)\lambda_m(F^TF)$.

Obviously, $V(t)$ is decreasing and the coupling weights $G_{ij}(t)$ $((i,j)$ and $(j,i) \in \mathcal{E})$ are bounded. Consequently, $\lim_{t\to+\infty} V(t)$ exists and is a nonnegative real number. Since $G_{ij}(t)$ $((i,j)$ and $(j,i) \in \mathcal{E})$ is monotonically increasing (see (10.5)), we have

$$\lim_{t\to+\infty} G_{ij}(t) = G_{ij}^* \in \mathbb{R} > 0, \ \ (i,j) \text{ and } (j,i) \in \mathcal{E}.$$

According to the definition of $V(t)$, we can derive that $\lim_{t\to+\infty} \sum_{i=1}^{N} \xi_i \int_\Omega e_i^T (x,t)e_i(x,t)dx$ exists and is a non-negative real number. In the following, we prove

$$\lim_{t\to+\infty} \sum_{i=1}^{N} \xi_i \int_\Omega e_i^T(x,t)e_i(x,t)dx = 0.$$

If this is not true, we have

$$\lim_{t\to+\infty} \sum_{i=1}^{N} \xi_i \int_\Omega e_i^T(x,t)e_i(x,t)dx = \mu > 0.$$

Then, there obviously exists a real number $T > 0$ such that

$$\sum_{i=1}^{N} \xi_i \int_\Omega e_i^T(x,t)e_i(x,t)dx > \frac{\mu}{2} \tag{10.24}$$

for $t \geqslant T$.

By (10.23) and (10.24), we can obtain

$$\dot{V}(t) < -\frac{\gamma\mu}{2\lambda_M(\xi)}, \quad t \geqslant T. \tag{10.25}$$

From (10.25), we have

$$-V(T) \leqslant V(+\infty) - V(T) = \int_T^{+\infty} \dot{V}(t)dt < -\int_T^{+\infty} \frac{\gamma\mu}{2\lambda_M(\xi)}dt = -\infty.$$

This yields a contradiction, and so

$$\lim_{t\to+\infty} \sum_{i=1}^{N} \xi_i \int_\Omega e_i^T(x,t)e_i(x,t)dx = 0. \tag{10.26}$$

By (10.26), we can easily obtain

$$\lim_{t\to+\infty}\|z_i(\cdot,t)-\sum_{j=1}^{N}\xi_j z_j(\cdot,t)\|_2=\lim_{t\to+\infty}\|e_i(\cdot,t)\|_2=0,$$

where $i=1,2,\ldots,N$. Therefore, the CRDNNs (10.1) achieves synchronization under the adaptive law (10.5). The proof is completed.

From Theorems 10.6 and 10.8, we can obtain the following conclusion.

Theorem 10.9 *The CRDNNs* (10.1) *is synchronized under the adaptive law (10.5) if there exist matrices* $0<W\in\mathbb{R}^{nN\times nN}$, $Q\in\mathbb{R}^{nN\times pN}$, $F\in\mathbb{R}^{n\times n}$, $H\in\mathbb{R}^{n\times p}$ *and* $\kappa=(\kappa_{ij})_{N\times N}$ *such that*

$$\begin{pmatrix}\chi_3 & \chi_5\\ \chi_5^T & \chi_4\end{pmatrix}\leqslant 0, \tag{10.27}$$

where F *is a nonsingular matrix,* $M=(M_{ij})_{N\times N}$, $\chi_3=\xi\otimes\big(-\sum_{k=1}^{q}\frac{2}{l_k^2}D-2A+BB^T+\Theta\big)+c(M+\kappa)\otimes\Gamma+(I_N\otimes F^T)W(I_N\otimes F)$, $\chi_4=-Q^T(I_N\otimes H)-(I_N\otimes H^T)Q+(I_N\otimes H^T)W(I_N\otimes H)$, $\chi_5=\xi\otimes E-(I_N\otimes F^T)Q+(I_N\otimes F^T)W(I_N\otimes H)$, $\kappa_{ii}=-\sum_{\substack{j=1\\ j\neq i}}^{N}\kappa_{ij}$, $\kappa_{ij}=\kappa_{ji}(i\neq j)$ *are nonnegative constants, and* $\kappa_{ij}=0(i\neq j)$ *if and only if* $A_{ij}(t)=0$.

10.4 Passivity of CRDNNs with Undirected Topology and Adaptive Coupling Weights

In this section, we consider the case that the coupling matrix $G(t)$ is symmetric in network (10.1). Thus, we have

$$\begin{aligned}\frac{\partial z_i(x,t)}{\partial t}&=D\Delta z_i(x,t)-Az_i(x,t)+J+Bf(z_i(x,t))\\&\quad+c\sum_{j=1}^{N}G_{ij}(t)\Gamma z_j(x,t)+Eu_i(x,t),\end{aligned} \tag{10.28}$$

where $i=1,2,\ldots,N$. $G(t)=(G_{ij}(t))_{N\times N}$ is a time-varying matrix, where $G_{ij}(t)$ is defined as follows: if there exists a connection between neural network i and neural network $j(i\neq j)$ at time t, then $G_{ij}(t)=G_{ji}(t)>0$; otherwise, $G_{ij}(t)=G_{ji}(t)=0(i\neq j)$, and the diagonal elements of matrix $G(t)$ are defined by

$$G_{ii}(t)=-\sum_{\substack{j=1\\ j\neq i}}^{N}G_{ij}(t),\quad i=1,2,\ldots,N.$$

Similarly, we also assume CRDNNs (10.28) is connected and the topological structure of CRDNNs (10.28) is fixed. Use the following adaptive law:

$$\dot{G}_{ij}(t) = \begin{cases} \beta_{ij} \int_{\Omega} \big(z_i(x,t) - z_j(x,t)\big)^T \Gamma \big(z_i(x,t) \\ \quad -z_j(x,t)\big)dx, & \text{if } (i,j) \in \mathcal{E}, \\ -\sum_{\substack{l=1 \\ l \neq i}}^{N} \dot{G}_{il}(t), & \text{if } i = j, \\ 0, & \text{otherwise}, \end{cases} \tag{10.29}$$

where $\beta_{ij} = \beta_{ji}$ are positive constants.

Let $\bar{z}(x,t) = \frac{1}{N}\sum_{i=1}^{N} z_i(x,t)$ and $e_i(x,t) = z_i(x,t) - \bar{z}(x,t)$. Then, we have

$$\begin{aligned} \frac{\partial e_i(x,t)}{\partial t} &= D\Delta e_i(x,t) - Ae_i(x,t) + Bf(z_i(x,t)) - B\sum_{i=1}^{N}\frac{f(z_i(x,t))}{N} \\ &\quad + c\sum_{j=1}^{N} G_{ij}(t)\Gamma e_j(x,t) + Eu_i(x,t) - E\sum_{i=1}^{N}\frac{u_i(x,t)}{N}, \end{aligned} \tag{10.30}$$

where $i = 1, 2, \ldots, N$.

Similarly, the output vector $y_i(x,t)$ of system (10.30) is defined as

$$y_i(x,t) = Fe_i(x,t) + Hu_i(x,t), \quad i = 1, 2, \ldots, N, \tag{10.31}$$

where $F \in \mathbb{R}^{m \times n}$ and $H \in \mathbb{R}^{m \times p}$ are known matrices.

10.4.1 Passivity Analysis of CRDNNs

Theorem 10.10 *The system (10.30) is passive under the adaptive law (10.29) if there exists matrix $Q \in \mathbb{R}^{mN \times pN}$ such that*

$$Q^T(I_N \otimes H) + (I_N \otimes H^T)Q > 0. \tag{10.32}$$

Proof Define the following Lyapunov functional for system (10.30):

$$V_2(t) = \sum_{i=1}^{N}\int_{\Omega} e_i^T(x,t)e_i(x,t)dx + c\sum_{i=1}^{N}\sum_{(i,j)\in\mathcal{E}}\frac{(G_{ij}(t) - \hat{\kappa}_{ij})^2}{2\beta_{ij}}, \tag{10.33}$$

where $\hat{\kappa}_{ij} = \hat{\kappa}_{ji}(i \neq j)$ are nonnegative constants, and $\hat{\kappa}_{ij} = 0(i \neq j)$ if and only if $G_{ij}(t) = 0$.

Calculating the time derivative of $V_2(t)$ along the trajectory of system (10.30), we can get

$$\begin{aligned}\dot{V}_2(t) = {} & 2\sum_{i=1}^{N}\int_{\Omega} e_i^T(x,t)\Big[D\Delta e_i(x,t) - Ae_i(x,t) + Bf(z_i(x,t)) - Bf(\bar{z}(x,t)) \\ & + Bf(\bar{z}(x,t)) - B\sum_{i=1}^{N}\frac{f(z_i(x,t))}{N} + c\sum_{j=1}^{N} G_{ij}(t)\Gamma e_j(x,t) + Eu_i(x,t) \\ & - E\sum_{i=1}^{N}\frac{u_i(x,t)}{N}\Big]dx + c\sum_{i=1}^{N}\sum_{(i,j)\in\mathcal{E}}(G_{ij}(t) - \hat{\kappa}_{ij})\int_{\Omega}\big(e_i(x,t) \\ & - e_j(x,t)\big)^T\Gamma\big(e_i(x,t) - e_j(x,t)\big)dx. \end{aligned} \tag{10.34}$$

Define the matrix $\hat{\kappa} = (\hat{\kappa}_{ij})_{N\times N}$, where $\hat{\kappa}_{ii} = -\sum_{\substack{j=1\\ j\neq i}}^{N}\hat{\kappa}_{ij}$, $i = 1, 2, \ldots, N$ Then, we have

$$\begin{aligned}&\sum_{i=1}^{N}\sum_{(i,j)\in\mathcal{E}}(G_{ij}(t) - \hat{\kappa}_{ij})\big(e_i(x,t) - e_j(x,t)\big)^T\Gamma\big(e_i(x,t) - e_j(x,t)\big) \\ &= -2\sum_{i=1}^{N}\sum_{j=1}^{N}(G_{ij}(t) - \hat{\kappa}_{ij})e_i^T(x,t)\Gamma e_j(x,t). \end{aligned}\tag{10.35}$$

By (10.13), (10.34) and (10.35), one obtains

$$\begin{aligned}\dot{V}_2(t) \leqslant {} & \sum_{i=1}^{N}\int_{\Omega} e_i^T(x,t)\Big(-\sum_{k=1}^{q}\frac{2}{l_k^2}D - 2A + BB^T + \Theta\Big)e_i(x,t)dx \\ & + 2c\sum_{i=1}^{N}\sum_{j=1}^{N}\hat{\kappa}_{ij}\int_{\Omega} e_i^T(x,t)\Gamma e_j(x,t)dx \\ & + 2\sum_{i=1}^{N}\int_{\Omega} e_i^T(x,t)Eu_i(x,t)dx. \end{aligned}\tag{10.36}$$

From (10.32) and (10.36), one has

$$\begin{aligned}&\dot{V}_2(t) - 2\int_{\Omega} y^T(x,t)Qu(x,t)dx \\ &\leqslant \sum_{i=1}^{N}\int_{\Omega} e_i^T(x,t)\Big(-\sum_{k=1}^{q}\frac{2}{l_k^2}D - 2A + BB^T + \Theta\Big)e_i(x,t)dx \end{aligned}$$

$$
\begin{aligned}
&+ 2c\sum_{i=1}^{N}\sum_{j=1}^{N}\hat{\kappa}_{ij}\int_{\Omega} e_i^T(x,t)\Gamma e_j(x,t)dx \\
&- 2\int_{\Omega} e^T(x,t)(I_N\otimes F^T)Qu(x,t)dx \\
&+ 2\sum_{i=1}^{N}\int_{\Omega} e_i^T(x,t)Eu_i(x,t)dx \\
&- \int_{\Omega} u^T(x,t)\Psi u(x,t)dx \\
\leqslant &\int_{\Omega} e^T(x,t)\Big[I_N\otimes\Big(-\sum_{k=1}^{q}\frac{2}{l_k^2}D-2A+BB^T+\Theta\Big) \\
&+ \Xi+(2c\hat{\kappa})\otimes\Gamma\Big]e(x,t)dx,
\end{aligned}
$$

where $\Xi=\big(I_N\otimes E-(I_N\otimes F^T)Q\big)\Psi^{-1}\big(I_N\otimes E^T-Q^T(I_N\otimes F)\big)$, $\Psi=Q^T(I_N\otimes H)+(I_N\otimes H^T)Q$.

On the other hand, there obviously exists a unitary matrix $\phi=(\phi_1,\phi_2,\ldots,\phi_N)\in\mathbb{R}^{N\times N}$ such that $\phi^T\hat{\kappa}\phi=\Lambda$ with $\Lambda=\mathrm{diag}(\lambda_1,\lambda_2,\ldots,\lambda_N)$. $\lambda_i, i=1,2,\ldots,N$, are the eigenvalues of $\hat{\kappa}$ and $0=\lambda_1>\lambda_2\geqslant\lambda_3\geqslant\cdots\geqslant\lambda_N$. Let $\eta(x,t)=(\eta_1^T(x,t),\eta_2^T(x,t),\ldots,\eta_N^T(x,t))^T=(\phi^T\otimes I_n)e(x,t)$. Since $\phi_1=\frac{1}{\sqrt{N}}(1,1,\ldots,1)^T$, one has $\eta_1(t)=(\phi_1^T\otimes I_n)e(x,t)=0$. Then, we can obtain

$$
\begin{aligned}
&\dot{V}_2(t)-2\int_{\Omega} y^T(x,t)Qu(x,t)dx \\
\leqslant &\int_{\Omega} e^T(x,t)\Big[I_N\otimes\Big(-\sum_{k=1}^{q}\frac{2}{l_k^2}D-2A+BB^T+\Theta\Big)+\Xi\Big]e(x,t)dx \\
&+ 2c\int_{\Omega} e^T(x,t)(\phi\otimes I_n)(\Lambda\otimes\Gamma)(\phi^T\otimes I_n)e(x,t)dx \\
\leqslant &\int_{\Omega} e^T(x,t)\Big[I_N\otimes\Big(-\sum_{k=1}^{q}\frac{2}{l_k^2}D-2A+BB^T+\Theta\Big)+\Xi\Big]e(x,t)dx \\
&+ 2c\lambda_2\int_{\Omega}\eta^T(x,t)(I_N\otimes\Gamma)\eta(x,t)dx \\
= &\int_{\Omega} e^T(x,t)\Big[I_N\otimes\Big(-\sum_{k=1}^{q}\frac{2}{l_k^2}D-2A+BB^T+\Theta\Big) \\
&+ \Xi+2c\lambda_2(I_N\otimes\Gamma)\Big]e(x,t)dx.
\end{aligned}
\tag{10.37}
$$

By selecting $\hat{\kappa}_{ij}$ sufficiently large such that

$$
\lambda_M\Big(-\sum_{k=1}^{q}\frac{2}{l_k^2}D-2A+BB^T+\Theta\Big)+\lambda_M(\Xi)+2c\lambda_2\lambda_m(\Gamma)+1\leqslant 0,
$$

one obtains

$$\dot{V}_2(t) - 2\int_{\Omega} y^T(x,t)Qu(x,t)dx \leqslant -\int_{\Omega} e^T(x,t)e(x,t)dx.$$

Therefore,

$$\dot{V}_2(t) \leqslant 2\int_{\Omega} y^T(x,t)Qu(x,t)dx. \tag{10.38}$$

From (10.38), we have

$$\int_{t_0}^{t_p}\int_{\Omega} y^T(x,t)Qu(x,t)dxdt \geqslant \hat{V}(t_p) - \hat{V}(t_0)$$

for any $t_p, t_0 \in \mathbb{R}^+$ and $t_p \geqslant t_0$, where $\hat{V}(t) = \frac{V_2(t)}{2}$. The proof is completed.

By a minor modification of the proof of Theorem 10.10, we can easily get the following conclusions.

Theorem 10.11 *The system (10.30) is input-strictly passive under the adaptive law (10.29) if there exist matrices $0 < W \in \mathbb{R}^{pN\times pN}$ and $Q \in \mathbb{R}^{mN\times pN}$ such that*

$$Q^T(I_N \otimes H) + (I_N \otimes H^T)Q - W > 0. \tag{10.39}$$

Theorem 10.12 *The system (10.30) is output-strictly passive under the adaptive law (10.29) if there exist matrices $0 < W \in \mathbb{R}^{mN\times mN}$ and $Q \in \mathbb{R}^{mN\times pN}$ such that*

$$\Psi - (I_N \otimes H^T)W(I_N \otimes H) > 0, \tag{10.40}$$

where $\Psi = Q^T(I_N \otimes H) + (I_N \otimes H^T)Q$.

10.4.2 The Relationship Between Output Strict Passivity and Synchronization of CRDNNs

Theorem 10.13 *The CRDNNs (10.28) is synchronized under the adaptive law (10.29) if F is a nonsingular matrix and system (10.30) under the adaptive law (10.29) is output-strictly passive with respect to storage function*

$$\hat{V}(t) = \frac{1}{2}\sum_{i=1}^{N}\int_{\Omega} e_i^T(x,t)e_i(x,t)dx + c\sum_{i=1}^{N}\sum_{(i,j)\in\mathcal{E}}\frac{(G_{ij}(t) - \hat{\kappa}_{ij})^2}{4\beta_{ij}}, \tag{10.41}$$

where $\hat{\kappa}_{ij} = \hat{\kappa}_{ji}$ are positive constants.

Proof Following similar arguments as in the proof of Theorem 10.8, we can obtain the desired result immediately.

From Theorems 10.12 and 10.13, we can obtain the following conclusion.

Theorem 10.14 *The CRDNNs (10.28) is synchronized under the adaptive law (10.29) if there exist matrices* $0 < W \in \mathbb{R}^{nN\times nN}$, $Q \in \mathbb{R}^{nN\times pN}$ *and* $H \in \mathbb{R}^{n\times p}$ *such that*

$$\Psi - (I_N \otimes H^T)W(I_N \otimes H) > 0,$$

where $\Psi = Q^T(I_N \otimes H) + (I_N \otimes H^T)Q$.

Remark 10.15 The passivity theory has long been a nice tool for analyzing the synchronization of complex dynamical networks. However, for passive CRDNNs with different dimensions of input and output, the synchronization criterion has not yet been established. In this chapter, two sufficient conditions for synchronization of the output-strictly passive CRDNNs (10.1) with adaptive coupling weights are obtained (see Theorems 10.8 and 10.13).

10.5 Numerical Examples

In this section, we give two examples and their simulations to show the effectiveness of the above-obtained theoretical results.

Example 10.16 A single reaction-diffusion neural network with Dirichlet boundary conditions is described by the following PDEs:

$$\begin{aligned}\frac{\partial w_i(x,t)}{\partial t} &= d_i\frac{\partial^2 w_i(x,t)}{\partial x^2} - a_i w_i(x,t) + J_i + \sum_{j=1}^{3} b_{ij}f_j(w_j(x,t)),\\ w_i(x,t) &= 0, \quad (x,t) \in \partial\Omega \times [0,+\infty),\end{aligned} \tag{10.42}$$

where $i = 1,2,3$, $\Omega = \{x \mid -0.5 < x < 0.5\}$, $f_j(\xi) = \frac{|\xi+1|-|\xi-1|}{4}$, $d_1 = 0.4$, $d_2 = 0.5$, $d_3 = 0.3$, $a_1 = 0.5$, $a_2 = 0.6$, $a_3 = 0.8$, $J_1 = 0.2$, $J_2 = 0.3$, $J_3 = 0.1$, and the matrix $B = (b_{ij})_{3\times 3}$ is chosen as

$$B = \begin{pmatrix} 0.4 & 0.2 & 0.5 \\ 0.7 & 0.2 & 0.3 \\ 0.3 & 0.1 & 0.6 \end{pmatrix}.$$

Obviously, $f_j(\cdot)(j = 1,2,3)$ satisfies the Lipschitz condition with $\rho_j = 0.5$.

Consider a directed CRDNNs consisting of 5 such RDNNs (10.42), which is characterized by

$$\frac{\partial z_i(x,t)}{\partial t} = D\frac{\partial^2 z_i(x,t)}{\partial x^2} - Az_i(x,t) + J + Bf(z_i(x,t)) + c\sum_{j=1}^{5} G_{ij}(t)\Gamma z_j(x,t) + Eu_i(x,t), \tag{10.43}$$

where

$$c = 0.6,$$
$$\Gamma = \mathrm{diag}(0.8, 0.9, 0.7),$$
$$E = \begin{pmatrix} 0.4 & 0.7 \\ 0.3 & 0.8 \\ 0.6 & 0.5 \end{pmatrix},$$
$$G(0) = \begin{pmatrix} -1 & 0.4 & 0 & 0.2 & 0.4 \\ 0.2 & -0.3 & 0.1 & 0 & 0 \\ 0 & 0.2 & -0.4 & 0.2 & 0 \\ 0.2 & 0 & 0.1 & -0.5 & 0.2 \\ 0.1 & 0 & 0 & 0.3 & -0.4 \end{pmatrix}.$$

By utilizing the MATLAB function NULL, we can find the positive vector $\theta = (0.5, 1, 0.5, 1, 1)^T \in \mathbb{R}^5$ satisfying $\theta^T G(0) = 0$.

Choose

$$F = \begin{pmatrix} 0.8 & 0 & 0 \\ 0 & 0.5 & 0 \\ 0 & 0 & 0.6 \end{pmatrix}, \quad H = \begin{pmatrix} 0.4 & 0.6 \\ 0.3 & 0.7 \\ 0.2 & 0.5 \end{pmatrix},$$

and use the YALMIP Toolbox of MATLAB. There exist matrices

$$W = I_5 \otimes \begin{pmatrix} 0.3815 & 0.0208 & -0.0073 \\ 0.0208 & 1.3610 & -0.1369 \\ -0.0073 & -0.1369 & 0.7422 \end{pmatrix} > 0,$$
$$Q = I_5 \otimes \begin{pmatrix} 0.4294 & 0.3827 \\ 0.4458 & 1.4361 \\ 0.2368 & 0.5213 \end{pmatrix},$$
$$\kappa = \begin{pmatrix} -5.4787 & 1.9707 & 0 & 1.6518 & 1.8562 \\ 1.9707 & -4.0340 & 2.0633 & 0 & 0 \\ 0 & 2.0633 & -4.1094 & 2.0461 & 0 \\ 1.6518 & 0 & 2.0461 & -5.4390 & 1.7410 \\ 1.8562 & 0 & 0 & 1.7410 & -3.5972 \end{pmatrix},$$

such that the condition (10.20) holds. According to Theorem 10.6, the system (10.9) with above given parameters is output-strictly passive under the adaptive law (10.5). The simulation results are shown in Figs. 10.1 and 10.2.

On the other hand, it is easy to verify that condition (10.27) in Theorem 10.9 is also satisfied. According to Theorem 10.9, CRDNNs (10.43) is synchronized under the adaptive law (10.5). The simulation results are shown in Figs. 10.1 and 10.2.

Example 10.17 Consider an undirected CRDNNs consisting of 5 such RDNNs (10.42), which is characterized by

$$\frac{\partial z_i(x,t)}{\partial t} = D\frac{\partial^2 z_i(x,t)}{\partial x^2} - Az_i(x,t) + J + Bf(z_i(x,t)) + c\sum_{j=1}^{5} G_{ij}(t)\Gamma z_j(x,t) + Eu_i(x,t), \tag{10.44}$$

where

$$\begin{aligned} c &= 0.3, \\ \Gamma &= \mathrm{diag}(0.5, 0.6, 0.4), \\ E &= \begin{pmatrix} 0.5 & 0.3 \\ 0.4 & 0.6 \\ 0.2 & 0.7 \end{pmatrix}, \\ G(0) &= \begin{pmatrix} -0.5 & 0.2 & 0 & 0.2 & 0.1 \\ 0.2 & -0.3 & 0.1 & 0 & 0 \\ 0 & 0.1 & -0.2 & 0.1 & 0 \\ 0.2 & 0 & 0.1 & -0.5 & 0.2 \\ 0.1 & 0 & 0 & 0.2 & -0.3 \end{pmatrix}. \end{aligned}$$

We take

$$F = \begin{pmatrix} 0.2 & 0 & 0 \\ 0 & 0.3 & 0 \\ 0 & 0 & 0.4 \end{pmatrix}, \quad H = \begin{pmatrix} 0.1 & 0.2 \\ 0.3 & 0.5 \\ 0.4 & 0.2 \end{pmatrix}.$$

By using the YALMIP Toolbox of MATLAB, we can find the following matrices:

$$\begin{aligned} W &= I_5 \otimes \begin{pmatrix} 2.9223 & 0 & 0 \\ 0 & 2.9223 & 0 \\ 0 & 0 & 2.9223 \end{pmatrix} > 0, \\ Q &= I_5 \otimes \begin{pmatrix} -0.9200 & 1.9854 \\ -1.1921 & 4.1796 \\ 5.7267 & -2.7179 \end{pmatrix}, \end{aligned}$$

satisfying (10.40). According to Theorem 10.12, the system (10.30) with above given parameters is output-strictly passive under the adaptive law (10.29). The simulation results are shown in Figs. 10.3 and 10.4.

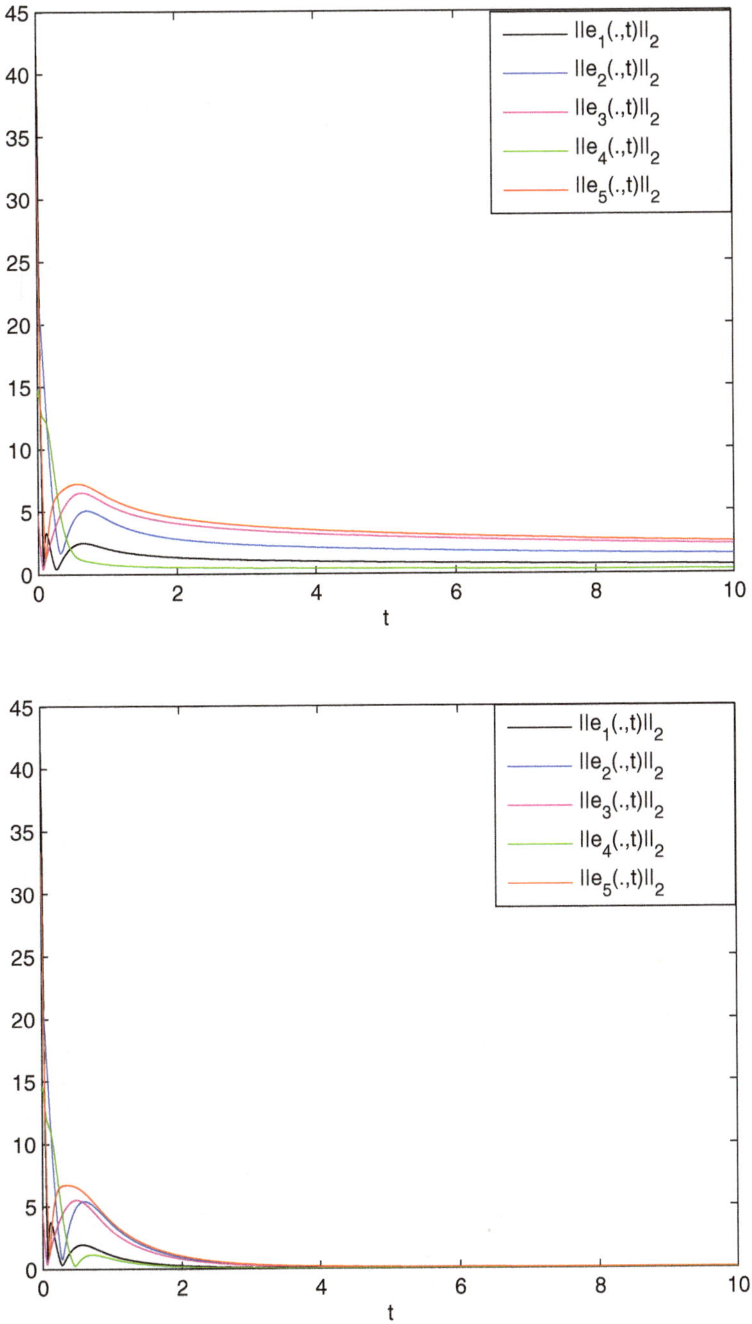

Fig. 10.1 The change processes of $\|e_i(\cdot, t)\|_2$ for directed network (10.43), where $z_{i1}(x, 0) = 24 \times i \times \cos(\pi x)$, $z_{i2}(x, 0) = 12 \times i \times \cos(\pi x)$, $z_{i3}(x, 0) = 3 \times i \times \cos(\pi x)$, $i = 1, 2, \ldots, 5$. In the first subfigure (from *top* to *bottom*), $u_{i1}(x, t) = 10 \times i \times e^{-0.01t} \cos(\pi x)$, $u_{i2}(x, t) = 20 \times i \times e^{-0.01t} \cos(\pi x)$; In the second subfigure, $u_{i1}(x, t) = u_{i2}(x, t) = 0$

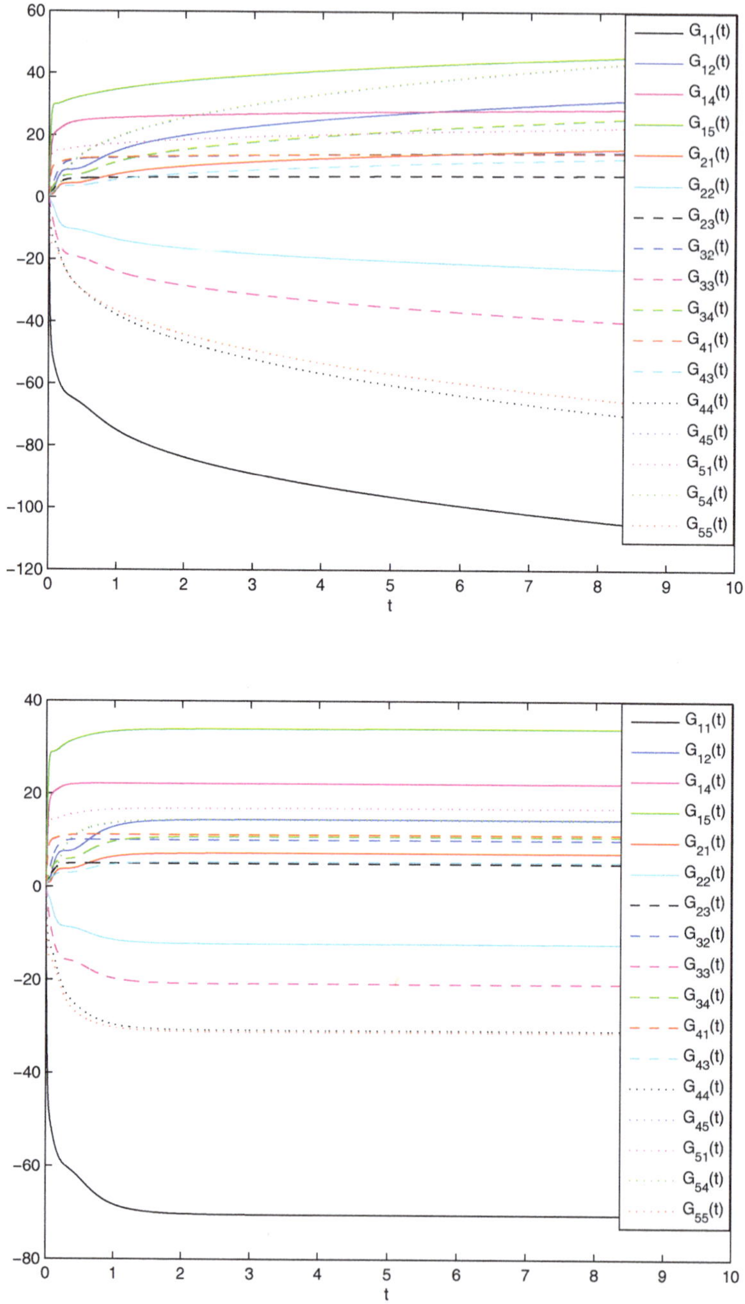

Fig. 10.2 The change processes of coupling weights for directed network (10.43), where $z_{i1}(x, 0) = 24 \times i \times \cos(\pi x)$, $z_{i2}(x, 0) = 12 \times i \times \cos(\pi x)$, $z_{i3}(x, 0) = 3 \times i \times \cos(\pi x)$, $i = 1, 2, \ldots, 5$. In the first subfigure (from *top* to *bottom*), $u_{i1}(x, t) = 10 \times i \times e^{-0.01t} \cos(\pi x)$, $u_{i2}(x, t) = 20 \times i \times e^{-0.01t} \cos(\pi x)$; In the second subfigure, $u_{i1}(x, t) = u_{i2}(x, t) = 0$

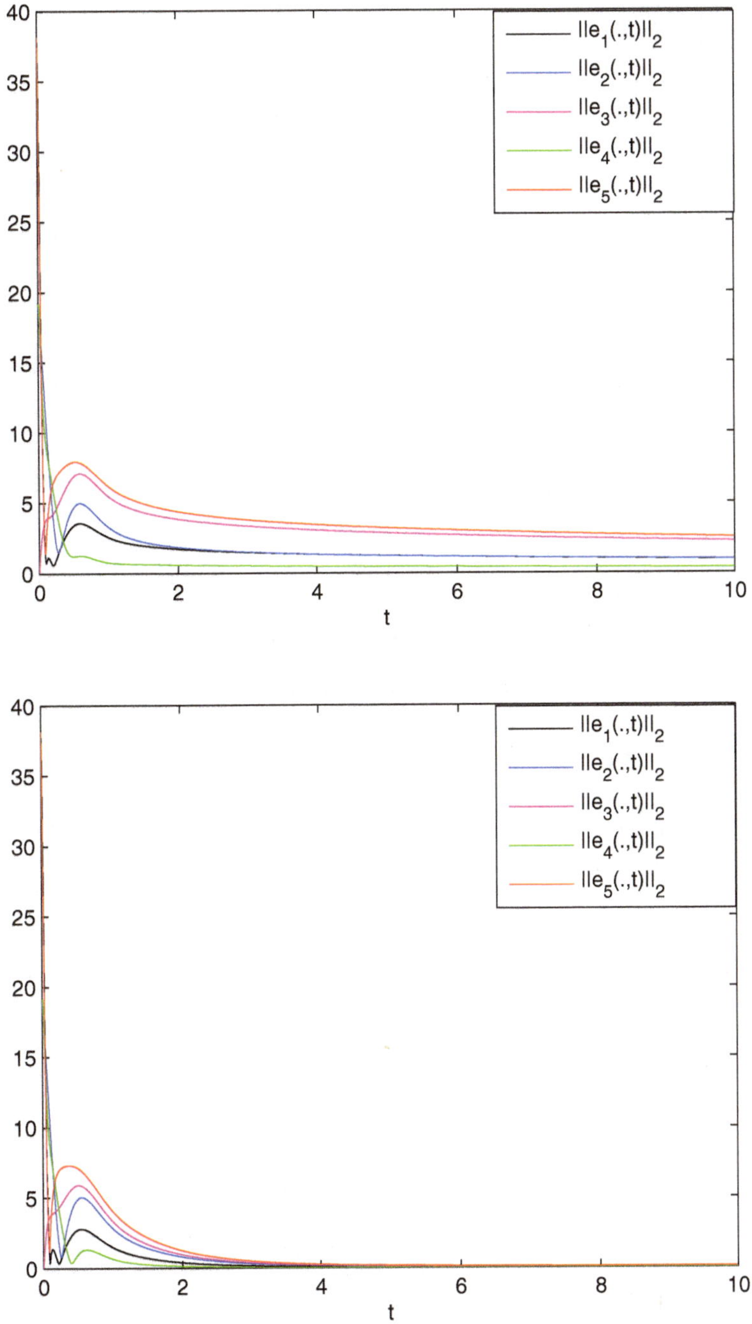

Fig. 10.3 The change processes of $\|e_i(\cdot, t)\|_2$ for undirected network (10.44), where $z_{i1}(x, 0) = 24 \times i \times \cos(\pi x)$, $z_{i2}(x, 0) = 12 \times i \times \cos(\pi x)$, $z_{i3}(x, 0) = 3 \times i \times \cos(\pi x)$, $i = 1, 2, \ldots, 5$. In the first subfigure (from *top* to *bottom*), $u_{i1}(x, t) = 10 \times i \times e^{-0.01t} \cos(\pi x)$, $u_{i2}(x, t) = 20 \times i \times e^{-0.01t} \cos(\pi x)$; In the second subfigure, $u_{i1}(x, t) = u_{i2}(x, t) = 0$

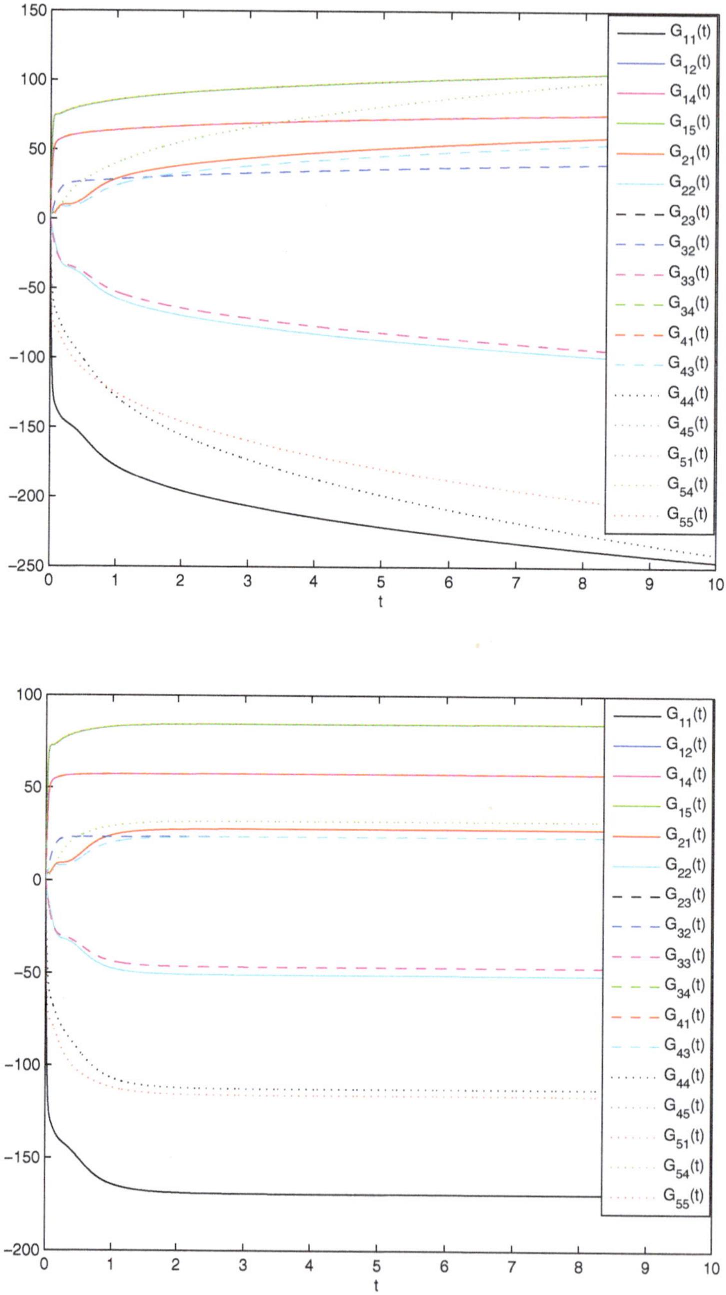

Fig. 10.4 The change processes of coupling weights for undirected network (10.44), where $z_{i1}(x,0) = 24\times i\times\cos(\pi x)$, $z_{i2}(x,0) = 12\times i\times\cos(\pi x)$, $z_{i3}(x,0) = 3\times i\times\cos(\pi x)$, $i = 1, 2, \ldots, 5$. In the first subfigure (from *top* to *bottom*), $u_{i1}(x,t) = 10\times i\times e^{-0.01t}\cos(\pi x)$, $u_{i2}(x,t) = 20\times i\times e^{-0.01t}\cos(\pi x)$; In the second subfigure, $u_{i1}(x,t) = u_{i2}(x,t) = 0$

On the other hand, it is easy to verify that the condition of Theorem 10.14 is also satisfied. According to Theorem 10.14, CRDNNs (10.44) is synchronized under the adaptive law (10.29). The simulation results are shown in Figs. 10.3 and 10.4.

From Figs. 10.1, 10.2, 10.3 and 10.4, we clearly see that coupling weights asymptotically converge to some finite values when CRDNNs (10.43) (or (10.44)) is synchronized under the adaptive law (10.5) (or (10.29)).

10.6 Conclusion

In this chapter, two kinds of passivity problems for CRDNNs have been investigated, that is, the cases with directed topology and with undirected topology. On the one hand, some adaptive strategies to tune the coupling weights among network nodes have been proposed, ensuring the passivity of the CRDNNs. On the other hand, several network synchronization criteria have been established by utilizing the relationship between synchronization and output strict passivity. Two numerical examples have been provided to verify the correctness and effectiveness of the theoretical results.

References

1. Z. Ma, G. Sun, D. Liu, X. Xing, Dissipativity analysis for discrete-time fuzzy neural networks with leakage and time-varying delays. Neurocomputing **175**, 579–584 (2016)
2. W. Zhang, Y. Tang, W.K. Wong, Q. Miao, Stochastic stability of delayed neural networks with local impulsive effects. IEEE Trans. Neural Networks Learn. Syst. **26**(10), 2336–2345 (2015)
3. Z. Wang, L. Liu, Q.-H. Shan, H. Zhang, Stability criteria for recurrent neural networks with time-varying delay based on secondary delay partitioning method. IEEE Trans. Neural Networks Learn. Syst. **26**(10), 2589–2595 (2015)
4. B. Zhang, J. Lam, S. Xu, Stability analysis of distributed delay neural networks based on relaxed Lyapunov-Krasovskii functionals. IEEE Trans. Neural Networks Learn. Syst. **26**(7), 1480–1492 (2015)
5. J. Lu, J. Zhong, C. Huang, J. Cao, On pinning controllability of boolean control networks. IEEE Trans. Autom. Control **61**(6), 1658–1663 (2016)
6. J. Lu, Z. Wang, J. Cao, D.W.C. Ho, J. Kurths, Pinning impulsive stabilization of nonlinear dynamical networks with time-varying delay. Int. J. Bifurc. Chaos **22**(07), 137–139 (2012)
7. W.-H. Chen, X. Lu, W.X. Zheng, Impulsive stabilization and impulsive synchronization of discrete-time delayed neural networks. IEEE Trans. Neural Networks Learn. Syst. **26**(4), 734–748 (2015)
8. J. Lian, J. Wang, Passivity of switched recurrent neural networks with time-varying delays. IEEE Trans. Neural Networks Learn. Syst. **26**(2), 357–366 (2015)
9. W. Gong, J. Liang, J. Cao, Global μ-stability of complex-valued delayed neural networks with leakage delay. Neurocomputing **168**, 135–144 (2015)
10. Q. Song, Z. Zhao, Stability criterion of complex-valued neural networks with both leakage delay and time-varying delays on time scales. Neurocomputing **171**, 179–184 (2015)
11. J. Cao, R. Rakkiyappan, K. Maheswari, A. Chandrasekar, Exponential H_∞ filtering analysis for discrete-time switched neural networks with random delays using sojourn probabilities. Sci. China Technol. Sci. **59**(3), 387–402 (2016)
12. T. Huang, Y. Huang, C. Li, Stability of periodic solution in fuzzy BAM neural networks with finite distributed delays. Neurocomputing **71**(16–18), 3064–3069 (2008)
13. T. Huang, Exponential stability of fuzzy cellular neural networks with distributed delay. Phys. Lett. A **351**(1–2), 48–52 (2006)
14. Z. Zeng, T. Huang, W.X. Zheng, Multistability of recurrent neural networks with time-varying delays and the piecewise linear activation function. IEEE Trans. Neural Networks **21**(8), 1371–1377 (2010)
15. S. Arik, An improved robust stability result for uncertain neural networks with multiple time delays. Neural Networks **54**, 1–10 (2014)

J.-L. Wang et al., *Analysis and Control of Coupled Neural Networks with Reaction-Diffusion Terms*, DOI 10.1007/978-981-10-4907-1

16. O. Faydasicok, S. Arik, Robust stability analysis of a class of neural networks with discrete time delays. Neural Networks **29**, 52–59 (2012)
17. J. Cao, Y. Wan, Matrix measure strategies for stability and synchronization of inertial BAM neural network with time delays. Neural Networks **53**, 165–172 (2014)
18. Z.-G. Wu, P. Shi, H. Su, J. Chu, Stability analysis for discrete-time Markovian jump neural networks with mixed time-delays. Expert Syst. Appl. **39**(6), 6174–6181 (2012)
19. R. Rakkiyappan, P. Balasubramaniam, On exponential stability results for fuzzy impulsive neural networks. Fuzzy Sets Syst. **161**(13), 1823–1835 (2010)
20. Z.-G. Wu, P. Shi, H. Su, J. Chu, Stochastic synchronization of Markovian jump neural networks with time-varying delay using sampled data. IEEE Trans. Cybern. **43**(6), 1796–1806 (2013)
21. Z.-G. Wu, P. Shi, H. Su, J. Chu, Exponential synchronization of neural networks with discrete and distributed delays under time-varying sampling. IEEE Trans. Neural Networks Learn. Syst. **23**(9), 1368–1376 (2012)
22. J. Cao, L. Li, Cluster synchronization in an array of hybrid coupled neural networks with delay. Neural Networks **22**(4), 335–342 (2009)
23. J. Cao, Z. Wang, Y. Sun, Synchronization in an array of linearly stochastically coupled networks with time delays. Phys. A **385**(2), 718–728 (2007)
24. Q. Song, J. Cao, Dynamics of bidirectional associative memory networks with distributed delays and reaction-diffusion terms. Nonlinear Anal. Real World Appl. **8**(1), 345–361 (2007)
25. R. Li, J. Cao, Stability analysis of reaction-diffusion uncertain memristive neural networks with time-varying delays and leakage term. Appl. Math. Comput. **278**, 54–69 (2016)
26. Y. Wang, J. Cao, Exponential stability of stochastic higher-order BAM neural networks with reaction-diffusion terms and mixed time-varying delays. Neurocomputing **119**(16), 192–200 (2013)
27. Y. Wang, J. Cao, Synchronization of a class of delayed neural networks with reaction diffusion terms. Phys. Lett. A **369**(3), 201–211 (2007)
28. P. Balasubramaniam, C. Vidhya, Global asymptotic stability of stochastic BAM neural networks with distributed delays and reaction-diffusion terms. J. Comput. Appl. Math. **234**(12), 3458–3466 (2010)
29. J.-L. Wang, H.-N. Wu, L. Guo, Passivity and stability analysis of reaction-diffusion neural networks with Dirichlet boundary conditions. IEEE Trans. Neural Networks **22**(12), 2105–2116 (2011)
30. C. Hu, H. Jiang, Z. Teng, Impulsive control and synchronization for delayed neural networks with reaction-diffusion terms. IEEE Trans. Neural Networks **21**(1), 67–81 (2010)
31. J.G. Lu, Robust global exponential stability for interval reaction-diffusion Hopfield neural networks with distributed delays. IEEE Trans. Circuits Syst. II Express. Briefs **54**(12), 1115–1119 (2007)
32. Q. Ma, G. Feng, S. Xu, Delay-dependent stability criteria for reaction-diffusion neural networks with time-varying delays. IEEE Trans. Cybern. **43**(6), 1913–1920 (2013)
33. Q. Song, Z. Wang, Dynamical behaviors of fuzzy reaction-diffusion periodic cellular neural networks with variable coefficients and delays. Appl. Math. Model. **33**(9), 3533–3545 (2009)
34. S.H. Strogatz, Exploring complex networks. Nature **410**(6825), 268–276 (2001)
35. R. Albert, A.L. Barabsi, Statistical mechanics of complex networks. Rev. Modern Phys. **74**, 47–97 (2002)
36. R. Cheng, M. Peng, Y. Weibin, Pinning synchronization of delayed complex dynamical networks with nonlinear coupling. Phys. A **413**, 426–431 (2014)
37. T.H. Lee, J.H. Park, H.Y. Jung, S.-M. Lee, O.-M. Kwon, Synchronization of a delayed complex dynamical network with free coupling matrix. Nonlinear Dyn. **69**(3), 1081–1090 (2012)
38. W. Wang, L. Li, H. Peng, J. Xiao, Y. Yang, Stochastic synchronization of complex network via a novel adaptive nonlinear controller. Nonlinear Dyn. **76**(1), 591–598 (2014)
39. Y. Wang, T. Li, Synchronization of fractional order complex dynamical networks. Phys. A **428**, 1–12 (2015)
40. J. Liang, Z. Wang, X. Liu, Exponential synchronization of stochastic delayed discrete-time complex networks. Nonlinear Dyn. **53**(1), 153–165 (2008)

41. W. Yu, P. DeLellis, G. Chen, M. di Bernardo, J. Kurths, Distributed adaptive control of synchronization in complex networks. IEEE Trans. Autom. Control **57**(8), 2153–2158 (2012)
42. J.-L. Wang, H.-N. Wu, Adaptive output synchronization of complex delayed dynamical networks with output coupling. Neurocomputing **142**, 174–181 (2014)
43. J. Yao, H.O. Wang, Z.-H. Guan, W. Xu, Passive stability and synchronization of complex spatio-temporal switching networks with time delays. Automatica **45**(7), 1721–1728 (2009)
44. X. Yang, J. Cao, J. Lu, Stochastic synchronization of complex networks with nonidentical nodes via hybrid adaptive and impulsive control. IEEE Trans. Circuits Syst. I Regul. Pap. **59**(2), 371–384 (2012)
45. Q. Song, J. Cao, On pinning synchronization of directed and undirected complex dynamical networks. IEEE Trans. Circuits Syst. Part I Regul. Pap. **57**(3), 672–680 (2010)
46. W. Yu, J. Cao, J. Lü, Global synchronization of linearly hybrid coupled networks with time-varying delay. SIAM J. Appl. Dyn. Syst. **7**(1), 108–133 (2008)
47. W. Yu, G. Chen, J. Lü, On pinning synchronization of complex dynamical networks. Automatica **45**(2), 429–435 (2009)
48. J. Zhao, D.J. Hill, T. Liu, Synchronization of complex dynamical networks with switching topology: A switched system point of view. Automatica **45**(11), 2502–2511 (2009)
49. T. Liu, D.J. Hill, J. Zhao, Synchronization of dynamical networks by network control. IEEE Trans. Autom. Control **57**(6), 1574–1580 (2012)
50. J. Yao, Z.H. Guan, D.J. Hill, Passivity-based control and synchronization of general complex dynamical networks. Automatica **45**(9), 2107–2113 (2009)
51. B. Liu, D.J. Hill, Impulsive consensus for complex dynamical networks with nonidentical nodes and coupling time-delays. SIAM J. Control Optim. **49**(2), 315–338 (2011)
52. B. Liu, X. Liu, G. Chen, H. Wang, Robust impulsive synchronization of uncertain dynamical networks. IEEE Trans. Circuits Syst. I Regul. Pap. **52**(7), 1431–1441 (2005)
53. J.-L. Wang, H.-N. Wu, Synchronization criteria for impulsive complex dynamical networks with time-varying delay. Nonlinear Dyn. **70**(1), 13–24 (2012)
54. A. Hu, Z. Xu, Pinning a complex dynamical network via impulsive control. Phys. Lett. A **374**(2), 186–190 (2009)
55. W. Lu, T. Chen, Synchronisation in complex networks of coupled systems with directed topologies. Int. J. Syst. Sci. **40**(9), 909–921 (2009)
56. B. Liu, W. Lu, T. Chen, Synchronization in complex networks with stochastically switching coupling structures. IEEE Trans. Autom. Control **57**(3), 754–760 (2012)
57. M. Chen, Synchronization in complex dynamical networks with random sensor delay. IEEE Trans. Circuits Syst. II Express. Briefs **57**(1), 46–50 (2010)
58. W. Yu, J. Cao, G. Chen, J. Lü, J. Han, W. Wei, Local synchronization of a complex network model. IEEE Trans. Syst. Man Cybern. Part B (Cybern.) **39**(1), 230–241 (2009)
59. Z. Fei, H. Gao, W.X. Zheng, New synchronization stability of complex networks with an interval time-varying coupling delay. IEEE Trans. Circuits Syst. II Express. Briefs **56**(6), 499–503 (2009)
60. Y. Tang, H. Gao, W. Zou, J. Kurths, Distributed synchronization in networks of agent systems with nonlinearities and random switchings. IEEE Trans. Cybern. **43**(1), 358–370 (2013)
61. Y. Liu, Z. Wang, J. Liang, X. Liu, Synchronization and state estimation for discrete-time complex networks with distributed delays. IEEE Trans. Syst. Man Cybern. Part B (Cybern.) **38**(5), 1314–1325 (2008)
62. B. Shen, Z. Wang, X. Liu, Sampled-data synchronization control of dynamical networks with stochastic sampling. IEEE Trans. Autom. Control **57**(10), 2644–2650 (2012)
63. J.-L. Wang, H.-N. Wu, Local and global exponential output synchronization of complex delayed dynamical networks. Nonlinear Dyn. **67**(1), 497–504 (2012)
64. Y.-W. Wang, J.-W. Xiao, C. Wen, Z.-H. Guan, Synchronization of continuous dynamical networks with discrete-time communications. IEEE Trans. Neural Networks **22**(12), 1979–1986 (2011)
65. J. Zhao, D.J. Hill, T. Liu, Global bounded synchronization of general dynamical networks with nonidentical nodes. IEEE Trans. Autom. Control **57**(10), 2656–2662 (2012)

66. J. Lü, G. Chen, A time-varying complex dynamical network model and its controlled synchronization criteria. IEEE Trans. Autom. Control **50**(6), 841–846 (2005)
67. J. Zhou, J.-A. Lu, J. Lü, Adaptive synchronization of an uncertain complex dynamical network. IEEE Trans. Autom. Control **51**(4), 652–656 (2006)
68. J. Zhou, J.-A. Lu, J. Lü, Pinning adaptive synchronization of a general complex dynamical network. Automatica **44**(4), 996–1003 (2008)
69. J.-J.E. Slotine, W. Wang, *A Study of Synchronization and Group Cooperation Using Partial Contraction Theory* (Springer, Berlin, 2005)
70. L. Scardovi, M. Arcak, E.D. Sontag, Synchronization of interconnected systems with applications to biochemical networks: An input-output approach. IEEE Trans. Autom. Control **55**(6), 1367–1379 (2010)
71. M. Arcak, Certifying spatially uniform behavior in reaction-diffusion PDE and compartmental ODE systems. Automatica **47**(6), 1219–1229 (2011)
72. L. Fang, Y.-H. Zhao, Passivity analysis of complex delayed dynamical networks with output coupling, in *Proceedings of the 33rd Chinese Control Conference* (2014), pp. 2794–2798
73. J.-L. Wang, H.-N. Wu, Z.-C. Yang, Passivity analysis of impulsive complex networks. Int. J. Autom. Comput. **8**(4), 484–489 (2011)
74. J.-L. Wang, Z.-C. Yang, H.-N. Wu, Passivity analysis of complex dynamical networks with multiple time-varying delays. J. Eng. Math. **74**(1), 175–188 (2012)
75. Z. Ye, H. Ji, H. Zhang, Passivity analysis of Markovian switching complex dynamic networks with multiple time-varying delays and stochastic perturbations. Chaos, Solitons Fractals **83**, 147–157 (2016)
76. F.C. Hoppensteadt, E.M. Izhikevich, Pattern recognition via synchronization in phase-locked loop neural networks. IEEE Trans. Neural Networks **11**(3), 734–738 (2000)
77. Y. Zhang, Z. He, A secure communication scheme based on cellular neural network, in *1997 IEEE International Conference on Intelligent Processing Systems* (1997), pp. 521–524
78. A.A. Ukhtomsky, *Collected Works* (Nauka, Leningrad, 1978)
79. C.M. Gray, Synchronous oscillations in neuronal systems: Mechanisms and functions. J. Comput. Neurosci. **1**(1), 11–38 (1994)
80. J. Feng, S. Wang, Z. Wang, Stochastic synchronization in an array of neural networks with hybrid nonlinear coupling. Neurocomputing **74**(18), 3808–3815 (2011)
81. H. Wang, Q. Song, Synchronization for an array of coupled stochastic discrete-time neural networks with mixed delays. Neurocomputing **74**(10), 1572–1584 (2011)
82. Y. Tang, J.-A. Fang, Adaptive synchronization in an array of chaotic neural networks with mixed delays and jumping stochastically hybrid coupling. Commun. Nonlinear Sci. Numer. Simul. **14**(9–10), 3615–3628 (2009)
83. Y. Tang, J.-A. Fang, Robust synchronization in an array of fuzzy delayed cellular neural networks with stochastically hybrid coupling. Neurocomputing **72**(13–15), 3253–3262 (2009)
84. Y. Tang, Z. Wang, J.-A. Fang, Controller design for synchronization of an array of delayed neural networks using a controllable probabilistic PSO. Inf. Sci. **181**(20), 4715–4732 (2011)
85. J. Cao, P. Li, W. Wang, Global synchronization in arrays of delayed neural networks with constant and delayed coupling. Phys. Lett. A **353**(4), 318–325 (2006)
86. W. Wang, J. Cao, Synchronization in an array of linearly coupled networks with time-varying delay. Phys. A **366**, 197–211 (2006)
87. W. Yu, J. Cao, W. Lu, Synchronization control of switched linearly coupled neural networks with delay. Neurocomputing **73**(4), 858–866 (2010)
88. J. Lu, D.W.C. Ho, M. Liu, Globally exponential synchronization in an array of asymmetric coupled neural networks. Phys. Lett. A **369**(5–6), 444–451 (2007)
89. K. Yuan, Robust synchronization in arrays of coupled networks with delay and mixed coupling. Neurocomputing **72**(4–6), 1026–1031 (2009)
90. T. Li, A. Song, S. Fei, Synchronization control for arrays of coupled discrete-time delayed Cohen-Grossberg neural networks. Neurocomputing **74**(1–3), 197–204 (2010)
91. N. Li, J. Cao, Passivity and robust synchronisation of switched interval coupled neural networks with time delay. Int. J. Syst. Sci. **47**(12), 2827–2836 (2016)

92. P.-C. Wei, J.-L. Wang, Y.-L. Huang, B.-B. Xu, S.-Y. Ren, Impulsive control for the synchronization of coupled neural networks with reaction-diffusion terms. Neurocomputing **207**, 539–547 (2016)
93. J.-L. Wang, H.-N. Wu, T. Huang, S.-Y. Ren, J. Wu, Passivity analysis of coupled reaction-diffusion neural networks with Dirichlet boundary conditions. IEEE Trans. Syst. Man Cybern. Syst. **PP**(99), 1–12 (2016)
94. J.-L. Wang, H.-N. Wu, T. Huang, S.-Y. Ren, J. Wu, Passivity of directed and undirected complex dynamical networks with adaptive coupling weights. IEEE Trans. Neural Networks Learn. Syst. **PP**(99), 1–13 (2016)
95. J.-L. Wang, H.-N. Wu, T. Huang, Passivity-based synchronization of a class of complex dynamical networks with time-varying delay. Automatica **56**, 105–112 (2015)
96. J.-L. Wang, H.-N. Wu, T. Huang, S.-Y. Ren, Passivity and synchronization of linearly coupled reaction-diffusion neural networks with adaptive coupling. IEEE Trans. Cybern. **45**(9), 1942–1952 (2015)
97. X. Liu, Synchronization of linearly coupled neural networks with reaction-diffusion terms and unbounded time delays. Neurocomputing **73**(13–15), 2681–2688 (2010)
98. K. Wang, Z. Teng, H. Jiang, Adaptive synchronization in an array of linearly coupled neural networks with reaction-diffusion terms and time delays. Commun. Nonlinear Sci. Numer. Simul. **17**(10), 3866–3875 (2012)
99. J.-L. Wang, H.-N. Wu, L. Guo, Novel adaptive strategies for synchronization of linearly coupled neural networks with reaction-diffusion terms. IEEE Trans. Neural Networks Learn. Syst. **25**(2), 429–440 (2014)
100. J.-L. Wang, H.-N. Wu, Synchronization and adaptive control of an array of linearly coupled reaction-diffusion neural networks with hybrid coupling. IEEE Trans. Cybern. **44**(8), 1350–1361 (2014)
101. X. Yang, J. Cao, Z. Yang, Synchronization of coupled reaction-diffusion neural networks with time-varying delays via pinning-impulsive controller. SIAM J. Control Optim. **51**(5), 3486–3510 (2013)
102. J.-L. Wang, H.-N. Wu, T. Huang, S.-Y. Ren, J. Wu, Pinning control for synchronization of coupled reaction-diffusion neural networks with directed topologies. IEEE Trans. Syst. Man Cybern. Syst. **46**(8), 1109–1120 (2016)
103. J.-L. Wang, H.-N. Wu, T. Huang, S.-Y. Ren, Pinning control strategies for synchronization of linearly coupled neural networks with reaction-diffusion terms. IEEE Trans. Neural Networks Learn. Syst. **27**(4), 749–761 (2016)
104. R.A. Brualdi, H.J. Ryser, *Combinatorial Matrix Theory* (Cambridge University Press, UK, 1991)
105. J.C. Willems, Dissipative dynamical systems part I: General theory. Arch. Ration. Mech. Anal. **45**(5), 321–351 (1972)
106. D.J. Hill, P.J. Moylan, The stability of nonlinear dissipative systems. IEEE Trans. Autom. Control **21**(5), 708–711 (1976)
107. C.I. Byrnes, A. Isidori, J.C. Willems, Passivity, feedback equivalence, and the global stabilization of minimum phase nonlinear systems. IEEE Trans. Autom. Control **36**(11), 1228–1240 (1991)
108. S.I. Niculescu, R. Lozano, On the passivity of linear delay systems. IEEE Trans. Autom. Control **46**(3), 460–464 (2001)
109. W. Yu, G. Chen, J. Lü, J. Kurths, Synchronization via pinning control on general complex networks. SIAM J. Control Optim. **51**(2), 1395–1416 (2013)
110. R. Lu, W. Yu, J. Lü, A. Xue, Synchronization on complex networks of networks. IEEE Trans. Neural Networks Learn. Syst. **25**(11), 2110–2118 (2014)
111. S. Boyd, L. El Ghaoui, E. Feron, V. Balakrishnan, *Linear Matrix Inequalities in System and Control Theory* (SIAM, Philadelphia, 1994)
112. C.P. Li, W.G. Sun, J. Kurths, Synchronization of complex dynamical networks with time delays. Phys. A **361**(1), 24–34 (2006)

113. W. Yu, G. Chen, M. Cao, J. Kurths, Second-order consensus for multiagent systems with directed topologies and nonlinear dynamics. IEEE Trans. Syst. Man Cybern. Part B **40**(3), 881–891 (2010)
114. W. Lu, T. Chen, New approach to synchronization analysis of linearly coupled ordinary differential systems. Physica D **213**(2), 214–230 (2006)
115. T. Chen, X. Liu, W. Lu, Pinning complex networks by a single controller. IEEE Trans. Circuits Syst. I Regul. Pap. **54**(6), 1317–1326 (2007)
116. J.-L. Wang, H.-N. Wu, L. Guo, Stability analysis of reaction-diffusion Cohen-Grossberg neural networks under impulsive control. Neurocomputing **106**(6), 21–30 (2013)
117. J.G. Lu, Global exponential stability and periodicity of reaction-diffusion delayed recurrent neural networks with Dirichlet boundary conditions. Chaos, Solitons Fractals **35**(1), 116–125 (2008)
118. J.-L. Wang, H.-N. Wu, L. Guo, Pinning control of spatially and temporally complex dynamical networks with time-varying delays. Nonlinear Dyn. **70**(2), 1657–1674 (2012)
119. Y.-W. Wang, H.O. Wang, J.-W. Xiao, Z.-H. Guan, Synchronization of complex dynamical networks under recoverable attacks. Automatica **46**(1), 197–203 (2010)
120. Y. Xu, W. Zhou, J.-A. Fang, W. Sun, Adaptive synchronization of the complex dynamical network with non-derivative and derivative coupling. Phys. Lett. A **374**(15–16), 1673–1677 (2010)
121. Y. Dai, Y. Cai, X. Xu, Synchronization criteria for complex dynamical networks with neutral-type coupling delay. Phys. A **387**(18), 4673–4682 (2008)
122. K. Wu, B.-S. Chen, Synchronization of partial differential systems via diffusion coupling. IEEE Trans. Circuits Syst. I Regul. Pap. **59**(11), 2655–2668 (2012)
123. B. Dubey, B. Das, J. Hussain, A predator-prey interaction model with self and cross-diffusion. Ecol. Model. **141**(1), 67–76 (2001)
124. L.-M. Zhu, A.-L. Wang, Y.-J. Liu, B. Wang, Stationary patterns of a predator-prey model with spatial effect. Appl. Math. Comput. **216**(12), 3620–3626 (2010)
125. Y. Tang, H. Gao, J. Lu, J. Kurths, Pinning distributed synchronization of stochastic dynamical networks: A mixed optimization approach. IEEE Trans. Neural Networks Learn. Syst. **25**(10), 1804–1815 (2014)
126. Y. Tang, H. Gao, J. Kurths, Distributed robust synchronization of dynamical networks with stochastic coupling. IEEE Trans. Circuits Syst. I Regul. Pap. **61**(5), 1508–1519 (2014)
127. Y. Tang, F. Qian, H. Gao, J. Kurths, Synchronization in complex networks and its application-a survey of recent advances and challenges. Annu. Rev. Control **38**(2), 184–198 (2014)
128. D. Wu, S. Zhu, X. Luo, L. Wu, Effects of adaptive coupling on stochastic resonance of small-world networks. Phys. Rev. E Stat. Nonlinear Soft Matter Phys. **84**, 021102 (2011)
129. X.-Z. Jin, G.-H. Yang, Adaptive synchronization of a class of uncertain complex networks against network deterioration. IEEE Trans. Circuits Syst. I Regul. Pap. **58**(6), 1396–1409 (2011)
130. S. Larsson, V. Thome, *Partial Differential Equations with Numerical Methods* (Springer Science & Business Media, 2008)
131. L. Xiang, J.J.H. Zhu, On pinning synchronization of general coupled networks. Nonlinear Dyn. **64**(4), 339–348 (2011)
132. F. Sorrentino, M. di Bernardo, F. Garofalo, G. Chen, Controllability of complex networks via pinning. Phys. Rev. E **75**(4 Pt 2), 77–85 (2007)
133. M. Porfiri, M. di Bernardo, Criteria for global pinning-controllability of complex networks. Automatica **44**(12), 3100–3106 (2008)
134. S. Liu, X. Li, W. Jiang, Y.-Z. Fan, Adaptive synchronization in complex dynamical networks with coupling delays for general graphs. Appl. Math. Comput. **219**(1), 83–87 (2012)
135. X. Guo, J. Li, A new synchronization algorithm for delayed complex dynamical networks via adaptive control approach. Commun. Nonlinear Sci. Numer. Simul. **17**(11), 4395–4403 (2012)
136. Y. Chai, L. Chen, R. Wu, J. Sun, Adaptive pinning synchronization in fractional-order complex dynamical networks. Phys. A **391**(22), 5746–5758 (2012)

137. X.-Z. Jin, G.-H. Yang, W.-W. Che, Adaptive pinning control of deteriorated nonlinear coupling networks with circuit realization. IEEE Trans. Neural Networks Learn. Syst. **23**(9), 1345–1355 (2012)
138. P. De Lellis, M. di Bernardo, F. Garofalo, Synchronization of complex networks through local adaptive coupling. Chaos **18**(3), 037110 (2008)
139. P. De Lellis, M. di Bernardo, F. Garofalo, Novel decentralized adaptive strategies for the synchronization of complex networks. Automatica **45**(5), 1312–1318 (2009)
140. P. DeLellis, F. Garofalo, M. Porfiri et al., Evolution of complex networks via edge snapping. IEEE Trans. Circuits Syst. I Regul. Pap. **57**(8), 2132–2143 (2010)
141. C.L.P. Chen, Y.-J. Liu, G.-X. Wen, Fuzzy neural network-based adaptive control for a class of uncertain nonlinear stochastic systems. IEEE Trans. Cybern. **44**(5), 583–593 (2014)
142. C.L.P. Chen, G.-X. Wen, Y.-J. Liu, F.-Y. Wang, Adaptive consensus control for a class of nonlinear multiagent time-delay systems using neural networks. IEEE Trans. Neural Networks Learn. Syst. **25**(6), 1217–1226 (2014)
143. Y.-J. Liu, C.L.P. Chen, G.-X. Wen, S. Tong, Adaptive neural output feedback tracking control for a class of uncertain discrete-time nonlinear systems. IEEE Trans. Neural Networks **22**(7), 1162–1167 (2011)
144. Y.-J. Liu, S. Tong, C.L.P. Chen, Chen, Adaptive fuzzy control via observer design for uncertain nonlinear systems with unmodeled dynamics. IEEE Trans. Fuzzy Syst. **21**(2), 275–288 (2013)
145. Y.-J. Liu, L. Tang, S. Tong, C.L.P. Chen, Adaptive NN controller design for a class of nonlinear MIMO discrete-time systems. IEEE Trans. Neural Networks Learn. Syst. **26**(5), 1007–1018 (2015)
146. Y. Xia, Z. Yang, M. Han, Synchronization schemes for coupled identical Yang-Yang type fuzzy cellular neural networks. Commun. Nonlinear Sci. Numer. Simul. **14**(9–10), 3645–3659 (2009)
147. Z.M. Ge, S.Y. Li, Fuzzy modeling and synchronization of chaotic quantum cellular neural networks nano system via a novel fuzzy model and its implementation on electronic circuits. J. Comput. Theor. Nanosci. **7**(11), 2453–2462 (2010)
148. D. Malagarriga, M.A. Garcia-Vellisca, A.E.P. Villa, J.M. Buldu, J. Garcia-Ojalvo, A.J. Pons, Synchronization-based computation through networks of coupled oscillators. Front. Comput. Neurosci. **9**, 97 (2015)
149. J. Hu, J. Cao, Q. Cheng, *Adaptive Pinning Synchronization of Coupled Inertial Delayed Neural Networks* (Springer International Publishing, Cham, 2014), pp. 35–42
150. X. Yang, Q. Song, J. Liang, B. He, Finite-time synchronization of coupled discontinuous neural networks with mixed delays and nonidentical perturbations. J. Franklin Inst. **352**(10), 4382–4406 (2015)
151. G. Wang, Q. Yin, Y. Shen, Exponential synchronization of coupled fuzzy neural networks with disturbances and mixed time-delays. Neurocomputing **106**(6), 77–85 (2013)
152. X. Wang, C. Li, T. Huang, L. Chen, Impulsive exponential synchronization of randomly coupled neural networks with Markovian jumping and mixed model-dependent time delays. Neural Networks **60**, 25–32 (2014)
153. Q. Song, J. Cao, F. Liu, Pinning synchronization of linearly coupled delayed neural networks. Math. Comput. Simul. **86**(86), 39–51 (2012)
154. M.-J. Park, O.-M. Kwon, J.H. Park, S.-M. Lee, E.-J. Cha, Synchronization criteria for coupled neural networks with interval time-varying delays and leakage delay. Appl. Math. Comput. **218**(12), 6762–6775 (2012)
155. J. Wang, H. Zhang, Z. Wang, H. Liang, Stochastic synchronization for Markovian coupled neural networks with partial information on transition probabilities. Neurocomputing **149**, 983–992 (2015)
156. M.J. Park, O.M. Kwon, J.H. Park, S.M. Lee, E.J. Cha, On synchronization criterion for coupled discrete-time neural networks with interval time-varying delays. Neurocomputing **99**(1), 188–196 (2013)
157. C. Zheng, J. Cao, Robust synchronization of coupled neural networks with mixed delays and uncertain parameters by intermittent pinning control. Neurocomputing **141**(4), 153–159 (2014)

158. X. Yang, C. Huang, Z. Yang, Stochastic synchronization of reaction-diffusion neural networks under general impulsive controller with mixed delays. Abstr. Appl. Anal. **137–138**, 2012 (2012)
159. J. Hu, Z. Wang, B. Shen, H. Gao, Quantised recursive filtering for a class of nonlinear systems with multiplicative noises and missing measurements. Int. J. Control **86**(4), 650–663 (2013)
160. J. Hu, D. Chen, J. Du, State estimation for a class of discrete nonlinear systems with randomly occurring uncertainties and distributed sensor delays. Int. J. Gen. Syst. **43**(3–4), 387–401 (2014)
161. J. Hu, Z. Wang, B. Shen, H. Gao, Gain-constrained recursive filtering with stochastic nonlinearities and probabilistic sensor delays. IEEE Trans. Signal Process. **61**(5), 1230–1238 (2013)
162. J. Hu, Z. Wang, S. Liu, H. Gao, A variance-constrained approach to recursive state estimation for time-varying complex networks with missing measurements. Automatica **64**, 155–162 (2016)
163. Y. Xu, W. Zhou, J.-A. Fang, C. Xie, D. Tong, Finite-time synchronization of the complex dynamical network with non-derivative and derivative coupling. Neurocomputing **173**, 1356–1361 (2016)
164. Z. Li, J.-A. Fang, Q. Miao, G. He, Exponential synchronization of impulsive discrete-time complex networks with time-varying delay. Neurocomputing **157**(2), 335–343 (2014)
165. Z. Tang, J. Feng, Y. Zhao, Global synchronization of nonlinear coupled complex dynamical networks with information exchanges at discrete-time. Neurocomputing **151**(3), 1486–1494 (2015)
166. J.-L. Wang, H.-N. Wu, Stability analysis of impulsive parabolic complex networks. Chaos Solitons Fractals **44**(44), 1020–1034 (2011)
167. J.-L. Wang, H.-N. Wu, Robust stability and robust passivity of parabolic complex networks with parametric uncertainties and time-varying delays. Neurocomputing **87**(87), 26–32 (2012)
168. H.-Q. Wang, B. Chen, C. Lin, Adaptive neural tracking control for a class of stochastic nonlinear systems. Int. J. Robust Nonlinear Control **24**(7), 1262–1280 (2014)
169. H.-Q. Wang, B. Chen, C. Lin, Direct adaptive neural tracking control for a class of stochastic pure-feedback nonlinear systems with unknown dead-zone. Int. J. Adapt. Control Signal Process. **27**(4), 302–322 (2013)
170. B. Chen, X. Liu, K. Liu, C. Lin, Novel adaptive neural control design for nonlinear MIMO time-delay systems. Automatica **45**(6), 1554–1560 (2009)
171. J. Lu, J. Cao, Adaptive synchronization of uncertain dynamical networks with delayed coupling. Nonlinear Dyn. **53**(1–2), 107–115 (2008)
172. L. Wang, H.P. Dai, H. Dong, Y.H. Shen, Y.X. Sun, Adaptive synchronization of weighted complex dynamical networks with coupling time-varying delays. Phys. Lett. A **372**(20), 3632–3639 (2008)
173. J. Cao, G. Chen, P. Li, Global synchronization in an array of delayed neural networks with hybrid coupling. IEEE Trans. Syst. Man Cybern. Part B (Cybern.) **38**(2), 488–498 (2008)
174. J. Lu, D.W.C. Ho, Globally exponential synchronization and synchronizability for general dynamical networks. IEEE Trans. Syst. Man Cybern. Part B Cybern. **40**(2), 350–361 (2010)
175. J. Liang, Z. Wang, Y. Liu, X. Liu, Global synchronization control of general delayed discrete-time networks with stochastic coupling and disturbances. IEEE Trans. Syst. Man Cybern. Part B Cybern. **38**(4), 1073–1083 (2008)
176. V. Belevitch, *Classical Network Synthesis* (Van Nostrand, New York, 1968)
177. D.J. Hill, P.J. Moylan, Stability results for nonlinear feedback systems. Automatica **13**(4), 377–382 (1977)
178. L. Xie, M. Fu, H. Li, Passivity analysis and passification for uncertain signal processing systems. IEEE Trans. Signal Process. **46**(9), 2394–2403 (1998)
179. C.W. Wu, Synchronization in arrays of coupled nonlinear systems: Passivity circle criterion and observer design, in *IEEE International Symposium on Circuits and Systems* (2001), pp. 1257–1261

180. M. Arcak, Passivity as a design tool for group coordination. IEEE Trans. Autom. Control **52**(8), 1380–1390 (2007)
181. N. Chopra, Output synchronization on strongly connected graphs. IEEE Trans. Autom. Control **57**(11), 2896–2901 (2012)
182. N. Chopra, M.W. Spong, *Passivity-Based Control of Multi-Agent Systems* (Springer, Berlin, 2006)
183. J.C. Willems, Dissipative dynamical systems part II. Linear systems with quadratic supply rates. Arch. Ration. Mech. Anal. **45**(5), 352–393 (1972)
184. I.-A.F. Ihle, M. Arcak, T.I. Fossen, Passivity-based designs for synchronized path-following. Automatica **43**(9), 1508–1518 (2007)
185. J.T.-Y. Wen, M. Arcak, A unifying passivity framework for network flow control. IEEE Trans. Autom. Control **49**(2), 162–174 (2004)
186. X. Fan, T. Alpcan, M. Arcak, J.T.-Y. Wen, T. Başar, A passivity approach to game-theoretic CDMA power control. Automatica **42**(11), 1837–1847 (2006)
187. M.Y. Ayad, M. Becherif, A. Henni, A. Aboubou, M. Wack, S. Laghrouche, Passivity-based control applied to DC hybrid power source using fuel cell and supercapacitors. Energy Convers. Manag. **51**(7), 1468–1475 (2010)
188. M. Hu, J. Cao, Y. Yang, A. Hu, Passivity analysis for switched generalized neural networks with time-varying delay and uncertain output. IMA J. Math. Control Inf. **30**(3), 407–422 (2013)
189. Q. Song, J. Cao, Passivity of uncertain neural networks with both leakage delay and time-varying delay. Nonlinear Dyn. **67**(2), 1695–1707 (2012)
190. Z.G. Wu, P. Shi, H. Su, J. Chu, Passivity analysis for discrete-time stochastic Markovian jump neural networks with mixed time delays. IEEE Trans. Neural Networks **22**(10), 1566–1575 (2011)
191. H.R. Pota, P.J. Moylan, Stability of locally dissipative interconnected systems. IEEE Trans. Autom. Control **38**(2), 308–312 (1993)
192. J. Zhao, D.J. Hill, Dissipativity theory for switched systems. IEEE Trans. Autom. Control **53**(4), 941–953 (2008)
193. G.L. Santosuosso, Passivity of nonlinear systems with input-output feedthrough. Automatica **33**(4), 693–697 (1997)
194. L.O. Chua, Passivity and complexity. IEEE Trans. Circuits Syst. I Fundam. Theory Appl. **46**(1), 71–82 (1999)
195. W. Yu, Passive equivalence of chaos in Lorenz system. IEEE Trans. Circuits Syst. I Fundam. Theory Appl. **46**(7), 876–878 (1999)
196. G. Calcev, R. Gorez, M. De Neyer, Passivity approach to fuzzy control systems. Automatica **34**(3), 339–344 (1998)
197. L. Su, H. Shen, Mixed H_∞ passive synchronization for complex dynamical networks with sampled-data control. Appl. Math. Comput. **259**(C), 931–942 (2015)
198. Y. Liu, J. Zhao, Generalized output synchronization of dynamical networks using output quasi-passivity. IEEE Trans. Circuits Syst. I Regul. Pap. **59**(6), 1290–1298 (2012)
199. Y. Han, P.J. Antsaklis, Output synchronization of networked passive systems with event-driven communication. IEEE Trans. Autom. Control **59**(3), 750–756 (2012)
200. E. Steur, I. Tyukin, H. Nijmeijer, Semi-passivity and synchronization of diffusively coupled neuronal oscillators. Physica D **238**(21), 2119–2128 (2009)
201. H. Li, J. Wang, H.-K. Lam, Q. Zhou, H. Du, Adaptive sliding mode control for interval type-2 fuzzy systems. IEEE Trans. Syst. Man Cybern. Syst. **46**(12), 1654–1663 (2016)
202. H. Li, J. Wang, P. Shi, Output-feedback based sliding mode control for fuzzy systems with actuator saturation. IEEE Trans. Fuzzy Syst. **PP**(99), 1 (2016)
203. X.-H. Chang, G.-H. Yang, New results on output feedback H_∞ control for linear discrete-time systems. IEEE Trans. Autom. Control **59**(5), 1355–1359 (2014)
204. Z. Tang, J.H. Park, T.H. Lee, Dynamic output-feedback-based H_∞ design for networked control systems with multipath packet dropouts. Appl. Math. Comput. **275**, 121–133 (2016)

205. Q. Zhou, P. Shi, S. Xu, H. Li, Observer-based adaptive neural network control for nonlinear stochastic systems with time delay. IEEE Trans. Neural Networks Learn. Syst. **24**(1), 71–80 (2013)
206. Q. Zhou, P. Shi, Y. Tian, M. Wang, Approximation-based adaptive tracking control for MIMO nonlinear systems with input saturation. IEEE Trans. Cybern. **45**(10), 2119–2128 (2015)
207. K. Mathiyalagan, T.H. Lee, J.H. Park, R. Sakthivel, Robust passivity based resilient H_∞ control for networked control systems with random gain fluctuations. Int. J. Robust Nonlinear Control **26**(3), 426–444 (2016)
208. T.H. Lee, Q. Ma, S. Xu, J.H. Park, Pinning control for cluster synchronisation of complex dynamical networks with semi-Markovian jump topology. Int. J. Control **88**(6), 1223–1235 (2015)
209. H. Shen, J.H. Park, Z.G. Wu, Z. Zhang, Finite-time H_∞ synchronization for complex networks with semi-Markov jump topology. Commun. Nonlinear Sci. Numer. Simul. **24**(1–3), 40–51 (2015)
210. H. Li, Z. Chen, L. Wu, H.-K. Lam, H. Du, Event-triggered fault detection of nonlinear networked systems. IEEE Trans. Cybern. **PP**(99), 1–12 (2016)

Papier | Fördert gute Waldnutzung
FSC® C083411

Kolibri 360 GmbH
Ferdinand-Jühlke-Straße 7
99095 Erfurt, Deutschland
produktsicherheit@kolibri360.de